P9-ARF-369

Time and the Dancing Image

Also by Deborah Jowitt

Dance Beat (Marcel Dekker, 1977)
The Dance in Mind (David Godine, 1985)

TIME *and the* DANCING IMAGE

Deborah Jowitt

William Morrow and Company, Inc. | *New York*

Library of Congress Cataloging-in-Publication Data

Jowitt, Deborah.
Time and the dancing image / by Deborah Jowitt.
p. cm.
Bibliography: p.
Includes index.
ISBN 0-688-04910-9
1. Dancing—History. 2. Dancing—Anthropological aspects.
3. Dancing—Social aspects. I. Title.
GV1601.J68 1988 87-34171
793.3'2—dc19 CIP

Printed in the United States of America

First Edition

1 2 3 4 5 6 7 8 9 10

BOOK DESIGN BY VICTORIA HARTMAN

This book is dedicated to the memory of three people: my great aunt, Susan T. Smith, for many years the head librarian of the Berkeley Public Library, who saw to it that all the best children's books fell into my hands; my father, Anthony Jowitt, who loved words so much that he couldn't look one up in the dictionary without being waylaid by others; my mother, Doris Anderson, who was my tough and loving copy editor from the time I was ten, and who taught me almost everything I know about writing.

Preface

A dance critic, attending performances night after night, devises strategies for keeping eye and mind fresh. Some years ago, no doubt influenced by a longstanding addiction to *National Geographic,* I began to find it useful, on occasion, to blot out all expectations based on knowledge of styles or techniques. Instead I imagined myself an anthropologist skulking in ambush, observing the activities of members of a hitherto undiscovered tribe—trying to discern their customs and social hierarchy before I stepped out of the bushes and made myself known to them.

The game, idly begun, eventually generated this book. The approach turned out to have variants that were applicable to the study of the past. I found that I liked to think about how dancers may have looked onstage at a given point in time, what kind of men and women they've appeared to be, and how their training, skills, physiognomy, and—certainly in the case of dancer-choreographers—their view of life affected what was presented in the theater.

But dancers and choreographers don't exist in a vacuum, however cut off from life their rigorous daily discipline can make them appear to be. Those of us who write about dance sometimes find that in our anxiety to capture and chronicle a notoriously ephemeral art we do it an inadvertent disservice: we focus so intently on it that we sever it from the culture that spawned it and which it serves. Part of the pleasure I've had in this project has come in questioning how the social and cultural climate, discoveries in science, and developments in philosophy and the other arts may have influenced the domains created onstage and the character of those who inhabit them. What makes an elusive supernatural female such a meaningful symbol for audiences of the 1830's and a sleek, powerful acrobat so enticing to audiences of the 1980's? What happened between 1900 and 1930 to turn the image of the vanguard female dancer from that of a well-fleshed woman with flowing draperies and flowing gestures to that of a forceful, angular, deliberately unglamorous one who wasn't afraid to tackle serious and complex social issues? What caused the role of the male dancer to decline in importance during certain periods, and to become prominent during others? Trying to answer such questions has necessarily involved investigating which qualities a dancer must suppress, which emphasize, what new skills must be developed and what traditional ones discarded to effect a new image.

In non-Western dance, the image of the performer seems to change very slightly over the years (although one could argue that subtle changes simply may not be visible to the outsider). Some Asian and African styles are rooted in religion and cannot change radically without destroying the continuity of a spiritual tradition. In highly refined secular styles, great performers or choreographers may effect changes or a shift of priorities, like the extraordinarily popular Bando Tamasaburo V, whose delicate and almost naturalistic approach to the female roles of Japanese Kabuki theater may well be influencing the next generation of *onnegata* interpreters. However, a Kabuki actor is usually more concerned with perfecting a stylistic approach handed down in his own family than in innovation, and the Noh actor performs his ritualized soliloquy in much the same way as his counterpart of centuries ago. Balinese dancers still consider I Mario's dances to be innovative, even though he created them in the 1920's. Certainly modern slang and contemporary political references crop up in Balinese dance-dramas; steps cribbed from Fred Astaire movies appear in the competition dances of male teams in South Africa; a Japanese pioneer, Tatsumi Hijikata, created a style, Butoh, which refers not only to German modern dance and Noh drama, but to the sensibility of contemporary Japan. Nevertheless, in African and Asian culture, it is always a great honor to be part of a tradition—even a new tradition like Butoh—while in Europe and America, at least since the early twentieth century, artists have tended to prize originality and contemporaneity.

In the West, dancing was not much admired by the dominant Catholic Church and never—with rare exceptions, like the dance that little boys known as "Los Seises" perform on certain holy days in the Cathedral of Seville—figured in services. Western theatrical dancing developed out of an uneven mix of social dancing, party entertainment, street performance, and court spectacle. Because of this, it has always been responsive to current trends. At its most profound, like the other arts, it reflects aspects of the current world picture; at its most superficial, it acknowledges the current fashions.

If Diaghilev's Les Ballets Russes was an instant hit with the public, and Merce Cunningham was not, the reason may be that some changes are easier to accept than others: however innovative Mikhail Fokine's ballets for Diaghilev's company were in terms of dance style, many of them fed his early-twentieth-century contemporaries' interest in violence, exoticism, and lust, while the dances that Cunningham began to make in the 1950's undermined our inherited ideas about what constituted form and order and an easy evening at the theater.

The so-called "modern dance" tends to progress by violent redefinitions: choreographers reject those images that mirror the past and then start over. The more codified, less idiosyncratic field of ballet advances by absorbing innovations into an existing vocabulary of movement. But however changes occur, the dancer's image has been subject to many alterations since the beginning of the nineteenth century in response to the immense social, political, scientific, and technological upheavals that have characterized the period.

The shifts in perspective affect both the role dancers may play and the roles they are *perceived* as playing. For millions of spectators, Rudolf Nureyev recharged the image of the male dancer, not just because of his virtuosity and brooding Tartar good looks, but because of the daring, impetuosity, and courage that his defection from Soviet Russia necessitated ("That Nureyev—a real *mensch!*," a man told me when he learned that I was a dance critic).

The body and its agenda may alter too. For example, during the 1950's in modern dance concerts, it was rare for women to touch each other—beyond the holding of hands that signified girlish playfulness or a folk-dance circle, the embrace that said "mother-and-child," and the grabbing and twisting that meant antagonism. For men to touch each other, except in athletic contests or war or bluff camaraderie, was almost as uncommon. After the Women's Liberation and Gay Liberation movements began in earnest, we started to see far more casual physical contact between people of the same gender—both onstage and off. On one level, of course, this affirmed that homosexuality existed, but it also affirmed that there were complex ways—having nothing to do with sex—in which women and women,

men and men, men and women might support each other and bear one another's weight. Although the sight of a woman lifting a man continued to draw laughter in conservative circles for some time, quite a few choreographers—Senta Driver, for one—made such an act a choreographic issue during the 1970's. Now it provokes few snickers and raises almost no eyebrows.

In a trenchant article entitled "The Aesthetic Interpretation of Dance History," John Chapman complains that dance historians too often view the past with the eyes of the present, unaware that they are even doing so. He points out, for example, that just because the eighteenth-century choreographer and theoretician Jean Georges Noverre spoke out in favor of modeling art on "nature," there is no reason to suppose that his ballets or the style of performing them would have struck us as natural, or to view him as the direct ancestor of Mikhail Fokine or Isadora Duncan or any other choreographer who pleads for an avoidance of artifice. Noverre's view of "nature" in art—elegant, subtle, ordered—was far more like that of his near contemporary Alexander Pope than it was like the view these twentieth-century choreographers took of it.

Implicit in any restructuring of history is the dangerous notion that the present is fulfilling the past, that dancers are better today than they were, say, fifty years ago. But the important point is not whether, if Anna Pavlova were set dancing before us, we would consider her technique inferior to that of a member of the New York City Ballet's corps today; the point is what she meant to her contemporaries, what her own sources may have been, and her influence on those who saw her.

Trying to view the dancers of the past as products of their age, to see them as their contemporaries saw them, is a challenge. Inevitably the distancing effect of time seduces us into making connections and evaluations that those living during the period discussed might not have been able to make. Where they saw trees, we see a wood, and trying to keep our sights on both perspectives can be vexing. When I write of dancers who are my contemporaries, the problem is reversed: it is easier to note a phenomenon than to see it in its context.

The fact that this book was intended as a collection of essays on the historical aspects of style, rather than as a comprehensive history or a series of brief biographies, has facilitated certain options in structure—giving me the freedom, for example, to focus on Fokine primarily in terms of his role in delineating an exotic and sexually charged image of dancers, or to mention Marius Petipa's work in Imperial Russia in three different contexts: the supernatural females of nineteenth-century ballet, Orientalism, and dance symphonism. Because I am dealing primarily with the invention or crystallization of an image, many important choreographers are mentioned briefly, if at all, and the whole field of vaudeville and musical-comedy styles doesn't figure. On the other hand, a group like Mummenschanz, whose work most people would consider mime rather than dance, *is* discussed, because disguise is an aspect of theatricality that I was interested in thinking about.

I regret that I couldn't hymn by name more of those dancers who do not create their own stage image single-handedly, but serve primarily as instruments for others. Because it is only through the motion of bodies that choreographic genius can be realized. And through all these nimble likenesses of ourselves, we see aspects of our behavior and our world that, left to ourselves, we might never recognize.

Acknowledgments

As a critic, I have been used to working alone. Little, except erudite lobby chitchat, a program, a press release, and an occasional book, mediates between me and the performance to be reviewed. Working on this project has involved an influx of ideas from others and has been from the start six years ago a highly sociable venture. If I were to thank all those who have dropped a useful name or book title in my ear, these acknowledgments would cover many pages.

I owe an enormous debt of gratitude to the curators and staff (past and present) of the Dance Collection of the Library and Museum of the Performing Arts at Lincoln Center—with special thanks to curator emeritus Genevieve Oswald, present acting curator Dorothy Lourdou, Monica Moseley, Nancy Shawcross, Winifred Messe, Lacy McDearmon, Karen Nickeson, and Henry Wisneski. They found what I could not find and drew my attention to items I didn't know existed. I would also like to thank Derek Damon and the other pages who set up tapes and films, took my call slips, and lugged out large, heavy collections of mounted photographs for me day after day; Tina Brooks even lessened my guilt by doing it with a smile.

I am extremely grateful to the photographers and their assistants who not only made their work available to me but gave me valuable advice and help: Arnold Eagle, Barbara Morgan and her daughter-in-law Liliane Morgan, Peter Moore and Barbara Moore, Carol Greunke of the Waldman Archives, Martha Swope, and James Klosty and my colleague at the *Village Voice,* Lois Greenfield, who literally opened their files to me. I would also like to thank Victoria Edwards at Sovfoto, the Leo Castelli Gallery, Alwin

Nikolais, the Trisha Brown Company, the Twyla Tharp Dance Foundation, Meredith Monk/the House, American Ballet Theatre, the Cunningham Dance Foundation, and ICM Artists. Ivor Guest and Parmenia Migel Ekstrom graciously allowed me to reproduce lithographs in their collections, Erna Schulz Lane and Barbara Cronk gave me the use of photographs they owned, and Fredrika Blair loaned me a negative of a precious Isadora Duncan photograph in her possession.

Joseph Reed Petticrew merits special thanks for his skill and patience in making photographic prints of over sixty pictures in the Dance Collection. Musicologist Linda Roesner kindly translated German text for me.

The dance community seems to me an amazingly generous one. The people listed below shared their knowledge, their memories, their memorabilia with me. Some answered a few important questions; others gave me hours of their time. Thanks to Trisha Brown, Carolyn Brown, Helen Priest Rogers, Martha Hill, Bessie Schönberg, Bertram Ross, Erna Lane, David Vaughan, Michael Bloom, Sally Sommer, Hortense Bunsick Fried, Edith Segal, Harriet Roeder, Louise Kloepper, Eve Gentry, Nancy Ruyter, Sophie Maslow, Lily Mehlman Rady, Jane Dudley, Barbara Morgan, Frieda Flier Maddow, Dorothy Bird, Margaretta Mitchell.

Many dance scholars who passed through classes I have taught in the department of performance studies and the dance department of NYU's Tisch School of the Arts have not only given me tangible assistance, they have stimulated me with their ideas and sparked my research by their own. Especially deserving of mention are Ellen Graff, Judy Burns, Ann Daly, Margaret Drewal, Robert Johnson, Martie Fellom, and Kristin Jackson.

Bonnie Bird allowed me to take some of her classes in early Graham technique, given for Senta Driver's company, Harry, in November 1981; these were useful, exciting, and invigoratingly painful. An inspiring series of classes with Annabelle Gamson helped me to feel something of Isadora Duncan in my own body.

I have been lucky to have had the support of a remarkable group of friends and colleagues who read chapters at various stages of development and gave me superb criticism, advice, and the encouragement that I needed: Sally Banes, Selma Jeanne Cohen, Barbara Davis, Toni Dorfman, John Mueller, Laura Shapiro, Suzanne Shelton, Marcia B. Siegel, and Burt Supree.

My gratitude also goes to those who ushered this book from idea to printed page: David White (who set a few things in motion), Robert Cornfield, and the people at William Morrow—in particular book design director Maria Epes, designer Victoria Hartman, copy editor extraordinaire Sonia Greenbaum, chief copy editor Deborah Weiss, editorial assistant Julia Kent, and senior editor Maria Guarnaschelli, who believed in this book from the beginning.

Last, but hardly least, I would like to express my love and gratitude to my husband, Murray Ralph, and my son, Toby Ralph, who put up with disorder and my calamities for much of the time this book was being written, and who gave me in return their full confidence, support, and love.

Contents

List of Illustrations

Note: Illustrations followed by an asterisk are reproduced through the courtesy of the Dance Collection of the New York Public Library, Lenox, Astor, and Tilden Foundations.

·1·

Flesh and Spirit: The Uneasy Balance

Marie Taglioni and Antonio Guerra in *L'Ombre.* Lithograph by Joseph Bouvier.

Marie Taglioni and Joseph Mazilier in the first scene of *La Sylphide* by Filippo Taglioni. Lithograph from a painting by Francois Lepaulle.

IN PURSUIT OF THE SYLPH

The painting is familiar to ballet-lovers. A young Scotsman slumbers in an armchair before the fire. Beside him, one hand laid lightly on his kilted thigh, kneels a woman wearing a pale, diaphanous dress—a wreath of flowers on her head, pearls at her throat and wrist. A small set of filmy wings sprouts from the region of her shoulder blades. Chin on hand, she gazes pensively out of the picture at us. Behind her, through a window of the gloomy baronial hall, shimmers an alluring vista of mist and light and foliage: her domain.

We know what happens next. She will dance with innocent buoyancy—bounding across the room, hovering on the tips of her toes. He will awaken and be enraptured by her. Eventually the man—his name is James—will forsake his hearth, family, friends, and betrothed to follow this creature into the forest. But, not content simply to dance with her and her bevy of airborne sisters, unable to subsist on dew and gauzy flights, he will strike a bargain with the forces of darkness—a witch he once thrust from his fireside—and attempt to secure the sylph with a magic scarf. While he watches aghast, her little wings will fall pitifully to the ground, and she will expire. As her comrades bear her aloft, a wedding procession will pass through the forest: the nuptials of James's former fiancée and his former rival. Unable to keep the desires of the spirit and those of the flesh in harmonious balance, the hero will have lost everything.

The painting by François Lepaulle, widely circulated in lithograph form, might well be called *The Poet and His Muse.* You can imagine it included in a popular class of pictures of artists being inspired, along with Delacroix's portrait of Michelangelo brooding in his studio, so aptly does it symbolize the Romantics' view of the artist as a solitary visionary and art not as profession, but as vocation—something to which one was called by mysterious forces. But whether the winged female is a muse, or an ideal to be victimized by a poet's impetuosity, or an enchantress, like Keats's "Belle Dame Sans Merci," who saps the reason of sensitive young men, she is indisputably one of the most famous ballerinas of her epoch.

The figures in the painting are Marie Taglioni and Joseph Mazilier in the opening pose of *La Sylphide,* choreographed by Marie's father, Filippo, and first performed at the Paris Opéra in May 1832. And there's a serious bit of poetic license in Lepaulle's much-copied picture: the sylph is barefoot. Taglioni's slippers were a vital part of the illusion she created. In those satin shoes, laced to her ankles with ribbons, with a high vamp, stiffening sewn inside, and a slim line of protective darning running over the tip, she could stand on her toes . . . no, not stand: a sylph never rests. She could poise on one toe for a few seconds; she could run with swift little steps that seemed to the enraptured observers scarcely to touch the stage floor.

La Sylphide is the work that officially brought ballet into the Romantic fold. Its scenario, by the acclaimed tenor Adolphe Nourrit, was said to be loosely based on *Trilby, or The Imp of Argyll* (1822), Charles Nodier's fantastic tale of a male sprite who woos a Scottish fisherman's wife.* Seven years before Nourrit devised his ballet plot, however, Victor Hugo had already clarified one fact: sylphs who visit important poets are female. In the poem addressed to "Trilby," he salutes *his* fairy and *his* sylph:

Qui me visitent sans bruit,
Et m'apportent, empressées,
Sur leurs ailes nuancées,
Le jour de douces pensées
Et de doux rêves la nuit.†

---

*Choreographer Louis Henry's widow later claimed that Filippo Taglioni had stolen steps and a few other ideas from her husband's *La Silfide,* which was first performed at La Scala in 1828. Although the plots differ considerably, I wouldn't be surprised, given Taglioni's reputation for plagiarism, although I *am* surprised by the charge: nineteenth-century balletmakers did a lot of "borrowing," and seldom did anyone make a fuss.

†Who visit me soundlessly,/ And, hastening, bring,/ On their tinted wings,/ Sweet thoughts by day,/ And sweet dreams at night.

Nourrit, a man of culture and poetic temperament, understood the potency of such images. He also knew that important *dancers* were female. Having performed with one of Paris's latest sensations, Marie Taglioni, in the opera-ballet *Le Dieu et la Bayadère* in 1830 and in Meyerbeer's opera *Robert le Diable* in 1831, he was able to capitalize on her outstanding ethereality.

His ballet touched off a vogue for unearthly heroines in Paris, London, Milan, Stuttgart, Vienna, St. Petersburg—wherever ballet flourished—and engendered a spate of works, such as Taglioni's *La Fille du Danube* (1836) and *L'Ombre* (1839); the Jean Coralli–Jules Perrot *Giselle* (1841); Perrot's *Alma, ou La Fille du Feu* (1842), *Ondine, ou La Naiade* (1843), *Eoline, ou La Dryade* (1845), and many more in which mortal men and fairy women fell hopelessly in love. Later in the century, when Parisian standards declined and Romanticism foundered in an excess of unreasonable spectacle and glamour, tales of bewitched or supernatural or dead maidens assumed new forms in Russia, with *La Bayadère* (1877) by Marius Petipa, and *Swan Lake* (first choreographed in 1877, but known to us in the Petipa-Ivanov version of 1895).

A number of practical inventions, decisions, and developments facilitated the emergence of the aerial ballerina and her ardent partner in these tales of gossamer and gloom. Less than a year after the July Revolution installed Louis Philippe on the French throne, the Opéra ceased to be court property. Dr. Louis Véron, as the new director of what was now a private enterprise with a government subsidy, wished quite naturally to make the Opéra's productions reflect both its new independence and the power of the bourgeoisie that had triumphed the previous July. He wanted that confident middle class in his audiences. They were already flocking to the boulevard theaters to see fairy spectacles and pantomimes, to see plays that laid on Gothic horror—their effects rendered more magical by improved stage lighting and machinery. The astute Dr. Véron could see that the public craved mystery and exoticism, that they would be thrilled to see on the Opéra stage the haunted German valleys and misty Scottish fens that they had long been reading about in ballads by Goethe or Heinrich Heine and in novels by Sir Walter Scott, to see vaporous, beckoning women—firing a man's imagination even as they chilled his flesh with long, pale fingers.

The magical *verismo* of these other worlds offered escape during a period of what must have seemed a dizzying succession of sweeping political changes, particularly in France. The present government's careful middle-of-the-road policies might as easily be swept away. Science was revealing more mysteries than it explained, and religion had lost much of its potency. On the one hand, instability and uncertainty as a condition of life; on the other, a complacent, plodding morality. No wonder that the Parisian public loved to see theater that made enigma and restlessness thrilling, but at the same time tamed it and contained it through theatrical conventions.

Up-to-date lighting equipment transformed the ballet stage into a fitting habitation for sylphs and other ethereal creatures. According to ballet historian Ivor Guest, one of Véron's first innovations at the Opéra was the installation of oil lamps with large reflectors to soften and diffuse the light. For the moonlit cloister act of *Robert le Diable,* he ordered the house lights extinguished. In short, he did everything that could intensify the atmosphere of light and shadow and heighten the effects of trapdoors, wires, veils, explosive powder, smoke machines, waterfalls, and other marvels of stage apparatus.

Gradual developments in ballet technique made possible what was generally considered a new style of dancing, one which was well suited to bring fantasies to life. From Taglioni and other dancers, as gifted if not as innovative, came the lightness and mobility that not only made fantasy flesh, and vice versa, but created a symbol of the unattainable far more profound than most of the ballet plots that made it possible.

Several scholars have wondered whether what we call Romantic ballet was perceived in its heyday as a vital part of the Romantic movement that flourished in painting, sculpture, music, and literature. Were the flittings of these dancers truly "Romantic" in the sense of challenging academic traditions? Was ballet not, they argue, a "juste milieu" phenomenon that, like some of the middle-of-the-road painting of the day, simply applied a patina of Romantic imagery to traditional theatrics and to the same dance techniques that served neoclassicism? They have observed that even the most ardent of balletomanes, Théophile Gautier, thought that "dancing is little adapted to render metaphysical themes . . ."

It is true that ballet choreographers that we consider Romantic exploited and developed a traditional vocabulary, but neither did an arch-Romantic like Byron deviate from traditional poetic forms. And the German poet Friedrich Schlegel viewed Romantic poetry as something that would open up ". . . a perspective upon an infinitely increasing classicism." In all fields of art, it was only pointless academicism that was to be resisted—like the approved genres of painting, and ballet's traditional classifications of male dancers according to physique as *danseur noble, demi-caractère,* or *caractère.*

The choreographers' choice of particular steps within the classical vocabulary and the freer way dancers performed them did indeed "increase" the range of classicism.

Certainly the general public did not attend the ballet for spiritual enlightenment, and it naturally lapped up spectacle and technical prowess. The *Petit Courrier des Dames* correspondent must have alarmed his Parisian readers when he described a sumptuous production of the colorful ballet *La Gitana* in St. Petersburg in 1838. There were, he exclaimed, 500 people in the last act's masked ball, 5,000 candles, and 120 chandeliers: "Is Europe saying the Opéra is no longer the first theater of the world for art and splendor?" When *Giselle* was first seen in London in 1841 in the form of a play—with dances, set to the original Adolphe Adam music, for those dangerous and alluring ghosts, the wilis—a poster advertised in huge letters what was obviously a major attraction: FIRST NIGHT OF THE REAL WATER!

Certainly the leaders of the Romantic movement in literature and painting—the fiercest balletomanes among them—looked down on ballet even as they delighted in it. It was, they understood, an excuse for watching pretty, lightly clad women disporting themselves. Yet everywhere their prose betrays deeper responses. Writing of *Giselle* in *Les Beautés de l'Opéra,* Gautier luxuriates in his description of the opening of Act II, when the heroine has become a wili: "And the rising moon that shows through the slashes of the leaves her sweet, sad, opaline visage, does her transparent whiteness not remind you of some young German girl who died of consumption while reading Novalis?" A contributor to the *Petit Courrier des Dames* compares the Danish ballerina Lucile Grahn both to Grecian maidens and the "enchantresses of German ballads" (certainly a simile in which classicism and Romanticism join hands). And if some have considered Gérard de Nerval insensitive for remarking that Giselle died of loving dancing too much, what more poetic fate could await anyone: the artist dying of excess devotion to art?

I think that the Romantic imagery in these ballets goes quite deep, and, whether spectators of the day realized it or not, the dark and mysterious currents within the plots, the edge of morbidity, the hallucinatory visions drew them to the ballet as much as did the acrobatic feats of dancers and their personal charm.

Everyone may have thought, with Gautier, that ballet was suited to express only passion and amorous pursuit, but it is passion darkened by the Romantic preoccupation with the dichotomy of flesh and spirit. Many of the ballets express the despairing notion that a perfect union between man and woman is possible only beyond the grave. Few supernatural ballets ended happily. In Filippo Taglioni's *La Fille du Danube* and August Bournonville's *Napoli,* the lovers are united on earth because they have proved their incorruptibility—never sullying their ideals, never swerving in their devotion to each other, no matter how many temptations or gorgeous look-alikes are strewn in their paths. Jules Perrot's *Ondine* originally had a happy ending of sorts: the water spirit couldn't exist on earth, and the hero married the girl next door as planned. Yet it seems significant that when Perrot re-created his work in Russia in 1851, he changed the ending: the hero drowned himself in the river and was finally united with his naiad in a watery apotheosis.

The heroes of nineteenth-century ballets behaved according to the Romantic ideal of the hero, of the artist. Customs wearied them, and they would brook no restrictions except those that they imposed themselves. Frequently they cast aside attractive women of the correct rank, amiability, and certified humanness to pursue their chosen sylph, undine, or even a fey and spiritual peasant girl (*Giselle*).

The crucial test to which the hero is often put exemplifies a highly Romantic dilemma. Will he be steadfast to his ideal, his true love, and not be taken in by a beguiling facsimile? When young Rudolph in *La Fille du Danube* becomes demented because his beloved (whom he takes to be mortal) has thrown herself into the river, his friends attempt to distract him by a veiled double who dances almost as well as does his darling. Weathering this deception, he plunges into the river and into a lively throng of veiled naiads, who, like the sea nymphs in *Ondine,* "dance their mazy fascinations around him." But his love has given him a posy—a talisman which helps him to distinguish between truth and illusion. In Bournonville's *Napoli* (1842), the faithful fisherman Gennaro, who has dived into the sea to die with his supposedly drowned Teresina, is confronted with a bevy of beguiling sea nymphs, among whom sports his reluctant, magicked sweetheart. The stage image is a familiar one: the lone man threading his way through swirling flocks of identically dressed females, looking searchingly at each one. It persists through *Swan Lake* and *The Sleeping Beauty* into Balanchine's ballets and such modern fantasies as the scene in the 1937 movie *Shall We Dance?* in which Fred Astaire dances perplexedly down a line of fetching women in Ginger Rogers masks.

True and false can confront the ballet hero in subtler forms too. The peri in the Orientalist fantasy of that name tests Achmet herself by taking over the body of a dead slave; even though he shows good taste in falling in love with her double, the peri doesn't make things easy for him. The tragedy of *Giselle* is often presented as arising from a nobleman's thoughtless dalliance with a peasant girl, but there is another possible interpretation: Albrecht and Giselle are soulmates, made for each other, but issues of class prevent him from recognizing this. Certainly the several great performances given in our time by Gelsey Kirkland and Mikhail Baryshnikov brought this to almost unbearably poignant life.

But although the themes of truth versus illusion, ideal versus real, that pervade so many of these ballets link them persuasively to Romanticism, the choice of subject was not the crux of the matter. According to Baudelaire, anything could be viewed "Romantically," and he cited as characteristics of such a vision, "intimacy, spirituality, color, aspiration toward infinity"—all of which distinguished Romantic ballet.

Realism, an instance of "intimacy" and a feature of much early Romantic painting, was an integral part of the Romantic ballet worlds too, especially those created by August Bournonville and Jules Perrot. Even in supernatural ballets, they prided themselves on the detail of their crowd scenes, the verisimilitude with which they evoked a Highland revel or a Naples dockside or a village festival. The admittedly theatricalized naturalism, with dances justified by parties or festivals, set off the spirit world where dancing was a given and a metaphor for the restlessness of spiritual longing.

The supernatural ballets also had connections with early nineteenth-century landscape painting. Although the popularity of this genre of painting reflected the desire to turn to nature as a constant in a bewildering world, the painter—imbuing nature with his own feelings—brought out its mysteriousness, its changeability. The play of light and shadow over a field could seem to suggest conflict among the powers of nature. The mysterious landscapes of Caspar David Friedrich are tranquil, yet disturbing; Hugh Honour has pointed out how Friedrich occasionally ". . . painted the foreground in great detail, but sank an immeasurable chasm between it and the distant, almost visionary, horizon, tantalizingly out of reach, creating an uneasy mood of yearning for the unattainable." And in Turner's sensuous vortexes of light, nature itself becomes a shimmering dreamworld.

The air, earth, fire, and water spirits of ballet awakened their lovers to the beauty of the natural world, to that landscape glinting beyond the window. In the second act of *La Sylphide,* the sylph flies to a treetop to bring down a nest for James's inspection and offers him spring water in her cupped hands. The scene must have seemed to contemporary spectators almost a visualization of Victor Hugo's popular poem "La Fée et la Péri," in which

Marie Taglioni in *La Sylphide,* Act II. Lithograph by J. H. Lynch from a drawing by Alfred Edward Chalon.

an Occidental fairy and an Oriental one vie for the soul of a dying child. The peri will show him fabled places; the fairy counters:

> *Si tu me suis, ombre ingénue,*
> *Je puis t'apprendre où va la nue,*
> *Te montrer d'où viennent les eaux;*
> *Viens, sois ma compagne nouvelle*
> *Si tu veux que je te révèle*
> *Ce que dit la voix des oiseaux.**

In considering how these ballets were perceived in their day, one must—as always—allow for individual sensibilities. When Hans Christian Andersen saw the "Ballet of the Nuns" in Act II of *Robert le Diable* at the Paris Opéra in 1833, he was overwhelmed by the thrilling atmosphere of death, misty female sensuality, forbidden pleasures, and religious blasphemy:

> By the hundred they rise from the graveyard and drift into the cloister. They seem not to touch the earth. Like vaporous images, they glide past one another. Suddenly their shrouds fall to the ground. They stand in all their voluptuous nakedness, and there begins a bacchanal like those that took place during their lifetimes, hidden within the walls of the convent.

But for every Andersen, there was undoubtedly a Fanny Appleton,† who, after mentioning in a letter the "diabolical music," the "terrible darkness," and the "strange dance," remarked that the members of the corps de ballet ". . . drop in like flakes of snow and are certainly very charming witches with their jaunty Parisian figures and most refined pirouettes."

Marie Taglioni's style of dancing, so enchanting to audiences, was the result, not just of her sensibility, but of changes in ballet technique. The manual of classical dancing produced in 1828 by the La Scala teacher and choreographer Carlo Blasis shows how the turnout of the hips had increased since the eighteenth century, making it possible for dancers to raise their legs higher, to execute more brilliant beats, to change directions more rapidly and more fluidly. Although dancing of the 1830's would probably not strike us as particularly expansive, it would have seemed to nineteenth-

---

*If you follow me, innocent spirit,/ I can teach you where the cloud goes,/ Show you where the waters rise;/ Come, be my new companion/ And I'll reveal to you/ What the birds are saying.

†Later Mrs. Henry Wadsworth Longfellow.

century balletomanes much freer and larger in scale than what they had seen around the turn of the century.

In this development, fashion played a role. When the *ancien régime* toppled, with it fell the ponderous, ornate, and constricting clothes that went with rank and power. The soft slippers and light, loose-fitting dresses and Grecian draperies that came with, or just after, the French Revolution enabled women dancers to increase the range of their movements significantly. By the time corsets returned, the dancers had already changed, and they never looked back. Even before *La Sylphide,* Taglioni was dancing onstage in simple light dresses similar to those her mother made for her to practice in. As historian Marian Hannah Winter has remarked, fashion freed dancers' thighs. Ecstatic reviews of the way Taglioni bent from side to side confirm that it made possible some freeing of the torso as well.

Pointework, so crucial to the image of the supernatural female, was not a new technique in 1832, although it wasn't standard equipment for all female dancers when Taglioni came to the Paris Opéra. (The 1830 edition of Blasis's manual doesn't even mention it.) Geneviève Gosselin, people remembered, had danced on her toes in 1815, maybe earlier, and "grotesque" (meaning acrobatic) dancers of both sexes did pointework. It was viewed more as a feat than anything else—and often seems to have been performed as one. When Taglioni saw Amalia Brugnoli go up on full pointe in Vienna in 1823, the year after her own debut there, she found the hoisting motions Brugnoli had to make with her arms distinctly unlovely.

Engravings of the little company that Filippo Taglioni assembled for the Opera in Stuttgart later in the 1820's show all the women on pointe. It is during those years that his daughter must have worked on perfecting her own approach—discovering ways to strengthen her feet and rise onto her toes without apparent effort.

Some of the impetus for pointework might have come from the elaborate "flying" techniques developed by Charles Didelot. His *Zéphire et Flore* astounded London balletgoers in 1796, St. Petersburg audiences in 1804, and Parisians in 1815 (when Mlle. Gosselin reputedly stood on her toes). It was still a favorite when Taglioni performed it in Paris with Jules Perrot in 1830. Didelot didn't simply fly down some heavenly personage on a cloud to cinch the plot. Any of his dancers might fly by means of individual wires and harnesses. They could take to the air, or be carefully lowered until only the tips of their toes touched the stage floor. For imaginative choreographers and ambitious dancers, it must have seemed natural to wish to echo that effect in passages where it wasn't possible to attach someone to a wire.

Circumstances, then, conspired to produce a style of dance and stage machinery ideal for supernatural subjects. Marie Taglioni's own attributes, her classes with Auguste Vestris in Paris, and her father's taste and inspired coaching defined that style. Not all dancers copied her purity, her coolness, or her deemphasis of athleticism, but they studied her delicate attack, her simplicity, the fluidity of her arms and torso. For example, it was remarked—as if it were a novelty—that she often held her long arms down, gently curved, instead of flourishing them about; one characteristic of the style that August Bournonville perpetuated in Copenhagen is just such a *port de bras* for jumps. It tends to make dancers look lighter and to focus the audience's attention on their nimble feet.

Dancers have always been praised for "lightness," but from the 1830's until late in the century, variants of the term saturated the metaphors and similes of writers on ballet. Light as weightlessness, light as luminosity; in English the same word serves both meanings. But even in languages where the words differ, the meanings intertwine in descriptions of ballerinas. *The Times* in London described Adèle Dumilâtre's dancing as being "so ethereal . . . that she almost looked transparent." The delight caused by airy and

Lucile Grahn in *Eoline, ou la Dryade,* by Jules Perrot. Lithograph by Edward Morton from a drawing by S. M. Joy.

seemingly effortless dancing, set off by mysterious lighting and gauzy, billowing skirts, seems related to the century's uneasiness about the flesh. Praising the lighting for Perrot's *Eoline,* Gautier raved that it gave the illusion that "Eoline is only the envelope, the transparent veil of a superior being, a goddess condemned by some fate to live among men." The "condemned" is telling, and it's interesting that it comes as if automatically from the pen of a cheerful hedonist like Gautier. Insubstantiality, then, is close to godliness.

Lightness in the sense of airiness complements the notion. The buoyancy of the female dancer helped her to embody a spiritual aspiration; the lightness of a male dancer suggested the hero's desires to transcend the limits of the flesh. It could also stand for the winging of his soul as he took on some of the qualities of the sought-after dream. Even the constant motion, the restlessness for which dancers were admired, can be seen as a dissatisfaction with present existence and a yearning for realms beyond. Too, in the Romantic era, the artist-as-rebel was a favored image. By their apparent denial of gravity, the sylphs and their kin prettily demonstrated their exemption from laws governing human behavior. So female dancers were enthusiastically compared to birds, butterflies, balloons, feathers, moonbeams, shadows, and criticized for showing too much vigor or attack. Male dancers, generally disprized during these years, succeeded the more they resembled the women in terms of style. It was "the aerial Perrot, Perrot the sylph, Perrot the male Taglioni." One English critic said flat out that "dancing excels in proportion exactly as it resembles flying."

The lithographers intensified the public's fantasies of supernatural heroines. The ballerinas—even when depicting real if exotic women—seem unstable, elusive. They hover on one improbably dainty toe, not in perfect equilibrium, but leaning slightly forward as if they're just passing through the pose. When shown in midair, their bodies are softly curved, legs barely apart—less as if they'd leapt than as if they'd been blown upward. Some of the poses may be artistic conventions rather than ballet reality—Carlotta Grisi in *La Péri* and Arthur St.-Léon in *La Esmeralda* are pictured by J. Brandard in exactly the same arched jump (a *jeté* surely; see page 53)—but the conventions were dictated by the artists' perceptions of the ballets. And as refinements of dance technique helped create the airborne images onstage, techniques in lithography developed as if in response. As Charles Rosen and Henry Zerner have pointed out, drawing on the surface of the stone made possible subtle nuances of tone and images that "vanish at the edges." The dancer becomes the glistening focal point in an evanescent and cloudy world.

Carlotta Grisi and Lucien Petipa negotiating the famous leap in Act I of *La Péri* by Jean Coralli (1843). Lithograph by Bertauts from a drawing by Marie-Alexandre Alophe.

She may have been an abstraction, but she was unmistakably female. In expressing the ambiguous tensions they felt between reality and spiritual longing, the ballet librettists and choreographers—almost without exception male*—revealed confused emotions in regard to women. On the one hand, almost all the stories were told from the hero's point of view; on the other, the ballerinas dominated the stage. Even in ballets where the hero dreamed the heroine, she was clearly superior to him, and freer than he was—enchanting, evasive, unrestricted by his codes, and able to drift about on her toes. Yet she flew into his arms of her own accord. In the thrilling, much-discussed dream sequence in *La Péri,* Carlotta Grisi leaped from the framed platform that contained "her world" and was bravely caught by Lu-

*Some of the exceptions are Lucile Grahn; Fanny Elssler's sister Therese; and Marie Taglioni, who choreographed *Le Papillon* in 1860 for her protégée, Emma Livry.

cien Petipa. The audience marveled over his strength, but was more excited by her daring, and the *Times* critic commented of some lifts in the same ballet: "She is supported by Petipa, but seems as if supported by air alone."

Only occasionally in these ballets does a male character obviously dominate a female one. Jules Perrot, a compelling performer as well as a brilliant choreographer, created intriguing demonic roles for himself in several ballets. As the Prince of the Gnomes in *Eoline,* he mesmerized the dryad heroine into an intoxicating and exhausting "Mazurka d'Extase." August Bournonville, working in Copenhagen relatively isolated from Parisian fads, refused to cater to the general dislike of male dancers; like Perrot, he was a good dancer and wanted to perform. He was outraged that Fanny Elssler evidently preferred her sister Therese to a male partner and would make her ballet lover sit out a *pas de deux* so that she could besport herself "with her tall sister, who always appeared as if she had just fallen from the clouds." The fairy worlds were, as a rule, unbalanced in their populations. Gautier explained that Myrtha, the Queen of the Wilis, had only female subjects, since men were ". . . too heavy, too stupid, too in love with their ugly hides to die such a pretty death."

These rather liberated creatures imagined by men stood opposed to the respectable middle-class wife and mother (also, to some extent, a male creation). In *La Sylphide,* James's betrothed, Effie, a woman of real weight and substance, with sensible shoes and domestic talents, is far less vivid than her rival. Despite their supposed purity, the supernatural creatures are femmes fatales, representing all that is erotically potent and compelling about women. They offer a double message, beckoning the hero both as the incarnation of an ideal and as a temptress luring him from the straight and narrow. It isn't for nothing that Erik Aschengreen called his fine monograph on Romantic ballet "The Beautiful Danger."

"Laughing terribly, irresistibly beautiful, the wilis dance in the moonlight . . .": this is how Heinrich Heine wrote of the betrayed maidens whose spirits dance unwary men to death. It is those wilis who came compellingly to life in *Giselle.* And of Ondine, Philarète Chasles cautions:

> Fear to fix your gaze on her: you will be bound by an invincible chain; you will follow her into her damp and marvelous palace, you will be chained with diamantine bonds at the bottom of this unknown and invisible world from whence you will never return.

Yet what Romantic hero worth his salt would hesitate?

Balletomanes were awed by the effort it took to appear effortless. In 1839, when Lucile Grahn first performed *La Sylphide* in Paris, the house held its breath to see if she would do Taglioni's "terrible pas" of Act II, which Elssler had cut: "C'était une question de vie ou de mort." And what applause when she triumphed in it! Because the preferred steps for spirits were bounding ones, the labor that went into becoming ethereal was considerable: the quantities of *ballottés, brisés, temps-levés, emboîtés, assemblés, cabrioles, sauts de basque,* and *ballonés* that packed the dances required strong ankles and good wind. It was thanks to Amalia Ferraris's "supple and sinewy foot" that she was able to "beat the *entrechat huit* to perfection."

In the days before the blocked pointe shoe was perfected, hovering on tiptoe required immense strength. Today's dancer, wearing a blocked and stiffened slipper, stands, in effect, on her toenails, on the very tip of her toe. The vertical equilibrium is so secure that, once up there, she is almost at rest. The ballerina of the mid-nineteenth century was stepping as high onto the toe pads as possible, and could stay there only by exertion of all her leg muscles and a tremendous lift in the body. Yet in 1846, in *Paquita,* Carlotta Grisi thrilled balletomanes by fancy hops "on the tip of the toe with a turn of dazzling vivacity . . ."

In addition, female dancers, along with the men, had to be skilled at balancing on the flat foot or half-toe for extended periods of time. The choreography of the day featured elaborate *adagio* sequences that are uncommon now except in the Bournonville repertory. Standing on one leg, the dancer would revolve smoothly, make one pretty pose metamorphose into another, bend forward or back—all without wobbling. Taglioni is said to have worked for two-hour stretches three times a day with her father while preparing for her debut. Classes, apparently, were grueling. Léopold Adice's syllabus of 1859 lists a barre in which exercises are performed one hundred times each. Bournonville dancers hoisted their legs in *grands battements* a total of 320 times. To be secure in *adagio,* a dancer might work at holding one leg in the air for one hundred counts, as Marie Taglioni did. Louise Fitzjames had her maid stand on her hips to increase turnout, and Carlotta Grisi said sourly that those times Jules Perrot stood on her hips while she lay face down on the floor with legs spread were the erotic high points of their liaison. Beginners forced their turnout by standing in a box with braces that could be adjusted via a series of grooves. (No wonder Marie Taglioni was easily able to pass off one of her pregnancies as knee trouble.)

In this drawing from Édouard de Beaumont's *Album Comique: l'Opéra au XIX Siècle,* the ballet master tells the young woman using the turnout machine that to chase fortune successfully, a dancer has to start by breaking her legs.

The prose devoted to ballerinas during the heyday of Romanticism makes it clear that spectators found the paradox of a real, and probably available, woman playing an incorporeal nymph a titillating one. Gautier lingers lovingly over descriptions of ballerinas' knees, ankles, breasts, noses, and chins. Fanny Cerrito "knows how to curve and soften her plump arms like the handles of an ancient Greek vase." Classical allusion aside, the sentence is unabashedly sensual.

Besides the enraptured accounts, a beguiling detritus of poems, lithographs, curios, and sheet music attest to the fervor dancers inspired. The *galop* from Act I of *La Fille du Danube* "invaded every ballroom," says critic André Levinson, and the sheet music of the "Bridal Waltz" from *A Folk Tale* (1854) could be found in the piano benches of Danish households. Portraits of the most famous ballerinas of the day assumed curious forms. A 1984 exhibit presented by the Theater Collection of the Austrian National Library contained the following mementos of the wildly acclaimed Fanny Elssler: a porcelain cast of her left hand, her right foot sculpted in marble, portrait medallions naming her "Terpsichore's Darling," images of Fanny and her sister Therese as sylphides painted on a cup and on a pipe, an amber and meerschaum cigarette holder and case in the shape of Fanny's leg. There was a tiny, winged enamel Fanny on a ladies' watch pin, a porcelain Fanny in costume for her famous (and very much of this world) Spanish "Cachucha" kneeling on an inkwell, a Fanny in metal relief dancing Perrot's *La Esmeralda* on the side of a lamp base. In terms of the souvenirs and verbiage they generated, dancers—female ones, at any rate—were the rock stars of the nineteenth-century bourgeoisie.

For the unearthly heroines of the ballets, physical union with a mortal posed usually fatal danger. Remember what happened to the sylph? Alma, the "daughter of fire," a statue who comes to life by day, faces a dilemma even more vexing: if she falls in love, she will become a statue forever. (Seldom was an audience more tantalized by the prospect of a lovely idol falling off her pedestal.)

By these standards, the women dancers of the nineteenth century were not very sylphish offstage. Salaries for all but the stars were not high; a well-to-do protector, not hard to come by, was considered by many to be a necessity. Members of the Jockey Club frequented the green room and backstage areas of the Paris Opéra; at Her Majesty's Theatre, the bloods

"Rats d'Opéra," one of the lithographs in Gustave Doré's *La Ménagerie Parisienne* (1854)

of London could obtain seats in the "omnibus boxes" on the sides of the stage—the better to ogle, and perhaps to pinch and pass messages. Benjamin Lumley, the director of this theater when *Ondine* was premiered there, related that backstage one evening when Fanny Cerrito was dancing the lovely "Pas de l'Ombre," frolicking on the seashore with her newly acquired shadow, Adeline Plunkett aimed a kick at Elisa Scheffer, her rival for the favors of the Earl of Pembroke, missed, broke the cord holding the "moon" lamp, and temporarily extinguished Cerrito's dance.

Sometimes, of course, dancers married other dancers or formed liaisons with them—as did Jules Perrot and Carlotta Grisi, Arthur St.-Léon and Fanny Cerrito, Fanny Elssler and (briefly) Anton Stuhlmüller. However, these relationships in no way exempted female performers from the solicitations of others. Young corps dancers were particularly anxious to secure wealthy protectors or husbands. Out of their meager salaries—often made even smaller by fines levied for various infractions—they had to pay for their daily classes, obtain practice clothes, scheme for advancement. Many came from poor families. In *Les Petits Mystères de l'Opéra,* Albéric Second's satirical look at the backstage world of the Paris Opéra, one *petit rat* wears a capacious pocket under her sylph costume, into which she packs useful objects she's picked up, including a pack of cards, five or six cigar butts, a squeezed half lemon, some cheese, a scrap of soap, and a necklace. Furthermore, she says, the bulging pocket gives her a "Spanish shape" pleasing to the gentlemen in the stalls. The same girl relates how the dancing master, Cellarius, lures *coryphées*—who rank above the girls of the quadrilles—to come to his place the three times a week when they're not performing and partner gentlemen who are ostensibly learning to waltz: five francs to dance with a chair, ten to dance with a *figurante* at the Opéra. Supper on the town afterward, where a girl can gorge . . .

Given the lack of birth control, it's not surprising that many female dancers became mothers. Ballerinas often danced well into their pregnancies. Sometimes their offspring accompanied them on their numerous tours or guest appearances. More often, the babies were brought up by grandparents or aunts or friends. "Well, Fanny, send the brat to me," Elssler's English friend Harriet Grote wrote cheerfully, when Fanny decided not to take her seven-year-old Therese to America. It was four years before Fanny retrieved her daughter.

The nineteenth-century female dancer would probably not have struck us as looking ethereal, considering her diet, childbearing, and the kind of muscles she had to develop. Fanny Elssler, to judge from her pink satin and black lace "Cachucha" costume, was a woman of medium height with a trim, but not tiny waist and a full, curving bust. A sylph could hardly be ethereal enough in her dancing, but the woman who played her could be too ethereal to suit public taste. Gautier couldn't abide shoulder blades that stuck out ("two bony triangles that resemble the roots of a torn-off wing"—the analogy is revealing). Poor Louise Fitzjames was constantly criticized for her thinness. A caricaturist presented her as a dancing asparagus, and Gautier said that she wasn't "even substantial enough to play the part of a shadow."

Being substantial enough to play a shadow . . . It might be considered the mission of the Romantic ballerina.

Most of the supernatural ballets of the early nineteenth century—along with a host of other ballets of the period—have perished. The only works in this genre that can be experienced today are *Giselle* and the Danish August Bournonville's version of *La Sylphide,** his *Napoli,* and *A Folk Tale* (in the last two, the heroines were not supernatural, but were temporarily in thrall to supernatural forces). All have been altered to some degree, and the *Giselle* we see today was largely rechoreographed by Marius Petipa in 1884. The poetic image of a mortal man lured by a filmy female vision into a magical world didn't perish with the decay of Romanticism, however. Transformed by new ideas and new styles in dancing, it bloomed again in St. Petersburg, and years later in London and New York City. It is with us still.

In 1854 an Englishman recollected Taglioni in *La Gitana,* fifteen years before. At the end of her final solo, she alighted from a passage of leaps and paused on her pointes, looking back over her shoulder. "It was," he said, "like a deer standing with expanded nostril and neck uplifted to its loftiest height, at the first scent of his pursuers in the breeze. It was the very soul of swiftness embodied in a look! How can I describe it to you?"

Dancers of the nineteenth-century supernatural ballets were among the first to embody abstract qualities, which not all of the spectators who flocked to adore them recognized. These performers didn't *represent* Beauty or Music or Fecundity as had their counterparts in earlier centuries, yet their light, fleeting, ardent dancing could suggest something larger than their stage personas and more ineffable than the roles they played.

*Pierre Lacotte mounted a controversial reconstruction of Taglioni's *La Sylphide* at the Paris Opéra in 1972, with all the women on pointe.

Marie Taglioni in *Le Dieu et la Bayadère.* Lithograph by Nathaniel Currier from a drawing by Napoleon Sarony.

HEROISM IN THE HAREM

She was Persian perhaps, or Indian, or Moorish. Although she was almost certainly in thrall to some man—a denizen of a pasha's harem or a temple dancer subject to a high priest's whims—she fought heroically for her chastity and her right to choose a mate. When ruse or defiance failed, her own death became the ultimate weapon and bargaining point. Spirited, provocatively beautiful enough to put that chastity in jeopardy, she usually won in the end—not only the hand of her lover in marriage (or, if all else failed, union beyond the grave in a stunning apotheosis), but the hearts of all spectators. Whether *kadine* or *bayadère,* she wore beribboned cloth slippers and danced on her toes. Her costume was a full-skirted dress with a nipped-in waist and a bodice that bared her bosom and shoulders; for whatever else she might be pretending to be, she was, above all, a ballerina of the nineteenth century, the sylph's more accessible cousin.

The ballet landscapes she roamed were the domain of pageantry, opulence, mystery, barbaric customs, and ravishing women. With the advent of Romanticism, choreographers and ballet librettists—like artists in other fields—turned to the East, much as they did to supernatural worlds. The Romantic temper bred images more seductive than those in the ballet dramas of the early nineteenth century and far removed from the decorative Orientalism of the eighteenth century: the nobility-in-turbans dances—as in Jean Philippe Rameau's *Les Indes Galantes*—or cavorting Siamese "grotesques" or the porcelain figures come to life in Jean Georges Noverre's *Les Fêtes Chinoises*.

Authenticity was never an issue. Although Westerners have been fascinated by the East for centuries, what one writer has referred to as "Europe's collective daydream of the Orient" has not always depended on a familiarity with Eastern customs or an understanding of Eastern philosophies. More often "The Orient" has designated a pleasure garden for the imagination—an Orient restructured to fire European longings and justify European conquest.

During the first part of the nineteenth century, the East on which these visions were based rarely included distant China and Japan; it was a nearer, more comprehensible Orient that clustered around the Mediterranean and flowed enticingly into the Caucasus, Persia, and India. The Orient began just east of Venice: present-day Yugoslavia, Albania, and Greece were all part of an exotic Ottoman empire. Spain, with its strains of Moorish culture and its Gypsies, was as "Eastern" as Morocco and Algiers.

The public's interest in the ancient cultures of Egypt and Syria was roused by the findings of archaeologists and other scholars who accompanied Napoleon's armies into those countries in 1798. Later, Greece's heroic struggles against the Turks stirred European hearts and engendered a rash of art and literature. By the middle of the nineteenth century, most of the major works of Asian literature had been translated into various European languages; among these were Kalidasa's play *Shakuntala* and *The One Thousand and One Nights*—the latter to be ransacked by composers, playwrights, choreographers, and vaudeville impresarios for years to come.

"Never," wrote Victor Hugo in the preface to his 1829 volume of poetry, *Les Orientales,* "have so many intellects excavated the great abyss of Asia." In some, the riches of the East excited ambitions for empire; in others, tales of exotic splendor and curious customs spiced escapist dreams; for the nineteenth-century artists, excavating the Orient meant digging into their own psyches. Eastern mysticism held a possibility of spiritual regeneration to those soured on Christianity. To those put off by increasing industrialization, the East offered visions of unspoiled nature, of vital, sensual temperaments free of the constraints of Western civilization. In the wildness, the willfulness, the strangeness, the cruelty of what they saw or read about, they recognized the contours of their own interior landscapes and, in turn, projected onto it their inchoate longings. As Raymond Schwab has written, "The more superficial used the Orient primarily for costumes, the more profound as fancy dress for the soul."

Some, like Goethe or Hugo, were armchair voyagers, devouring the profusion of travel literature; others, like Flaubert, Byron, Delacroix, and de Nerval traveled in the East, adjusting the discrepancies between the real

Orient and their private visions of it. "For me," wrote de Nerval ruefully to his friend Théophile Gautier, "a lotus is only a kind of onion." Yet even while seeming to submit to the East, they possessed and transformed it as thoroughly as did the colonizers. Delacroix filled sketchbooks with beautifully detailed drawings of Moroccan buildings and people, but a great anecdotal canvas like *The Massacre at Chios* is wholly imagined, invested with his own vision—both personal and political—of disorder, random violence, and loss of focus.

The public who filled the theaters and read the books and stared at the paintings were thrilled by this exotic East. Hugo's *Les Orientales* went through fourteen editions in a month. Byron's epic poem *The Corsair,* the author's publisher was pleased to inform him, sold ten thousand copies the first day it was printed, in 1814.

The fictional East was a cultural marketplace, in which artists borrowed freely from each other. Cultivated people of the day could easily become expert in its lore. Looking at the several paintings that Delacroix based on Byron's long poem of revenge *The Giaour,* they would have understood the import behind, say, a scene in which two enraged men in billowing robes rein their plunging horses close to each other the better to wield their daggers. This public would have known how closely Albert's ballet based on *The Corsair* followed Byron's scenario and how Joseph Mazilier's version diverged from it. Those who saw Carlo Blasis's pre-Romantic ballet *Mokanna* at La Scala or Jules Perrot's *Lallah Rookh, or The Rose of Lahore,* or heard Robert Schumann's *Paradise and the Peri* would undoubtedly have known that all three were drawn from episodes or tales in Thomas Moore's hugely popular Orientalist odyssey *Lallah Rookh.*

They judged these works more in relation to each other than to any real East. Some of those who saw the ballet *La Péri* (1843) at the Paris Opéra might even have noticed that one backdrop—depicting the view of Cairo from the hero's harem—was modeled on a painting by Georges Antoine Prosper Marilhat, but they wouldn't have questioned librettist Théophile Gautier's reasons for locating this tale of a Persian fairy in Egypt. As Thomas Moore's biographer Howard Mumford Jones said, speaking of Moore's *Lallah Rookh,* "What matter if to the general imagination India and Egypt, the Turks and the Parsees, the Bosphorus and the Vale of Cashmere were indistinguishable parts of a vague, rich universe of dreams?"

Le sérail! . . . Cette nuit il tressaillait de joie.
Au son des gais tambours, sur des tapis de soie,
Les sultanes dansaient sous son lambris sacré;

Et, tel qu'un roi couvert de ses joyaux de fête,
Superbe, il se montrait aux enfants du prophète,
De six mille têtes paré!

Livides, l'oeil éteint, de noirs cheveux chargées,
Ces têtes couronnaient, sur les créneaux rangées,
Les terrasses de rose et de jasmin en fleur!
Triste comme un ami, comme lui consolante,
La lune, astre des morts, sur leur pâleur sanglante
*Répandait sa douce pâleur.**

So begins the second part of Hugo's "Les Têtes de Sérail"—a poem lamenting a Greek defeat at the hands of the Turks. Luxury and remorseless cruelty, blood and flowers, life and death temporarily resolve in a rueful, yet voluptuous vision.

But such a vision could not be fully realized by nineteenth-century balletmakers. A certain prudery—both genuine and hypocritical—militated against it. "It is the fashion now to be virtuous and Christian; people have taken a turn for it," jeered Gautier in the preface to his sensational novel *Mademoiselle de Maupin*. In no theater would the public be likely to see such savage male grace as that depicted in Henri Regnault's painting of a black man, bloody sword in hand, standing over a decapitated body, or the voluptuous naked women being put to the sword in Delacroix's *The Death of Sardanapalus* (see page 120). Painted or described, such visions were tolerated; alive and moving onstage, they would have been threatening to morality.

Too, the earlier public expected to see the dancer's virtuosity in *ballon*, in the sensitive use of pointe technique that Marie Taglioni had made the new standard. At the Opéra, they expected some of the order and logic that had dominated the neoclassic ballets of the first quarter of the century.

---

*The harem! . . . Tonight it quivered with joy.
To the sound of gay drums, on silken carpets,
The sultanas danced beneath its inviolate roof;
Like a king superbly decked with festal jewels,
It displayed itself to the children of the prophet,
Adorned with six thousand heads!

Livid, eyes extinguished, heavy with black hair,
These heads, ranged along the battlements, crowned
The terraces flowering with rose and jasmine!
Sad as a friend, as consoling as he,
The moon, star of the dead, poured her pale sweetness
Over their bleeding pallor.

Later, in the St. Petersburg of Marius Petipa's heyday, mandatory displays of supported pointework and divertissements in classical style made realism-in-fantasy even less possible. Choreographic Orientalism, then, lay primarily in the subject matter and decor, in certain beguiling possibilities for dances and tableaux, and in the roles assumed by the dancers.

Choreographers doled out exoticism in judicious doses. The backdrops were often modeled after paintings by minor Orientalists, but the music was thoroughly Western—tuneful and buoyant.* While costumes for male dancers and some of the secondary women might be splendidly, if inauthentically, Oriental, the ballerina always wore her bouffant knee-length or calf-length tutu, garnished, perhaps, with fugitively "Eastern" trim. Carlotta Grisi's harem costume for *La Péri* (in which she played both the fairy of the title and Leila, the dead slave whose form the peri assumed) consisted of a full skirt and a pale blue bodice cut like an Indian dancer's *choli* to bare the midriff, but with the space between skirt and bodice filled in with flesh-colored net.

Carlotta Grisi in *La Péri*. Lithograph by John Brandard.

*Jules Perrot's *Lallah Rookh* seems to have been an exception. Although some of the music was written by Cesare Pugni, the middle section was set to Félicien David's *Le Desert,* which contained Middle Eastern motifs.

However, for all their charming *pas* and happily-ever-after plots, the ballets had the power to evoke for their audiences the darker, crueler, more sensual Orient of art and literature. In spite of Gautier's dismissal of ballet as a vehicle for serious themes ("It is not easy," he complained, "to write for legs."), the letter he wrote to Gérard de Nerval describing his ballet *La Péri* wafts in a vision worthy of Delacroix:

> In the interior of a harem with marble columns, mosaic pavements, fretted walls like lace, in the midst of perfume rising in clouds, jets of water falling like pearly dew, a handsome young man, rich as a prince of *The Thousand and One Nights,* nonchalantly dreams, his elbow deep in a lion's mane, his foot resting on the throat of one of those Abyssinians whose skins are always cold, even in the gust of the fiery desert wind: a kind of oriental Don Juan, his senses satiated, but not his desires.

Orientalism provided nineteenth-century ballet with a body of themes and motifs. You find them in both early Romantic works, like Filippo Taglioni's *The Revolt in the Harem* (1833), and ones made toward the end of the century, like Marius Petipa's forgotten *Zoraiya, or The Moorish Woman in Spain* (1881). Playwrights, librettists, and vaudeville producers appropriated some of the same fantasies. For almost a century, the stages of opera houses and popular theaters—in America as well as Europe—teemed with enslaved heroines, treacherous rivals, disguises, fateful talismans, lovers offering to sacrifice their lives or their purity for each other, intrigues, threats of hideous punishment, opium dreams, spectacular scenic effects, and, of course, dances galore.

New plots had to be concocted and old ones reshaped in order to make the harem a domain of action—a thrilling place as well as a picturesque one. Gautier's scenario for *Sacountala* (staged in 1858 by Marius Petipa's brother Lucien) dispensed with the religious elements that imbued Kalidasa's play, made the heroine a temple dancer, condensed the action of years into a few days, and added a vengeful female rival to thwart further the heroine's claims to the love of King Dushmata—thus transforming, as Gautier's biographer Edwin Binney remarks disapprovingly, ". . . a long psychological study into an agreeable fairy tale." The version of Byron's *The Corsair* concocted for Joseph Mazilier's ballet in 1856 is the one that has persisted, through many reworkings, into the twentieth century. This may be because Mazilier and his collaborator, Jules-Henri Vernoy de Saint-Georges, had the sense to see that if the heroine, Medora, is the beloved wife of Conrad the pirate (as she was in Byron's poem and in Albert's 1837 ballet) and plays little part in the action except to die at home of natural causes, the ballet public is being denied some potential thrills. (At the same time, it is of necessity denied the complex moral dilemmas Byron provided for the hero, since these don't translate easily into ballet terms.) In Mazilier's scenario, Medora is a comely slave girl with a flair for dancing who takes Conrad's fancy in the marketplace just as she is being sold to the lustful Seyd Pasha. Now, along with daring raids, treachery, capture, and escape, the audience can revel in the lovers' plight: will Medora yield to the Pasha to save Conrad from execution for piracy? Can Conrad accept such a sacrifice? Will the ruse of the heroic concubine Gulnare succeed? (She may. He can't. It does. And after a stupendous shipwreck, the audience is gifted with a happy ending.)

If the roles of harem woman and temple dancer were favored, it was not only because of the aura of glamour and intimations of promiscuity that surrounded them, but because of the extra piquancy the characters' status as chattels gave to their bravery and resourcefulness. The heroines of this strain of ballets are usually dominated, or about to be dominated, by a man whose rule is malign. As the ballet audience knew—with a frisson of horrified delight—Oriental girls who disobeyed their masters were likely to be burned alive or sewn up in a sack and thrown into the sea. The melodramatic complications in the heroine's pursuit of happiness dwarfed the hero's more manageable dilemma of love versus duty, or "my property versus your property"; he was a free man to start with—often a prince, a pirate, or a bandit, subject to few restraints beyond his own code of honor. And only in the Jean Coralli–Théophile Gautier *La Péri* (1843) and the Marius Petipa–S. N. Khudekov *La Bayadère* (1877) were the heroes the languid and perturbed dreamers we associate with the Romantic movement—men whom no reality can truly satisfy.

The courage and initiative that these Orientalist ballerinas display is remarkable. In the ballet *Brahma,* which Hippolyte Monplaisir created at La Scala in 1868 for Amalia Ferraris, there is a scene in which the heroine imperiously reveals herself as a priestess—shielding the sleeping Brahma from his enemies. In a critic's view of the great dramatic ballerina Virginia Zucchi as Padmana, "The fierce, expressive eyes that shine like rubies in the dark, the sculptured pose of the body, the imperiously regal gesture transport the spectator into the azure regions of the ideal, into a fairyland among the passionate sultanas of the *Thousand and One Nights*." The dreamed-up Oriental woman, fired by passion, was apparently capable of a daring beyond that of the proper European lady, and certainly beyond that of the real sultana, whiling away the long hours of her indolent and restricted life.

Consider, too, an ancestor of *Brahma, Le Dieu et la Bayadère* (known in England as *The Maid of Cashmere*). Concocted in Paris in 1830 by composer Daniel François Esprit Auber, the well-known playwright and librettist Eugène Scribe, and Filippo Taglioni, the work was designed to show off the seductive spirituality of Taglioni's daughter, Marie. The structure—an opera-ballet with a tenor for the hero—was hardly innovative, but the heroine, however victimized, did not behave like the acquiescent, sorrowing victims of early nineteenth-century ballets, and she played a more fiery role than the heroine of Goethe's ballad "Der Gott und die Bajadere," the source of the plot. The poem tells of the god Shiva, wandering the earth in disguise to learn the ways of men. A temple dancer (and hence, in the Western view, a prostitute) offers him a true and unselfish love, and when he dies—or seems to—she casts herself on his funeral pyre. By this act, she is redeemed from her moral waywardness and ascends to heaven in the arms of the god. The opera-ballet was the first new work staged at the Paris Opéra after the July Revolution of 1830, and, while it cannot be considered overtly political, it does show a heroine determined to resist a tyrannical government official—in this case a judge who abuses his power. Because the dancer, Zoloé, spurns the judge, mischievously pointing to a handsome stranger (now, for some reason, Brahma in disguise, rather than Shiva) as the kind of man she might fancy, the enraged judge sentences her to imprisonment. The Unknown heroically takes Zoloé's punishment on himself, and she, not to be outdone, offers herself to the judge in exchange for his freedom. (A mortal hero would refuse to accept her sacrifice, but Brahma is testing the power of the heroine's love, and the bargain is struck.)

Soldiers enter and offer a reward for information leading to the capture of a certain "stranger," who faces execution. Anyone found sheltering him will be punished, but when Zoloé realizes that the one they seek is the mysterious man she has already saved once, she unhesitatingly spirits him away and hides him in her hut, where she vows her love to him. When soldiers arrive, she points him toward a secret cave and will not reveal his whereabouts.

Cyril Beaumont's description of the end of the ballet is not an eyewitness account, but it captures the mixture of mystery, terror, spectacle, and heroism that the final apotheosis must have conveyed:

> The captain orders her to be seized and burnt. Soldiers chop the door into pieces and build a pyre on which Zoloé is placed. The wood is fired and immediately thunder rolls and lightning flashes. Zoloé is surrounded by leaping flames when the Unknown, now seen to be the God Brahma, shining like the sun, appears at her side and takes her in his arms. Together they rise toward the clouds where gleams the Indian paradise. The clouds close over them and the lovers pass from view.

This dramatic and heartrending spectacle must have been a success: it was performed at the Paris Opéra on and off for thirty-six years, and August Bournonville created a version of it for the Royal Opera House in Copenhagen.

The ballet heroines, with their exotic and perfumed names—Zoloé, Zulma, Gulnare, Medora, Nikiya, Niriti, Nisia, Leila, Padmana, Sacountala—were not only brave, they were endowed by their male creators with certain of the qualities usually reserved for Romantic heroes. The paradox undoubtedly arose from the fact that, although women were second-class citizens, ballerinas took precedence over male dancers in the esteem of public and critics. In acquiring time onstage, they acquired power as well. I don't mean to imply that these pretty, essentially simple heroines resembled the tormented and skeptical protagonist of Georges Sand's scandalous novel *Lélia* (1833), who dominated the men in her life and appropriated for herself all the characteristics of the Romantic hero. The ballet women did, however, fight for their liberty, assert the primacy of their desires, and refuse to be ruled by any restraints but their own will, even though, because they *were* women, their will was to submit themselves to the mate they chose, and "liberty" simply meant married happiness. The eponymous stolen-by-Gypsies-at-birth heroine of Mazilier's *Paquita* (1846) will not dance when the Gypsy chieftain, who virtually owns her, orders her to; she will do so when she understands that her dancing will please Lucien d'Hervilly and decides that she wishes to please him. Even in Filippo Taglioni's somewhat atypical *The Revolt in the Harem,* the heroine does not question the institution of the harem. She leads her fellow concubines in rebellion in order to protest the treachery that Mahomet, the king of Granada, has employed in annex-

Adèle Dumilâtre and Eugène Coralli in Act II of Joseph Mazilier's *La Gipsy.* Lithograph from a drawing by Achille Devéria.

ing her, and to assert her right to marry her betrothed. In the end, the liberated women (who've now learned how to shoot) agree to serve the king as troops—and presumably as paramours as well—if he'll treat them with respect.

To the male balletomane of the nineteenth century, the Orientalist heroine's blend of boldness and submissiveness was no paradox, but a compliment to his gender: that a woman so potent would inevitably submit to a man's domination added to the man's own strength. Nor was her courage seen as comparable to the manly variety. Lord Byron, traveling through Spain early in the century, was pleased by the demure appearance of Augustina, the Maid of Zaragoza, who, during a battle of the Napoleonic Wars, had reputedly stepped over the dead and dying of an entire artillery unit in order to fire a cannon and rouse the citizens. His lines on her in *Childe Harold* could scarcely reflect more accurately the acceptable contemporary view of ballet's heroic harem women and bandit queens and lady pirates:

> *Yet are Spain's maids no race of Amazons,*
> *But formed for all the witching arts of love;*
> *Though thus in arms they emulate her sons,*
> *And in the horrid phalanx dare to move,*
> *'Tis but the tender fierceness of the dove,*
> *Pecking the hand that hovers o'er her mate.*

The titillating blend of "masculine" strength and "feminine" voluptuousness found its most extreme expression in those ballets that required a woman to adopt male attire—*Le Diable Amoureux,* for instance, in which the pretty demon Urielle did her tempting disguised as a page. A woman in travesty, her curves revealed by tight breeches, was a feast for the eyes of the jaded sensualist. D'Albert, the voluptuary hero of Gautier's *Mademoiselle de Maupin,* inflamed with desire for the disguised heroine, "Théodore," apostrophizes hermaphroditism:

> This son of Hermes and Aphrodite is, in fact, one of the sweetest creations of Pagan genius. Nothing in the world can be imagined more ravishing than these two bodies, harmoniously blended together. . . . To an exclusive worshipper of form, can there be a more delightful uncertainty than that into which you are thrown by the sight of the back, the ambiguous loins, and the strong, delicate legs . . . ?

In order to give the influential subscribers in the boxes of the Opéra ample views of strong, delicate legs and not-so-ambiguous loins, management decreed that the Spanish Gypsies who flourished their capes about their partners in Mazilier's *Paquita* be female corps members in disguise, as were the sailors in the same choreographer's *Betty.* By 1870 in St.-Léon's *Coppélia,* this emasculated and seductive vision reached its apogee (or nadir, depending on your point of view): a woman (Eugénie Fiocre) played the youthful hero, Franz.

As noted in the previous chapter, the female dancer herself embodied two paradoxes: she was both a hardworking professional and a potential mistress, a poetic image and a panting, perspiring body. And just as ideas

Louise Fairbrother as Aladdin in *Open Sesame.* Lithograph by John Brandard after a drawing by J. W. Childe.

of strength and submissiveness married curiously in these ballets, so did the themes of female sensuality and female chastity. Even these lush, exotic fairy tales acknowledged that the imagined division between flesh and spirit had become a large and troubling chasm.

With a naïveté that is as touching as it is amusing, the hero of Félicien Champsaur's 1888 novel *L'Amant des danseuses* describes his emotions on following his ballerina-mistress (modeled after Virginia Zucchi) down the stairs:

> . . . I looked at her again from behind. Ah, her rump! I burned with the desire to spring at her, like a faun, to turn her round, to embrace her . . . It was a strange passion. Inwardly, I fought the brutal frenzy of a cloven-footed satyr for this splendid dancer, whom I respect as one of those rare beings in whose soul some unknown power has distilled the elixir of magnetism, and whom I venerate as I would a poem.

This kind of double-think pervades the ballets. In them, the heroine was, by implication, attainable, expected to gratify men's appetites; her "Oriental" nature alone held an erotic potential supposedly lacking in Western women. Yet the librettists—and, frequently, their heroes—persisted in viewing these heroines as virtuous maidens worthy of respect *and* as voluptuous chattels. Far from seeing anything odd about the harem girl who furiously defended her virtue unto death, the theatergoing public needed to have the Oriental-style sensuality controlled by familiar Christian moral codes.

Since most of these ballets have not been preserved, one can only guess that the degree of eroticism that was tolerated on the ballet stages of the nineteenth century would be considered reticent by today's standards. In 1985 Utah's Ballet West reconstructed a "lost" Oriental ballet that had been created in 1853 by Bournonville. The hero of *Abdallah,* a comedy-ballet with elements of *Aladdin, or The Wonderful Lamp,* is a humble shoemaker who, by means of a magic candelabrum, conjures up a dream harem complete with concubines. Yet—if the reconstruction is to be trusted—these appear to be anything but lascivious women. "Lounging" on a tier of marble steps, their tutus spread neatly around them and their pointed toes crossed, they resemble a decorous Danish garden. Their only concessions to the harem behavior of literature, painting, and fancy are their lack of shyness, the promptness with which they perform pretty, buoyant dances for their new master, and the eagerness with which they kneel at his feet or kiss his hand.

Admittedly, Bournonville, unlike many of his contemporaries, made no bones about prizing domesticity or seeing beauty in homey virtues, but the bulk of contemporary descriptions of ballerinas evokes charm and vivacity more often than sensuality. Here is Gautier, raving about his beloved Carlotta Grisi in *Paquita:*

> Never did foot so small support a more supple body, and never did castanets chatter more gaily at the tips of more agile fingers. How lightly she bounds and quickly escapes the two gipsies who are her partners, poor devils who think to be able to pinch her waist or kiss her hand! She darts away like an adder, smiling maliciously over her shoulder, and the pursuit begins all over again!

Of course, the degree of provocativeness varied from dancer to dancer. Marie Taglioni brought out the spirituality of every character that she played—her eroticism confined to a dainty teasing. With Fanny Elssler, it

Fanny Elssler in her famous "Cachucha." Lithograph by N. Sarony.

was otherwise. Yet the allure that she had for contemporary critics, the qualities that led Gautier to characterize her as a "pagan" dancer (as opposed to Taglioni, who was a "Christian" one) might be relative to the era. If recent reconstructions of her famous "Cachucha" from Jean Coralli's 1836 *Le Diable Boiteux* are accurate, the solo affords the dancer opportunities for intricate *tacqueterie*—the stamping of a heel, the tapping of a toe, the turn-in–turn-out of a foot to make the lace-edged skirt sway—for coquettish glances, supple *port de bras,* and a resiliently arching waist. It is a real Spanish dance and, as such, shows its Eastern influences. (Obviously the visit to the Paris Opéra in 1834 by four Spanish dancers of the Escuela Bolero had an influence on ballet choreography that the real "bayadères" who appeared in 1838 at the Théâtre des Varietés did not.) Today this "Cachucha" seems charming, flirtatious, mildly sensual; it certainly doesn't appear erotic to the degree that it did to Gautier ("What fire! What voluptuousness! What ardour!") or to Charles de Boigne, who wrote ecstatically:

> Those contortions, those movements of the hips, those arms which seem to seek and embrace an absent lover, that mouth crying out for a kiss, that thrilling, quivering, twisting body . . .

Elssler's pink satin dress with its black lace trim, its décolleté, its rounded bust and small waist still exists; her performance is unrecoverable. Either she performed the dance as no one would perform it today, or—more probable—the sensibility of her audience was such that the merest hint of heat set men ablaze. Perhaps in contrast to the more dignified dances of the late eighteenth and early nineteenth century, it *was* thrilling.

Eastern customs bowed to Western ideals of behavior in these and other theatrical works; Eastern dance styles and postures seem scarcely to have made a stand. We can suppose the bathing scenes in *The Revolt in the Harem* and *Le Corsaire* to have been inspired by Ingres's calmly seductive turbaned nudes, yet cannot imagine that they actually invoked them. We imagine instead some provocative frolicking, half hidden by the rims of *faux-marbre* bathing pools, "dressing" accomplished behind veils held by slaves. Looking at the handfuls of lithographs that are the most tangible evidence of scenes from these ballets, we wonder, for instance, just what kind of a dance was performed in *La Péri* by four "almays," their arms around each other's waists, wearing the kind of coat-over-trousers outfit that American dress reformer Amelia Bloomer unwittingly lent her name to. Does Achille

The *pas de quatre* from Act II, Scene 1, of *La Péri.* Lithograph from a drawing by Victor Coindre.

Devéria's drawing of Eugène Coralli and Adèle Dumilâtre dashing away together (see page 58) tell us about the choreography of *La Gipsy,* or only about its spirit?

Sometimes it was the props that set the tone—the veil dances and shawl dances that engendered such beguiling tableaux. In *Le Dieu et la Bayadère:* "The pink scarves of the dancing girls floated and hovered in the air in the most ingenious combinations, at times streaming in undulating folds, now hanging loosely, and at one moment, stretched out fan-wise with their ends gathered beneath Taglioni's foot so that she appeared like Venus emerging from the waves in her shell." In Perrot's *Lallah Rookh,* slave maidens formed elaborate tableaux around Fanny Cerrito, manipulating pink scarves (evidently pink was the popular hue) to suggest Hermes, The Shell, The Kiosks, The Cage, The Mirrors, The Harps, The Framed Picture, The Morning Breeze, The Stars, The Pine-Apples, The Car of The Rising Sun, The Butterflies, The Sun's Rays, and The Living Statue and Its Pedestal. This *pas symbolique* enumerated the heroine's virtues. We know of dances with pitchers in *Sacountala* and *La Bayadère.* The sheikh's sons in *Abdallah* perform a striking dance with swords, re-creating through their slashing and stabbing gestures, their leaps, spins, and crouches, the furious battle which—thank God for their prowess!—has just been won. Perhaps there were other such dances, although few choreographers in that century prized male dancing as Bournonville did.

Some dances boggle the imagination. The concubines turned militant of *The Revolt in the Harem* are performing dashing drills with lances when a spy sends for the guards; immediately, by virtue of a magic bouquet, the lances turn into lyres and the women mask their fierceness with dulcet steps.

One can be sure, however, that the dancing was couched in the familiar ballet vocabulary of the day. In Petipa's heyday, as Fokine later noted despairingly, "character dances" in ballets had a semblance of the national dances they referred to, but the principal dancers made all their utterances in the lingua franca of classicism. (The Spanish *pas* of mid-nineteenth-century France were an exception.) Attempts at "authentic" dances would have thwarted the public's expectations—particularly as far as their favorite ballerinas were concerned, and the ballerinas were careful to please. For Gautier's *La Péri,* based on his own story, "La Mille et Deuxième Nuit," the author had choreographer Jean Coralli concoct a "Pas de L'Abeille." Gautier had not yet visited Egypt, but had encountered the dance on his voyages through travelers' tales. Of course, Carlotta Grisi could not be expected to copy the harlots of Cairo, who, pretending that a bee had gotten into their clothing, contrived to strip themselves garment by garment, all the while dancing frantically. Grisi had only to drop a few outer garments, but apparently the unruly drama of the solo didn't suit her, and she soon substituted a more popular number. Gautier drily remarked that the sight of a peri dancing a cachucha with castanets was something that "our imagination had not foreseen."

Just as cultural and societal differences between the nineteenth and the twentieth centuries gradually altered the character of the unattainable female onstage, they altered—far more radically—the role and domain of the Orientalist heroine. The public in Paris and London that first thrilled to Diaghilev's Les Ballets Russes in 1909 was more comfortable with its libido than earlier generations of ballet fans had been, and could tolerate—indeed craved—images as sensual and cruel as any Delacroix or Byron could have devised. Furthermore, they were willing to exchange the pleasure of seeing familiar ballet steps for darkly convincing fantasies. By the early years of the twentieth century, audiences in both Europe and America were beginning to concede that the spiritual aspects of the East were also fit subjects for dancing, and to admire the young American Ruth St. Denis, who brought these to their attention.

Voyagers in the earlier ballet Orient, like armchair travelers, gave up none of the comforts of home. Perhaps to many of them the proximity of the dancers was spice enough. But the exotic settings and perilous plots gave the pretty, spirited dancing an edge and offered spectators a whiff of fragrance that seemed more intoxicating by far than that of their native daisies.

·2·

The Search for Motion

Isadora Duncan photographed in 1899 by Jacob Schloss, probably to publicize her performances with playwright Justin McCarthy of the *Rubaiyat* of Omar Khayyam

> It is far back, deep down the centuries that one's spirit passes when Isadora Duncan dances; back to the very morning of the world, when the greatness of soul found free expression in the body, when the rhythm of motion corresponded with the rhythm of sound, when the movements of the human body were one with the wind and the sea, when the gestures of a woman's arm was as the unfolding of a rose petal, the pressure of her foot upon the sod as the drifting of a leaf to earth.

Isadora Duncan inspired many such passages of fervid prose among her admiring contemporaries. From the beginning of her career, around the turn of the twentieth century, to her death in a bizarre automobile accident in 1927, those who saw her dance were less interested in writing about what she actually did than in presenting their bedizened responses to her.

Yet even as she evoked a fragrant never-never land, this American was creating a contemporary image for the stage dancer. Not a steely-legged virtuoso whipping off pirouettes, not a coquettish quasi-virgin, not a disembodied nymph, but a noble-spirited woman, bold, yet pliant—free to use her imagination and her body as she wished.

Artists in other fields were struck by her audacity: innovators in theater, such as Edward Gordon Craig, Konstantin Stanislavski, and Eleonora Duse; sculptors like Auguste Rodin and Émile-Antoine Bourdelle; poets, musicians, painters, dress designers. The artists who admired her were often the ones who shared her aversion to academic forms and her emphasis on expressiveness. But to them, as to a larger public, she articulated through dancing several important notions of her day. She evoked an idyllic "nature," even as developments in science and industry were shrinking the countryside, finally stripping poverty of its last veil, picturesqueness. She emphasized the connectedness of body and soul at a time when links between human beings, their work, and the land were being severed, and Victorian prudery shaped moral law. Like certain of her contemporaries in the Arts and Crafts movement—Gustav Stickley, for instance—she advocated simplicity and organic design when public taste decreed elaboration in both decoration and decorum. (A look at the exhibits of the Chicago Columbian Exposition of 1893 is enough to make you believe that the appetite for things carved, chased, embroidered, teased, and draped until no plain surface is visible must have reached some kind of apogee.) Finally, and perhaps most important, she made herself into an emblem of freedom—freedom not only from conventions of dance, but from conventional ideas about how women ought to dress and conduct their lives.

She profited—no doubt about it—from the fairly debased state of ballet in Western Europe and America around the turn of the century. The Ballets Russes of Serge Diaghilev didn't arrive to dazzle Europe with the boldness of its colors, music, dancing until 1909. The heyday of Romantic ballet had been over for at least sixty years. Few outside Russia had seen the high-order spectacles Marius Petipa was creating in St. Petersburg. The legacy of fine dramatic ballets that August Bournonville had left to the Danes was stored in Copenhagen. The public's appetite for spectacle and virtuosity was being satisfied to a stupefying degree by enterprising managers. Dancers, wrote George Bernard Shaw, come dispiritedly from another evening of ballet amid the smoke fumes at London's Alhambra Theatre, were ". . . still trying to give some freshness to the half-dozen *pas* of which every permutation has been worn to death any time these hundred years."

A spectator seeing Isadora Duncan perform at the peak of her power—arguably, sometime between 1902 and 1913—would have seen her alone onstage; although, after she founded her first school in 1904, her little girl students sometimes appeared to flock around her, to hold her hands and dance in a ring.* She might alternate her appearances with selections of important music—Schubert, Chopin, Brahms, Wagner—played by a pianist, or slip onto the stage to dance a portion of a symphonic work, then

*Later, various of her adopted daughters—Irma, Anna, Therese, Lisa, Margot, and Erica—might, on occasion, spell her onstage.

exit, leaving the orchestra to continue without her, as if she were a solo instrument.

Her settings were simple, but grand. The gathered blue-gray voile curtains she traveled with were very long so that they lay along the floor in foamy piles. People remember that when the light struck them in certain ways, you could see trees and clouds. She preferred lighting to bathe her in the roses and ambers of dawn and dusk, with an occasional flame red or deep blue for moodier pieces. White light, she detested; it didn't look natural to her. And she quite understandably disliked follow spots: what light in real life, she wanted to know, ever followed a person around?

She danced on a carpet—usually described as blue-gray, but sometimes as emerald green—which also traveled with her. It must have given additional softness and resilience to her footsteps, a stand-in, perhaps, for the grass that in her imagination carpeted her dancing ground. A cushion, too, for the shock of her (shockingly) bare feet.

The spectator would have seen a woman somewhat taller than medium height—five foot five or five foot six, people say—and rather voluptuous. (We know from her letters that she considered herself thin when she weighed 128 pounds and not distressingly fat when she weighed 143 pounds.) Her hair would be loosely caught back to frame the mild, sweet face. She'd be bare-legged under veils or a tunic of the lightest China silk or silk chiffon, little ragged wispy things that would float in your hand if you picked them up. The tunics she usually gathered in under her breasts and around her hips, after the fashion of one of the Graces in Botticelli's *Primavera,* a copy of which, she claimed, always hung in her family's various homes while she was growing up.

Perhaps the orchestra or the pianist would begin to play; or perhaps she would first walk quietly onto the stage and simply stand for a while, listening to the music, swaying slightly as if waiting for a wave to gather force inside her. Then she would embark on phrases built of simple, eloquent walks and runs, of buoyant waltz steps, of skips and leaps—not so much covering the stage with patterns as making of it a three-dimensional world in which invisible presences or aspects of the musical climate drew and repelled her. The phrases were elastic and expansive. Not stuffed with steps. There was plenty of time for the spectator's eye to take in the dancing figure, to follow her motions as she bounded into the air or fell to the floor, acknowledging the weight of her body, complying with gravity as no ballet dancer did. And the fullness of her gestures in space increased this sense of amplitude, of generosity.

Some of the gestures were abstract, like the lovely upward drift of the

Botticelli's *Primavera* was a source for costume designs, subject matter, and movement motifs in Duncan's early dances.

arms with which she so often ended her solos; some were pantomimic. Some sixty or seventy years after seeing Duncan perform her "Maidens of Chalkis" dance in Gluck's *Iphigenia in Aulis,* Marie Rambert remembered vividly how "she threw a ball and ran after it, bounced it, caught it in mid-air; or else she played with 'osselets,' reclining by the sea, leaning on one elbow and throwing up those little square bones from the inside of her hand to the outside . . . Of course she had neither a ball, nor the little bones—but you saw it all in her dance."

It must have been logical to accept her as a symbol of freedom. To her contemporaries, a small stock of knowledge about her life would have sharpened this perception. She talked and wrote frequently to deplore women's lack of liberty. She insisted on her right to bear children out of wedlock, subject to no man's control. She even danced while visibly pregnant, enhancing the image of herself as a fecund Demeter and giving Puritans a good shock. But a spectator who knew nothing about her could respond to the air of freedom she created onstage—in moments of serene repose, as well as in those wild maenad skippings with head flung back and strong throat exposed. Her gestures, instead of culminating in poses, flowed out into the space around her. Neither her clothing nor the conventions of dancing restricted her. Clearly, this was no Trilby, being manipulated by a shrewd manager or ballet master. She looked as if she were dancing out her own feelings in response to the music, making her own decisions.

She displayed at times, too, the artless vigor of a fine animal—an unusual image for a female and a dancer. A San Francisco lawyer told his daughter that Isadora gave him the feeling that she "could leap to the ceiling, but she never did," that she leaped like a coyote in the wilds, and "you always felt there was boundless exertion in back of her." The painter J. B. Yeats compared Duncan dancing onstage to a kitten playing by itself:

> We watched her as if we were each of us hidden in ambush. I don't wonder that at first New York rejected her—she stood still, she lay down, she walked about, she danced, she leaped, she disappeared and reappeared—all in curious sympathy with a great piece of classical music, and I sometimes did not know which I most enjoyed, her or the music.

The very qualities in Duncan's art that Yeats enjoyed were ones that galled many who saw her. Those who associated dancing with a mastery of ballet technique applied the words "amateur" and "dilettante" to her. Some spoke of the poverty of her dance vocabulary and attributed her popularity to a fluke of charisma. On first seeing her in Russia, young Vaslav Nijinsky, unlike many of his colleagues, did not fall under her spell. He told his sister point-blank, ". . . her performance is spontaneous and is not based on any school of dancing and so cannot be taught . . . it is not art."

There were many who were outraged instead of intrigued by the way she allied herself with major composers, daring to say that she was interpreting the soul of their music. When she danced to the "Liebestod" from Wagner's *Tristan* in New York in 1911, conductor Walter Damrosch felt it prudent to warn the audience that ". . . as there are probably a great many people here to whom the idea of giving pantomimic expression to the 'Liebestod' would be horrifying, I am putting it last on the program, so that those who do not wish to see it may leave." A critic might accept her glosses on Schubert's German Dances or waltzes by Chopin or Brahms, but recoil before the spectacle of Miss-Isadora-Duncan-from-California essaying Beethoven's Seventh Symphony: "The little figure leading its own life alongside it was a frustrating and upsetting experience—this immense art work accompanied by all that triviality." The reaction is understandable, but I don't think it would have occurred to Duncan to consider a symphony in terms of tonal variety or harmonic richness and thus hesitate to tackle it: she thought she heard Beethoven speaking and wished to respond. Her first experiences with great composers were intimate ones: when she was growing up, her mother played the piano decently and often. Beethoven and Chopin were revered members of the family.

When Duncan's lover, Gordon Craig, nicknamed her "Topsy," he may have been thinking of her ramshackle, "artistic" childhood. If she did, in some sense, "just grow" into an artist, the climate of her native California may have been partly responsible. Aesthetic movements flowered robustly in the warm air and sunshine. People could stage poetry readings, put on plays and pageants out of doors. They could wear Greek outfits as a sign of liberal thinking or artistic proclivities without the bother of donning long woolens under them—as a doctor–dress reformer gravely counseled British aesthetes to do. Poet Charles Keeler had himself photographed in a windblown toga, gazing searchingly into the distance.

Duncan's own father wrote poetry:

See, the centuried mist is breaking!
Lo, the free Hellenic shore!
Marathon-Platea tells us
Greece is living Greece once more.

It was understood that the writer was pondering an intaglio, but he might well have been looking across the San Francisco Bay to the hills of Berkeley, soon to be known as the "Athens of the Pacific."

Cultivated folk could see fine theater and hear fine music in San Francisco, a boomtown desirous of becoming a great city. The great Polish actress Helen Modjeska made her American debut there in 1877, the year Angela I. Duncan was born (the "Isadora" took precedence later). Sarah Bernhardt played San Francisco, and Edwin Booth and Adelina Patti, as well as Lottie Collins, the high-kicker from the London music halls, and many others. Duncan, as a girl, could have seen the theatrical dancing of the day in variety shows or musical plays—pert corseted females (some with scant training) marching about the stage to form a variety of eye-catching designs; skirt dancers with their spirited blend of clog dancing, ballet steps, and skirt maneuvers. She might have seen extravaganzas like

The Black Crook—a revival of the 1866 hit played San Francisco in 1893—in which ballet dancers spun and stalked about on the tips of their toes between the tableaux, the mist-and-fire stage effects, and the scraps of lurid, often inconsistent plot.

The four Duncan children—Augustin, Elizabeth, Raymond, and Isadora—were all mad for the theater. Encouraged by their mother and an aunt Augusta, who had hoped to go on the stage, they and their friends put on plays in the barn belonging to a house they rented in Oakland, and the four of them toured a high-toned family variety show up and down California. Small wonder that when Isadora left San Francisco in 1895 to seek her fortune, she sought—and found—work in a theatrical company, that of impresario-director Augustin Daly.

Poverty molded Isadora Duncan's character as profoundly as art did. In addition to writing poetry, her resourceful father, Joseph Duncan, at various times published a newspaper, ran an auction gallery, speculated in real estate, and was instrumental in setting up two banks, but around the time of Isadora's birth, it was discovered that, during a period of financial disasters when many California banks were forced to close, he'd been using bank funds to finance private stock speculations. Among other things. He could have gone to jail. Mrs. Duncan, angered as much by his philandering as by his financial peccadillos, divorced him.

In a series of rented lodgings in Oakland and later in San Francisco, always behind with the rent, she took in sewing and taught piano. (She also read her children, says Isadora, freethinker Robert Ingersoll's preachings against such Christian institutions as marriage.) The kids ran wild and grew up clannish and resourceful. It was in her childhood that Isadora developed a talent for scrounging, for coaxing money out of friends and credit out of merchants. She quit school young and embarked on a prodigious course of reading. She began teaching some kind of "expressive dancing" to neighborhood children. At fifteen, "Miss A. Dora Duncan," along with her sister Elizabeth, set herself up as a teacher of ballroom dancing—schottisches, waltzes, polkas, mazurkas, gavottes. Augustin and Raymond were recruited, no doubt to give female pupils a more delightful time, and by 1894 all four young Duncans were listed in a San Francisco directory as teachers of dancing.

Like many of her day, Duncan prized above all art that appeared to be a spontaneous expression of feeling, and so successfully did she create the illusion of spontaneity in her performing that many people supposed she was improvising. In her later years, she must have derived a certain mischievous pleasure from the bafflement of onlookers at "dress rehearsals" during which she simply sat and listened intently to the orchestra interpret the music she was to dance to—more willing to be thought an instinctive genius than observed as a sweating worker.

Evidently, she did sometimes improvise, and she allowed herself considerable leeway in performance. Yet, if the reconstructions of her dances can be relied on, she joined them to the music with firm craftsmanship; every time a musical phrase repeats, the dance steps associated with it appear again also.* Too, she did on occasion defend herself against the charges of mindless self-expression: ". . . even in nature you find sure, even rigid design. Natural dancing should only mean that the dance never goes against nature, not that anything is left to chance." She was a disorganized and pleasure-loving woman, but sentences here and there in her letters leave little doubt that she was a choreographer, a real one, who worked on her dances. "Dear," she wrote to Gordon Craig in 1906, "it's rather discouraging trying to work with one's body as an instrument . . . However I found something quite new for <u>me</u> this afternoon—only a little movement, but something I have never done before & may be the <u>key note</u> to a great deal." (This, incidentally, was written in the middle of a strenuous tour just three months after the birth of her daughter Deirdre.)

In keeping with her role as Romantic revolutionary, she also misted over the evolution of her style, disavowing any connections with existing dance traditions and promulgating the gospel of sudden inspiration. She created, in her various writings, pictures of herself being inspired. Duncan, standing for hours with her hands crossed over her solar plexus, finally deciding that it was the "central spring" of movement. Duncan standing, again for hours, before the Parthenon until her hands floated upward: "I had found my dance, and it was a prayer." Young Isadora dancing on the San Francisco beach below Cliff House, imitating the waves. She was, undoubtedly, a woman to whom transcendent moments came, and she did build her dancing persona on her own body and her own disposition—impetuous, mischievous, serious, generous, candid. However, she also drew from a variety

*During Duncan's lifetime, her dances were not filmed or notated. Her daughters—Lisa in Paris; Irma, Anna, and Therese in the United States—passed on to their students what they had learned or observed. Elizabeth Duncan's pupils also learned dances, and some—like Anita Zahn and Erna Schulz—went on to perform them. It is likely that details have eroded or been changed, and possible that the process of teaching may have tied down elements that were originally allowed to change from one performance to the next.

of sources any idea that would support a link between the perfectability of the body and the perfectability of the soul and launch her into motion as its messenger.

How, if at all, did she train to be a dancer? Late in her life, Duncan claimed that she had taken three ballet lessons as a child, hated their rigidity, and took no more. This may not be strictly true; there's some slender evidence that, as a young woman, she studied briefly with Marie Bonfanti in New York and with Katti Lanner in London. But all her life she railed against ballet training because it took account only of the body and because its gestures struck her as unnatural, even though she admired the great Russian dancers she met: Pavlova, Karsavina, Nijinsky.

So ballet played little, if any, role in her development. What did then? A daughter of Isadora's girlhood friend, Florence Treadwell (later Boynton), has said that she's sure her mother and Isadora used to go together to the Oakland *Turnverein,* where they could have taken part in folk and social dancing and in gymnastic classes. That's something Isadora's mother might have approved of. However disciplined the drills and procedures, gymnastics since the Civil War had subtly aligned itself with other movements concerned with the liberation of women—liberation, in this case, from the tyranny of corsets and heavy, tight clothing, from unbalanced diets, and a lack of fresh air and exercise. Gymnastics, when Isadora was growing up, was no longer a men-only, muscle-building activity. The "new gymnastics" stressed flexibility, coordination, balance. Its bending, twisting, and stretching exercises led to a more mobile torso. The jumps, done without ever putting the heels down, promoted lightness and strong ankles. (Later, Isadora taught just such a way of jumping to her pupils.) In a moral climate far more wholesome than that of a ballet class, girls as well as boys moved in vigorous rhythms, sometimes to musical accompaniment, even pairing up to make pleasing designs as they manipulated the *de rigueur* props: slim wands, small wooden rings, lightweight dumbbells, and Indian clubs.*

Duncan on occasion decried the mechanical aspect of gymnastics; nonetheless, some form of it served her as barrework serves the ballet dancer. In all the schools she founded, gymnastics began the day. To her lover,

*There were some differences between the gymnastics taught in public schools and colleges and what was taught in the *turnvereins.* Nevertheless, there was considerable similarity between the "new" American gymnastics and calisthenics, as set forth by such reformers as Dio Lewis, and the "new" approach to German gymnastics generated by Adolf Spiess.

Gordon Craig, from the icy darkness of St. Petersburg in January of 1908, she wrote, "I go each morning to a Swedish institute for gymnastics in order to keep alive." She knew that the body had to be primed for dancing, and some of her favorite authors supported her in this. Hadn't Plato in his "glorious *Republic*" decreed that music and gymnastics should be the foundation of education? Hadn't Goethe written that a certain amount of technical expertise was a prerequisite for art? The Duncan skip, with the free leg lifting to the back and then swinging front, is like a standard gymnastic exercise, set skimming in space and given artistic modeling by the floating arms and arching back.

By the middle of the 1880's, when Duncan would probably have been exposed to it, gymnastics, like almost every other area of American cultural life, was tinted by a new craze: Delsartism. In his native France, François Delsarte had developed an intelligent and systematic way of analyzing posture, gesture, and vocal expression by linking these with corresponding mental and spiritual states, intending his system to serve professional orators, actors, and singers. He died in 1871 without visiting America; he might have been astonished to see the offshoots of his teachings, carried there by disciples. Delsarte principles added a moral note to gymnastics, helped society ladies to acquire graceful gliding walks, toned up schoolchildren's recitations with appropriate gestures, brought new expressivity to "living pictures." Delsartism made thinking about the body not only advisable but fashionable. And spiritually uplifting into the bargain. Consider these words by the author of *The New Calisthenics: A Manual of Health and Beauty:*

> Get your slouching John and your shuffling Peter into an erect manly carriage, if you can, for three minutes, and see what the moral effect will be. Get that sneaking, lying boy to stand erect, throw up his head and say: "I am no knave; I am the King." He will be the King, while he says it at least; and you—perhaps he—will have learned a lesson.

The rising young salon artist Miss Isadora Duncan, interviewed in 1898 by the *New York Herald Sun,* made it clear that she admired Delsarte ("Delsarte, the master of all principles of flexibility, and lightness of body, should receive universal thanks for the bonds he has removed from our constrained members . . .") and that she had absorbed his message about the connection between movement and mental attitude.

It would have been hard for a bright, serious young person with theat-

A turn-of-the-century recitation of "Hark the Wedding Bells Are Ringing," enhanced by Delsartean gestures and Grecian attire

rical aspirations growing up in America in the 1880s and 1890s *not* to have been influenced by Delsarte. Duncan could have picked up some Delsartean theory grafted onto school gymnastics or elocution classes; she could have attended one of the "schools of expression" founded by Delsarte pupils (there was one in San Francisco by 1888, one in Oakland by 1892); she could have studied privately or read some of the Delsarte manuals flooding the market (she owned one, Craig said). She couldn't have read a treatise by Delsarte—he never got around to writing one—but she might well have read *The Delsarte System of Expression* by Genevieve Stebbins, a book that went through six editions between 1885 and 1902. There are enough similarities between Mrs. Stebbins's ideas and ones Duncan later expressed, between some of Stebbins's "drills" and exercises later taught in Duncan's schools to suggest that Isadora might have read this manual.

A would-be dancer could learn much from Stebbins's elucidation of Delsarte theory. She could be made aware of the moral function of art: François Delsarte considered that to value art for art's sake was as absurd as to value the telescope for the telescope's sake instead of for what it brought into focus. She could learn how the various parts of the body express emotion by studying Delsarte's charts—charts so bristling with trinities that one follower called Delsarte "Swedenborg geometrized," and G. B. Shaw suspected him of trying to found a "quack religion." Equipped with the knowledge that the body was divided into three zones, the head, the torso, and the limbs—corresponding to the three "essences" of human behavior: the mental, the moral, and the vital—and understanding that action occurred in three corresponding ways—away from the center (excentric), balanced (normal), and toward the center (concentric)—she could proceed to the myriad of permutations. She could learn from reading Stebbins about the "harmonic poise" derived from Greek statues and practice the rules for "artistic statue posing," watching in a mirror to see if she could indeed move from pose to pose in a fluid manner "as unaffected as the subtle evolutions of a serpent." Book propped on a table, she could follow some of the more strenuous exercises in "aesthetic gymnastics" Stebbins had devised:

> (a) Assume attitude of explosion [a lunge—"excentro-excentric"] . . . (b) throw body forward, striking floor on thigh of strong leg. Be careful to protect the face with the forearms as you throw torso to the floor.

Although Delsarte theory dealt primarily with "attitudes" and gestures, performed while standing in place, it was full of spatial implications. The performer's body was dynamically linked with points in three-dimensional space in terms of attraction and repulsion, fall and rise, tension and relaxation. So the system provided a splendid base for the expansion of expressive gesture into dancing. Duncan made her gestures travel across the stage by means of the simple walks, runs, and skips that came easily to a naturally athletic child, by the jigs and reels she learned from her Irish grandmother, and by the social dances she studied and taught. In those of her dances that have been reconstructed from the memories of her pupils—particularly the ones made during the first half of her career, like the lovely Brahms waltzes, the Schubert and Chopin pieces—the basic three-step pattern of the waltz appears in many guises. Pulsing in place, rushing forward, turning. Usually it's a very light and lilting thing, done mostly on the toes, but without a trace of stiffness. The robust hop-step-step-step of the polka she completely transformed—sharpening it sometimes by opposition in the arms and body, delivering it with triumphant force, even compressing it, on occasion, into 3/4 time.

After her girlish confidences of 1898, Isadora never credited Delsarte again, although others linked her with him. Certainly her dancing was "no art of Delsartean formulas," as Redfern Mason pointed out in the *San Francisco Chronicle* in 1918, but her gift for pantomime, her idealistic view of art, her Greekishness connected her with Delsarte in the minds of many spectators. Delsarte had given the public a framework in which to consider new forms of dance. Isadora stretched that framework; still, at the core of her art always lay the idea that a Delsartean gymnastics teacher had bluntly expressed in 1889: "Strength at the centre; freedom at the surface."

By the time Isadora Duncan gave those first recitals in New York, she had served a solid apprenticeship in the theater—about two years touring America and England with Augustin Daly's company in productions that ranged from a musical play, *The Geisha,* to bowdlerized Shakespeare. She'd done a little singing, small-part acting, dancing along the flitting-fairy line (in *A Midsummer Night's Dream*), and probably something like skirt dancing. Yet at the beginning of her career as a solo artist, she set about severing in audiences' minds the connection—and it was one that was taken for granted—between dance and entertainment, by allying herself with the work of sculptors, painters, composers, and writers whose reputations were unassailable. Much later in her career, she said grandly to a reporter, "I use my body as my medium just as the writer uses his words. Do not call me a dancer." Her quibble was not really, I think, with being a dancer, but with being what people thought a dancer was.

In the early solo programs, the art she chose to link herself with was, for the most part, trendy rather than great: the *Rubaiyat of Omar Khayyam,* Ethelbert Nevin's *Narcissus,* the *Idylls* of Theocritus (usually read in performance by Elizabeth or one of the brothers), Ophelia—a popular heroine with late-nineteenth-century painters. She danced her Botticelli *Primavera,* "scattering seeds as she goes, plucking the budding flower," to music by Johann Strauss.

Descriptions of Duncan around this time make it sound almost as if she were doing the kind of narrative "plastiques" Genevieve Stebbins had performed: "Dressed in a simple white frock, Miss Duncan took up a series of graceful poses, but passed from one to another so rapidly that the succession of postures resolved itself into a dance." The photographs Jacob Schloss took of her, probably early in 1899, suggest, endearingly, that she had not yet settled on the image she wished to project. In one, she stands, somewhat awkwardly, in fifth position on the tips of her toes; in another she smiles at the camera in a seductive back arch, caught perhaps in a skirt dancer's waltz turn; in a third, she offers herself to the camera with a simple, generous gesture, arms open, head falling back, which presages the later Isadora. It's illuminating to compare one of Schloss's pictures of her, in a "Mercury attitude," with one her brother Raymond took in Greece five years later. The pose is nearly the same, but the genteel nymph has become a true bacchante, the solo "artiste," a great dancer.

"The genteel nymph has become a true bacchante . . ."

Actually, the change seems to have taken place quite rapidly, with only a finishing luster applied by the air of living Greece, where she and her clan attempted to build a grandiose house-temple-theater complex until a search for water drained their (her) resources. The transformation began in London. (Like many American artists, Duncan had to go to Europe to build a reputation and she was always more appreciated there.) In 1900, at London's New Gallery, she gave three recitals sponsored by many prominent persons in the arts, among them the writers Henry James and Andrew Lang; the painters Sir Lawrence Alma-Tadema, William Holman Hunt, the only surviving founder of the Pre-Raphaelite Brotherhood, and Walter Crane, the designer and illustrator, prominent in the Arts and Crafts movement; musicologist Sir Hubert Parry and the music critic of *The Times,* J. Fuller Maitland. The Greek scholar Jane Ellen Harrison took over the Duncan siblings' role of reading Theocritus while Isadora danced. In one recital, all the dances were inspired by literary works, in another by paintings, in a third by music. From these performances, Duncan gleaned theories, images, sources of inspiration that nourished her for many years.

It's no wonder that Holman Hunt invited Duncan to perform at his house and to act as sponsor and adviser for the New Gallery concerts. Not only was she a charmingly frank young barbarian who doted upon famous men, she had the Pre-Raphaelitish moral fervor and zeal for social reform as well as a taste for quattrocento Italian painting. And how delicately she learned to bring to life those motifs beloved of Botticelli and of the by-then-deceased Edward Burne-Jones and Dante Gabriel Rossetti. It was not the realism of the English painters that drew her—the Renaissance domesticated—but their mythological and religious fantasies. And she must also have been attracted to the implied motion in some of these pictures, the ways in which the artists distilled anecdote into gesture. She too loved to create the vision of a chaste but voluptuous flight and amorous pursuit, thin draperies blown back by the rushing: Pan and Echo, Bacchus and Ariadne. You can see her motions figured in the Zephyr and hastening Chloris in the corner of Botticelli's *Primavera,* in the *Phyllis and Demophöon* of Burne-Jones—only in the latter painting, it's a youth who flees a pursuing maiden, his head turned back to regard her in fear and fascination.

Duncan continued to perform some of the "Dance Idylls" presented at the New Gallery for many years, long after she had stopped making dances rooted in painting. The activity within a picture frame showed her a way to fill the stage frame, playing all the roles if necessary; and that knowledge she kept and built upon. The New Gallery recitals also cured her of dancing illustrations of poetry, line by line, and upgraded her musical taste. She danced, apparently for the first time, and on Fuller Maitland's suggestion, to music by Chopin. She abandoned Strauss and performed her *Primavera* to Mendelssohn's *Spring Song* and, in another of the programs, to an air from Fabritio Caroso's sixteenth-century dance manual, ferreted out by musicologist Arnold Dolmetsch. At one of these recitals, she danced, perhaps for the first time, the minuet from Gluck's *Orpheus*. Throughout her life, she remained partial to Gluck's music, ignoring his eighteenth-century restraint, and to the Orpheus theme: the descent into darkness, the emergence into light, the loss of a beloved, the creation of an artist through suffering.

During Duncan's first few years as a concert artist, she seemed always to be searching for sources of motion—in museums, books, nature. She was trying to invent dancing powerful enough to match the vision she expressed in her best Whitmanesque rhetoric: "I see America dancing, beautiful, strong, with one foot poised on the highest point of the Rockies, her two hands stretched out from the Atlantic to the Pacific, her fine head tossed to the sky, her forehead shining with a crown of a million stars." It only seems odd at first that some of her sources for this heroic "Dance of the Future" were the vase paintings, bas-reliefs, and statues of ancient Greece.

Denying the charge—or compliment—by her contemporaries that she was trying to bring museum figures to life, she said that she was learning from them how to study nature. To her, as to Auguste Rodin and John Ruskin, the Greek statues, like the gods and goddesses they so often represented, embodied natural forces. It was these forces she was seeking when she browsed through the British Museum, the Louvre, museums in Berlin, Munich, Athens itself. She was drawn to figures in arrested motion, in poses suggesting action just completed or about to start: the Nike of Samothrace, the small Tanagra figures (out of whose poses she—or perhaps her protégée and adopted daughter Irma—did later develop phrases to edify her little students), the three headless Fates who once reposed on the east pediment of the Parthenon, a stooping knucklebone player, a frenzied maenad. They aided her to a calmness, a firmness, a force different from the fluid, weightless traceries of motion in the Pre-Raphaelite canvases. And they did supply her with gestures: postcards in a sizable collection she supposedly carried around with her show lunging warriors, reclining figures, processions of draped women, whose postures were echoed in her dancing and in the offstage persona that she displayed to photographers.

It was said that Duncan was trying to re-create the dance of the ancient Greeks. She denied this charge too. It was her brother Raymond who,

Greeks and Amazons from the Phigaleian frieze in the British Museum

Isadora Duncan, photographed by Elvira, Munich, 1904

much to her dismay, gave occasional recitals of two-dimensional, straight-off-the-vase stuff. And she went far beyond the many other professionals and amateurs who, since the beginning of the nineteenth century, had turned to pseudo-Grecian dancing and costume as an aesthetic gesture or a statement of freedom.

That so many writers saw in her their favorite museum figures vivified or an imagined antiquity brought to life reveals their own slant of mind. When Duncan began to develop her art, European cultural life had been tinged with Hellenism for more than a century; she fitted herself, most satisfyingly, into a trend. To link something with Greece automatically dignified it. In 1895 a Frenchman, Maurice Emmanuel, had written a treatise on *The Antique Greek Dance,* making scholarly (if often wrong-headed) attempts to reconstruct ancient dance steps by analyzing statues and the figures on vases; one of his aims was to link Greek dance with "modern dance," i.e. French ballet, by pointing out similarities and differences, thus aggrandizing a form temporarily in need of all the justification it could get.

Although Duncan went to Greece, the trip only fueled ideas she already had. As her rival Ruth St. Denis later pointed out, Greece was for Isadora "a state of mind." In this, she was like many of the cultivated people to whom Greek poetry, Greek philosophy, Greek architecture, Greek statues, bas-reliefs, and vases were of consuming interest. Few of them had been to Greece, not even the early German Hellenists Lessing, Winckelmann, and Goethe. In the Greece they imagined, disease, fear, brutality, exhaustion, hunger had no place. Neither did bad art. If there had been any, it hadn't survived. With the educated naïveté typical of the age, Gilbert Murray marveled that the citizens of Athens in the fifth century B.C. hadn't seemed to notice how extraordinarily splendid life in their enlightened society was.

The statues epitomized this distant grandeur, with their serenity, their impersonal gaze, their pallor. Perhaps most important, they afforded a way for people to look at and think about the body, while ignoring some of the realities that the nineteenth-century mind was not prepared to accept as beautiful. The statues were nude or, like Isadora Duncan, barely clad, but you could stare at them without feeling guilty, because the people they portrayed were so noble, so apparently beyond desire. Neither hair nor wrinkles grooved the smoothness of their marble torsos. To avid viewers, their voluptuous forms embodied intellectual and spiritual ideals. "Greek sensuousness, therefore," said Walter Pater, "does not fever the blood."

That was what Duncan believed too, and traded on. She exposed a good deal of flesh, but with such exalted innocence that the eye could ponder her unrebuked. "It was with a gesture full of chastity and grace," declared a Russian actor, that she would replace a breast that had fallen out of her chiton. The official position is clarified in these words written by a Dutch critic in 1905: "Everything was touched by pure beauty; sensuality was completely absent." In other words, for her contemporaries to find her dancing beautiful, they *had* to find it chaste.

So, in part, did she. She thought of her body the way an aesthete thought of a Greek statue: as a temple ennobled by the spirit and intellect it housed. She could refuse to wear the long chemise Cosima Wagner urged on her at Bayreuth, when she danced the "Bacchanale" in *Tannhäuser* in 1904, because she dismissed the idea that her body could inspire lust in an audience as unthinkable. Her knowledge of Delsarte fortified her in the position that it wasn't the body itself, but only certain movements of the body that could be thought of as shameless. When the Black Bottom and the Charleston became popular, she deplored them because they emphasized action in the hips—"the seat of the appetites."

It was not just by looking at Greek art and wandering about the Parthenon that Duncan developed her perspective on the uses of antiquity. Between 1900 and 1905, critical years in her development, she read Nietzsche. It was in his *The Birth of Tragedy from the Spirit of Music* that she was first exposed to the notion that the fierce Dionysian aspect of Greek tragedy and Greek religion was a necessary balance to the Apollonian aspect that entranced Nietzsche's—and most of Duncan's—contemporaries. As Nietzsche wondered whether dissonance in music might not serve to express Dionysian frenzy, Duncan, for her dancing Furies and maenads, experimented with such dissonant images as clawed hands, crouched body, the upward fling of the head that she noticed in moon-besotted dogs as well as in painted bacchantes. The Fury dances from her *Orpheus* (as reconstructed by Julia Levien, a pupil of Isadora's adopted daughter Irma, and performed by Annabelle Gamson) are quite terrifying. The dancer becomes a flock of creatures who prey on each other as well as on their destined victims. Now tormenting, now tormented, they snatch at the air as if it were human flesh, lift huge imaginary stones, flee and recoil, finally fall to the ground in what might be satiety or exhaustion, or both. These dances, and others, like the bold dances of the Scythian warriors from Gluck's *Iphigenia in Tauris,* contained images far darker and less polite than any that Duncan's beloved Gluck might have envisioned, as, guided by Nietzsche, she turned to the choruses of Greek tragedy for inspiration:

Isadora Duncan, sketched by José Clará, ca. 1910

Oh, they like a colt as he
Runs by the river,
A colt by his dam
When the heart of him sings,
With the keen limbs drawn
And the fleet foot aquiver
Away the bacchanal springs!

It may be mere coincidence that the copy of Euripides's *The Bacchae* that she used to travel with* has a bookmark placed at this passage, but it's easy to imagine her enthralled by the grand, free rush of movement it conjured up.

Nietzsche would probably have been astonished at the use to which Isadora put his theories about the chorus in the antique drama. He had asserted that

> . . . the scene, together with the action, was thought of only as a *vision,* that the only reality was just the chorus, which of itself generates the vision and celebrates it with the entire symbolism of dancing, music, and speech.

Duncan, therefore, decided that she would never play the protagonist in a potentially dramatic dance, but that—all alone—she would be the chorus: "When I have danced, I have tried always to be the Chorus . . . I have never once danced a solo." An undated program for a performance of *Orpheus* during her 1909 American tour indicates that she wished to be seen as the companions of Orpheus, as witnessing shades during the scene in Hades, as happy spirits in Elysium. Later, she added a dance of the Furies. If the nouns are plural, it is because she wasn't so much choosing to play the one-who-stands-for-many as trying to suggest an entire dancing throng, in which call and response become simply differing modes of a single plastic impulse, ". . . bound endlessly in one cadence."

This would have been a way of liberating herself from the more narrowly defined role-playing of the "Dance Idylls," a way of creating a more flexible scene. There are those who swear they saw her as Iphigenia going to the sacrifice, but it's clear that she was trying to suggest the action *as it might have been envisioned by the chorus,* thereby giving the picture a heroic impersonality and magnifying herself into something more than a solitary woman dancing for an audience. It could be said that in doing this, she presaged the abstractions of early modern dance, in which the dancer would eschew impersonation and equate her own persona with universal human feelings and drives.

Much has been made of the back-to-nature aspect of Duncan's work. In 1903 or 1904, an Englishman who saw her dance in Berlin named her a

*Part of the Irma Duncan Collection, now housed in the Dance Collection of the Library and Museum of the Performing Arts at Lincoln Center (a branch of the New York Public Library).

new Joshua, before whom ". . . the factory walls fell down, the festering slums and ugly places of London crumbled away . . ." However, what she sought in nature was what she was looking for in Renaissance paintings and Greek statues: laws of motion. And at the turn of the century, she saw the natural world from a perspective that wouldn't have been possible fifty years earlier. The harnessing of electricity gave her a new framework, a new vocabulary for dealing with nature, no matter how mystically and romantically she set about it.

"The true dance must be the transmission of the earth's energy through the body." Isadora was twenty-five or twenty-six when she jotted that down. Elsewhere, she described how, standing with her hands over her solar plexus, she envisioned herself receiving the "rays and vibrations" of the music in this "crater of motor power," which would then transmute the musical essence into movement and send it flowing out through her limbs.

She thought of herself as a dynamo.

She made audiences perceive her this way too, even though they would not have recognized the source. What they saw was how all the gestures of her limbs seemed to be impelled by a strong lift of her rib cage, by an expansion or folding in at the center of her body. It's what motivated her characteristic run, which Frederick Ashton, enraptured by her in 1921, described like this: "She had a wonderful way of running, in which she what I call left herself behind, and you felt the breeze running through her hair and everywhere else."

Did Isadora Duncan read treatises about electrical energy? Probably not (although she certainly read a surprising assortment of books). What she did do, in the summer of 1900, along with the rest of Europe, was stroll through the streets of the International Exposition in Paris. At night, colored lights played on the waters at the Château d'Eau in front of the brilliantly illuminated Palace of Electricity with its Hall of Dynamos. Like her fellow American Henry Adams, she might well have been roused to an almost religious contemplation of the great forty-foot dynamos humming with inexplicable power.

At the fair, too, she saw a compatriot, Loïe Fuller (later—briefly—her sponsor), performing her famous "Danses Lumineuses" in the Art Nouveau theater that had been built in the fairgrounds expressly for her. Fuller, manipulating yards of silk by means of wands concealed underneath her outsized skirts, illumined by her own ingenious inventions in lighting, functioned as a kind of motor. Even such potentially static images as her *Lily* were created through motion, kept in the air by motion; their fluid outlines would not have been visible except for the constant churning of her hidden arms and body.

Whether these particular sights influenced Duncan is, of course, a moot point, but excitement about the marvels of electricity was widespread—a fit subject for an impressionable young dancer's fantasies. Earlier, Genevieve Stebbins had written about the implications for movement of ideas expressed by John Tyndall, the British scientist and popular writer, teacher, and lecturer on physics, who had toured the United States in 1872–1873. Like Duncan, Stebbins had utilized analogies drawn from electrical theory to create the image of a charged and vibrant body. Indeed, it is startling to come across in one of Stebbins's books a passage that might explain Isadora's decision to use her middle name—a name which, she once explained to Gordon Craig, means "gift of Isis," or "child of Isis." Stebbins writes: "Osiris, the all-powerful god, gives light to Isis, who modifies it and transmits it by reflection to men."*

Isis/Isadora knew also that this luminosity was transmitted in waves, and with her customary untidy eloquence she linked these invisible undulations to those visible in wind and ocean and those sensed in her own interior tides. "Waves—love waves—," she wrote to Craig, "I've been writing about dance waves, sound waves, light waves—all the *same*—." What she had observed while letting off adolescent steam on a California beach, and fortified with lashings of science and Schopenhauer—and, perhaps, Stebbins's ideas about the undulatory nature of breathing—she built into a theory of dance capable of connecting her to the processes of the universe. That it also, in a sense, linked the machine with organic nature wasn't part of her plan and remained unperceived by her.

Waves pervade her dances—as line, pattern, gesture, and, in a deeper sense, as impetus. Even the pictures of Duncan in repose show her arranged in a series of serpentine curves—a single undulating line flowing through her head, torso, legs, like waves arrested. No wonder Rodin so admired her, and she him; it is in continuous curves that the viewer's eye travels over his twined figures washing up out of the marble.

In *The Three Graces* (a solo she expanded into a trio for her pupils), the

---

*Margaret Thompson Drewal has pointed out that Stebbins's and Duncan's references to Isis and their connecting of electricity and magnetism with the soul could both have been inspired by the two-volume *Isis Unveiled* by theosophist Helen Blavatsky (1877—the year of Duncan's birth). To Blavatsky and to Stebbins, Isis was not only a potent goddess, but a symbol of nature, ancient wisdom, and spiritual power—a role model, in a sense, for women wishing to affirm their instinctual connection with nature and their identity as spiritual beings.

three women, arms wreathed around each other as in Botticelli's *Primavera,* come toward the audience with light, pulsing steps so tiny that it takes them a long time, and a long swatch of the Schubert music, to reach the front of the stage. Once there, they break apart, make a few swirling, beckoning gestures to the space and each other, run to the back, and gently regroup in order to repeat their delicately tidal advance. In *Water Study,* another of Duncan's Schubert pieces, the dancer gazes down and ripples her arms and wrists toward the floor, as if calming or imitating waves that are gradually becoming more powerful. In a solo like her wildly popular *The Blue Danube,* the phrases were built on the impetus of a wave of water: the rush forward, the slight suspended pause, the retreat as if being sucked backward. The wavelike pattern in space is amplified and given weight by its dynamic relation to the contour of the tide—never even subject to growth and diminution.

These are the obvious examples, but in all of Duncan's work, the rising and sinking, tension and relaxation, ebb and flow acknowledged gravity as a force to strive against. This gave her dances, even the lightest and softest of them, a weight unlike that of ballet and a rhythmic physiognomy different from all other dancing contemporary with her.

Yet, despite her affirmation of gravity, she stressed the expansion rather than the contraction, the upward gesture rather than the downward one, the lightening of weight rather than the fall. Her solos often suggested that the performer could overcome something—fear, grief, gravity—to enter an altered state.

She was in love with Darwin's theory of evolution. One of the reasons she railed at ballet was that its gestures did not evolve naturally, as hers did, and her own work evolved during her lifetime to suit her aging body and battered optimism. She still performed some of her old solos after the tragic drowning of her two children in 1913 and the subsequent death of a third shortly after its birth; but the new ones she made were slow, dark, weighty—in some of them she barely moved—and the theories she had developed when she and the century were fresh expanded to fit them. In the poignant *Mother,* she didn't so much stand for a chorus of sorrowing women as for *every* mother who has lost a child. Spurred on by the struggles of France during World War I, she danced the *Marseillaise,* making herself the heroic symbol of a multitude. In Russia, she danced the serf's entry into freedom to Tchaikovsky's *Marche Slave,* and made pieces for her young pupils set to revolutionary songs, like *Warshavianka,* in which her idea about undulating forces served to depict successive waves of patriots—a new one always there to take the flag from the hand of a fallen comrade.

Isadora Duncan, photographed by Arnold Genthe in 1916

She added her contemporary Alexander Scriabin to the pantheon of composers who inspired her.

The scope and influence of Duncanism—as separate from Isadora's own performing—are slippery things to assess. Duncan's premature death, at fifty, happened in 1927, the year in which Martha Graham and Doris Humphrey presented their first certifiably "modern" dances in America. Mary Wigman, the leading force in Germany's *Ausdrucktanz,* had founded a school and company two years earlier. Before the onslaught of the modernists, Duncanism without Duncan, as a theater art, retreated into a backwater, although revivals staged by such Duncanites as Maria-Theresa (Therese), one of Duncan's six adopted daughters, and Julia Levien, pupil of another "daughter," Irma, created a new stir of interest in the 1970's when they coincided not only with the centenary of Duncan's birth, but with a contemporary back-to-nature urge.

Duncan herself had imagined the continuity of her teaching on a lofty scale. Her interpretations of Nietzsche and Darwin had led her to a vision of the "dancer of the future," as an emblem of an improved human species. It's no wonder that she felt it her mission to found schools. She made no

Isadora Duncan at Grunewald, surrounded by her little students

money from them. Indeed, the first of these, which she established at Grunewald, Germany, in 1904 when she was still in her twenties, was a constant drain on the revenues her then-flourishing career brought in.

Yet unresolved difficulties and inherent contradictions plagued the schools. They were grandiose in concept. Duncan had been inspired not only by Darwin and Nietzsche, but by reading *Émile,* by the eighteenth-century philosopher Jean-Jacques Rousseau, whose glorying in nature, reverence for music, and impatience with civilization's constraints endeared him to her. Rousseau had decreed that his imaginary pupil, Émile, must be with his tutor constantly, away from his parents and other influences, preferably in a pastoral place. In accord, Duncan founded a succession of schools that were like glorified orphanages or the state-supported ballet schools of Europe. No tuition. Parents, particularly working people or impoverished widows, were glad to give her their daughters—and it was always girls; a few boys came now and then, rarely stayed. Isadora thought in terms of women's bodies and cheerfully ignored Rousseau's opinion of women as second-class citizens, as she did Nietzsche's and Darwin's.

Duncan believed firmly in the power of beauty to mold character, but along with the Renaissance paintings, the Della Robbia plaques, the Donatello statues, and the Tanagra figurines that adorned her schools, there were also figures of Spartan girls engaged in strenuous exercise for the students to gaze upon. And, while she clearly agreed with Rousseau when he said of Émile's physical regimen, "I would make him the rival of the roebuck, rather than the dancer of the opera," her students were not little dilettantes.

A girl in training for the dance of the future had to do an hour of gymnastics every morning before breakfast, first running around the lawn in her little one-piece maillot if it was fair, then hanging on to a barre to do stretches, body bends, kneebends, leg swings—increasing the throw of the leg and the arch of the back until she could kick the back of her head. She had to do sit-ups, push-ups, and squat jumps. She had to learn to leap high over a taut rope and wide, across two ropes ("Let him learn to make jumps, now long, now high . . . ," Rousseau had advised).

After the limbering-up, the schoolwork, the singing, the music theory, the extra sessions with a Swedish gymnastics instructor, the children had their dancing lesson—with Isadora (oh joy!) if she was not off touring, otherwise with Elizabeth Duncan or, later, with Irma or one of the older girls. They learned the simple, noble gestures Isadora favored. They learned to walk and run and leap with the wind behind them. They studied se-

Erna Schulz, photographed by Arnold Genthe at the Vanderlip estate in Scarborough-on-Hudson, ca. 1917, when Elizabeth Duncan's school was located in Tarrytown, New York

quences like the "Tanagra Figures." In pairs, they did polkas, mazurkas, gavottes, the Duncan waltz with its upward lilt. They were encouraged to work alone in those dance rhythms, making of them something looser and freer. They composed dances or learned ones from Isadora or Elizabeth.

But the schools were a quixotic venture. Duncan couldn't, or wouldn't, spend a lot of time in any of them. Giving systematic instruction warred with her dedication to the spontaneity of dancing. And, of course, she had to make money. Occasionally she would appear to inspire everyone; the rest of the time she contented herself with thinking of all those little girls growing up in beauty, with composing reams of exercises to send to Elizabeth. It was her sister who ran all Duncan's schools except the one in Moscow—Elizabeth, a shadowy figure, whom Isadora loved and trusted; whom Irma in her memoirs put down as a martinet with a limp and no understanding of dance; whom others have remembered with admiration and some affection—one former student recalls that she was a beautiful waltzer despite her handicap.

Grunewald lasted about four years. Bellevue, founded in 1914 near Paris, had to be abandoned almost immediately when World War I began. The school Duncan established in Moscow in 1921 she left in the hands of Irma, while she herself danced to raise the funds that Lenin's government had promised, but could not, in the end, provide. Many potential sponsors looked askance at Duncan's life-style, and she forever scotched her chances of opening a school in America during her ill-starred American appearances in 1922—heroic, but ". . . too obviously high in flesh" for some people's tastes; lecturing audiences from the stage on the brave new world of Soviet Russia, accompanied by a belligerent and often drunken young husband, the poet Sergi Esenin. The schools that Elizabeth Duncan opened in her own name bounced back and forth between Germany and donated estates in Croton-on-Hudson, Yonkers, and Tarrytown, New York. In retrospect, it seems as if, much of the time, the Duncan "school" was a migratory band of children floating through the cities of the world.

Even more damaging to the future of Duncan's art than the schools' instability was her vacillation as to what she was training the students to *be*. Were the schools primarily "schools of life," as she said they were, or was she training dancers, as she sometimes seemed to be doing? During a four-year period that began in 1905, the Grunewald children appeared in over seventy performances, with and without Isadora. They were much admired:

> They appear one after the other, from opposite sides of the stage, bare-armed, bare-legged, in zephyr waving tunics . . . They move with seraphic lightness, deftness, grace, abandon, youth, and pleasure . . . They come on like rippling, light-tossed waves. It makes one happy—it makes one better—to see them.

But the loss of her own offspring turned Duncan's interest in children into an obsession. After those tragedies, her thoughts became fixed on a vision of teaching endless waves of children who would grow up to teach other children. One summer, she and her Russian pupils taught five hundred workers' children in Moscow's Red Stadium; they can't all have been beautiful, but the socialistic scale of the enterprise, their zeal, and her overwhelming loneliness sanctified them all. She resented it that her six best pupils, her adopted daughters, didn't share her enthusiastic plans for them as teachers of future generations. They thought they had been trained to be dancers, and that's what they wanted to be. It's hard to tell from the admonishing letters she wrote them when, as young women, they em-

Four of the "Isadorables": (from left) Therese, Irma, Lisa, and Anna. Photograph by Arnold Genthe.

barked on a tour of America whether she was jealous of them, whether she doubted their talents, or whether she simply didn't want them to grow up and leave her.

It's ironic that a woman so excited about evolution should have created an art that survived her only in a relatively ossified form. She bred disciples who were unable or unwilling to allow themselves the kind of development she had allowed herself. Those who came into contact with her were happy to promulgate her aesthetic exactly as they remembered it. Her private tragedies, her roving, untidy life, her early death all conspired to fix her direct legacy primarily onto the kinds of pieces she made for students or the lighter solos she deemed suitable to teach young girls. Her six "daughters," especially Anna, Irma, and Therese, received much praise from critics and spectators, but none surpassed her, or dreamed of it. In 1979 Therese

(Maria-Theresa), in her eighties and still dancing, said fervently that to be a flame of *her* fire was almost honor enough.

The new image of woman-as-dancer that Duncan had invented for herself assumed other forms—shadowy, most of them, or diluted, though pervasive. During her lifetime there were professionals (or amateur professionals like Lady Constance Stewart-Richardson, wafting through London drawing rooms and onto American stages) who, arguably, imitated the Duncan look: the Wiesenthal sisters of Vienna; Orchidée, billed as a "child of nature," who performed with Loïe Fuller's company in New York in 1909; and, particularly, the Canadian-born Maud Allan. When Duncan went to London in 1908, she found herself playing at the same time as Allan, whom much of London considered prettier and more musical than she. Allan, too, danced barefoot and in a Grecian tunic (except in her risqué *Vision of Salomé*); she danced to some of the same music Duncan used—Chopin piano pieces, Mendelssohn's *Spring Song;* she claimed that she had been turned from a promising career as a pianist, around 1903, by the sight of Botticelli's *Primavera* at the Uffizi (Gordon Craig made no bones of his opinion of Allan: "the first to try it & to find the robbery would pay").

Beyond direct imitation, there was the Duncan contagion. She can't, of course, be held responsible for all the pageants and Greek games in college theater and physical education departments, for the well-meaning girls in

Freshmen at Beloit College, Wisconsin, ca. 1928, in a generic "Greek" dance

bare feet and bunchy tunics tripping over midwestern grass, flourishing the scarves, garlands, and sacrificial bowls that were, by 1914, indispensable in displays of "natural" or "interpretive" dancing. It is, nevertheless, true that her dancing embodied the ideals of spontaneity and self-expression that many progressive educators favored.

Duncan influenced people's lives in ways she couldn't have conceived. Her girlhood friend Florence Treadwell Boynton may not have been the only one to create a style of living that Isadora, tripping from one expensive hotel to another, only thought about. Letting her long red hair hang loose, discarding conventional dress and most cooked food, Mrs. Boynton moved herself, her lawyer husband, and her eight children into a fantasy of a Greek temple she had constructed in the Berkeley hills. There, with only roll-down canvas curtains to protect them from the elements, the family practiced dance as a life art. Florence Boynton's daughter, Sulgwynn Quitzow, who until her death in 1983 at the age of eighty-two, was still at the Temple of the Winds, teaching Duncan dances to Bay Area girls, said of her mother's extraordinary venture, "Isadora was building a temple in Greece, and my mother decided she wanted a temple here. I guess she [Isadora] was the only one Mama ever wanted to keep up with."

Some of Duncan's influence is tricky to pin down. For instance, Marie Rambert admitted to having imitated, out of admiration, Duncan's "Danse des Scythes" from *Iphigenia* in her own early composition *Dance of Warriors,* but young Doris Humphrey, modernist-to-be, had not seen Duncan when she staged a *Greek Sacrificial Dance* and a dance set to Schubert's *Moment Musical* in Chicago, shortly after 1914, for pupils of her former teacher, Mary Wood Hinman. In photos the girls seem cast in the image of Duncan—maidens in tunics lifting imaginary vessels or arranged in willowy lines, each girl's hand on the shoulder of the girl in front, to flank a central figure who lifts both arms to the sky. But what Humphrey had been exposed to, through Hinman, were the plastique gestures and Eurthythmic studies of the Swiss composer and movement theorist Émile Jaques-Dalcroze. Dalcroze knew Duncan and admired her freedom and plasticity (while deploring her musical inexactness); in pictures, his skipping students could easily be mistaken for young Duncan dancers.

Duncan's most potent influences may have been, indirectly, on twentieth-century ballet and on modern dance. Because of her enormous success on her several visits to Russia, "Duncanism" or dancing "à la Duncan" became part of the ballet world's vocabulary. Lines were drawn among dancers and balletomanes: one faction admired her expressiveness, the other ridiculed the simplicity of her steps. Anna Pavlova, who found Duncan marvelous, studded the repertory for her constant tours with such bits of balletic Duncaniana as Mikhail Mordkin's *Bacchanale* and Ivan Clustine's *Orpheus* and *Dionysius.* (An extant film of one of Pavlova's solos shows her swirling her arms, arching extravagantly, and running and skipping with little hint of balletic turnout.) The restless and innovative choreographer Alexander Gorsky grafted Duncanisms onto classical ballet, but the young Mikhail Fokine was influenced in a subtler sense. "Duncan proved," he later said, "that all the primitive, plain, natural movements . . . are far better than all the richness of ballet technique, if to this technique must be sacrificed grace, beauty, expressiveness." Some of his contemporaries claim that it was Duncan who inspired him to look at ancient art, to choreograph to the music of Chopin, to use the dancers' arms and torsos in a freer way. Whether that is true or not, it is certain that she gave him a tremendous jolt into a path he may have been already groping for.*

George Balanchine, who saw Duncan in Russia much later (1921) thought her "awful . . . unbelievable—a drunken fat woman who for hours was rolling around like a pig." But Frederick Ashton, who saw her in London the same year, has written, "The way she used her hands and arms, the way she ran across a stage—these I have adopted in my own ballets." Marie Rambert, who was to guide such British choreographers as Ashton and Antony Tudor, had worshiped Duncan and studied with Dalcroze at Hellerau.

In America Ted Shawn helped spread the notion that Duncan's art had been a dead end, a matter of individual genius. It was he, he said, and his partner-wife, Ruth St. Denis, who had founded modern dance in America. What Shawn failed to mention were the ways in which he and St. Denis had been influenced by Duncan—he, in such flaming pieces as his *Revolutionary Étude* (1921) or the *Primavera* he created to Strauss music in 1917; she, in her Duncan-inspired, Dalcroze-based "music visualizations," like a dance she performed to a Chopin prelude in December 1917 (one month after she and Duncan both danced in San Francisco) and two lovely solos, to Brahms's Waltz No. 15, Opus 29, and Lizst's *Liebestraum* (both in 1921).

*In December 1904 (by the Russian calendar), Duncan gave her first performances in Russia. According to Fokine, sometime near the end of that year, he sent his famous manifesto against undue artificiality and unquestioned tradition in ballet to the director of the Imperial Theaters. Since no copy of the manifesto has ever been found in the archives, it is impossible to say which happened first. However, I can't imagine that anyone who has studied Duncan dancing could watch Fokine's *Les Sylphides* (the original version, *Chopiniana,* was created in 1907) without having certain gestures trigger kinetic memories.

These last dances are, in their ebb and flow and simple presentation, different in style from St. Denis's Orient-inspired vignettes, and quite Duncanesque.

Yet Duncan herself left something to choreographers of the next generation: the idea that the body itself, and not just the choreographic scenario, ought to reflect the creator's private response to the world, and could be altered to do so. When the renegade Denishawn alumni Martha Graham, Doris Humphrey, and Charles Weidman set out to present sterner truths than beauty, they worked them out on their own bodies. And in that, they were not like Ruth St. Denis and Ted Shawn, they were like Isadora Duncan.

·3·

Sphinxes, Slaves, and Goddesses

Margarita Petrovna Pustanovka (Froman) and Mikhail Mordkin in *Aziade,* 1917

Anna Pavlova and Mikhail Mordkin in Alexander Gorsky's production of *Daughter of the Pharaoh*

THE VEILS OF SALOME

Two Russian ballets on Egyptian themes, separated from each other by half a century, illustrate not only a metamorphosis in exotic roles, but a change in how dancers moved and presented themselves onstage. The ballets have quite a few things in common. Both Marius Petipa's *La Fille du Pharaon* (*Daughter of the Pharaoh,* 1862) and Mikhail Fokine's *Cléopâtre* (1909) were based on stories by Théophile Gautier; both were first performed on the stage of the Maryinsky Theater in St. Petersburg—at least, *Une Nuit d'Egypte,* the first draft of *Cléopâtre,* received its premiere there at a charity performance in 1908. Fokine, riffling through the theater's wardrobe for suitable costumes, even borrowed a few from *Daughter of the Pharaoh* (others, not surprisingly, came from *Aida*). The ineffable Anna Pavlova appeared in both ballets. Still two years away from graduation, she danced in the "Pas d'Almées" in the older ballet, and traveled to Moscow in 1906 to play the leading role in Alexander Gorsky's revised version. In *Une Nuit d'Egypte* she was the heroine, Veronika—the juvenile lead, really—and played the same role, now renamed Ta-Hor,* in *Cléopâtre* when it stunned audiences at its first performance by the company that Serge Diaghilev brought to Paris under the name of "Les Ballets Russes."

The differences between the two ballets, however, were more noteworthy than the similarities. Photographs of Pavlova in a *pas de deux* from both ballets clue us in. For *Daughter of the Pharaoh,* Pavlova's hair is swept up into a fashionable *fin de siècle* pompadour, and her jewelry is strictly

*This, confusingly, was the name of the *hero* of Petipa's ballet and a role Fokine himself had once played.

Anna Pavlova and Mikhail Fokine in Fokine's *Cléopâtre*

contemporary, unless you count the tiara in the form of a snake. Except in nondancing scenes, this Pharaoh's daughter wears pointe shoes and a tutu trimmed with Egyptian motifs. Her partner, Mikhail Mordkin, supports her in approved manner, even though the unusual freedom of her torso reminds us that Gorsky had been wildly excited by Isadora Duncan. In poses from *Cléopâtre,* on the other hand, Pavlova and her partner, Fokine, hold their bodies in profile after the fashion of ancient Egyptian art: in one photo, they kneel facing each other, each raising one palm as if in prayer; in another, he kneels and, holding her hands, supports her in an extravagant backbend. Although she wears pointe shoes, her hair is simply dressed, and her costume is designed to look like a skillfully draped piece of striped cloth. Under his striped kilt and Egyptian collar, Fokine wears dark leotards, although he tells us in his autobiography that at the first performance of *Une Nuit d'Egypte* he danced in bare skin and dark makeup—a radical step at the Imperial Ballet.

Daughter of the Pharaoh was one of Petipa's full-blown Oriental ballets—rife with adventure, peril, heroism, and rich in classical *pas d'action* and divertissements. Its prologue and three acts, based on Gautier's tale "Le Roman de la Momie," tells of an Englishman and his servant who, sheltering in an Egyptian tomb, enter the ancient past in an opium dream. Reembodied as Ta-Hor, the hero adventures with Princess Aspicia—saves her from a lion, in fact—but her father wishes to marry her to the King of Nubia. Her suicidal dive into the river Nile to escape the king's unwanted attentions occasions an elaborate underwater ballet. (Fokine expressed scorn for the fact that the Neva River was costumed as a boyar and danced like one, despite the scarcity of such figures in ancient Egypt.) The music was by the useful and tuneful Cesare Pugni. There was an "oriental" sword dance for the heroine, and *bayadères* are listed in the cast, but Petipa pooh-poohed the notion of re-creating Egyptian art in ballet. He had dutifully visited the Egyptian Museum in Berlin, but noted not unreasonably:

> . . . I surmised that the profile positions were the result of the insufficiently developed craftsmanship of the painters, and did not compel the artists in *The Daughter of the Pharaoh* to dance only in profile. Although their painters drew people in a certain way, the Egyptians certainly walked as we do, that is, in a straight direction.

Fokine's *Cléopâtre* had almost no plot to speak of. Like "Une Nuit de Cléopâtre," the Gautier story that was its inspiration, it concerns a young hunter, Amoun, who, not content with the beautiful girl he might marry, aspires to an impossible love—Cleopatra, Queen of Egypt, paramour of Antony. He shoots an arrow bearing the message "I love you," and it falls at her feet. Aroused from her boredom, Cleopatra offers the youth a night of carousing, to be followed at dawn by his death. He agrees. It happens.

In Gautier's story, which was written in 1845, action is subordinated to description. His words linger over gold and marble, evoke colored silks and the sheen and scent of flesh. Passion and death, languor and violence emerge from a wealth of exquisitely arranged details. The ballet, to judge from ecstatic descriptions of it, must have had a similar perfume and a more chilling sado-erotic edge. There was a dance for Amoun and his sweetheart, Ta-Hor, in which she played at being a deer and he mimed hunting her down. Later, Ta-Hor entertained Cleopatra's court with a snake dance—writhing her supple body in imitation of that most potent of erotic symbols.

If the style was not strictly Egyptian, neither was it classical. Few dancers wore pointe shoes. There was a scarf dance for two slaves (Tamara Karsavina and Vaslav Nijinsky), a dance of captive Jewish maidens, a wild quartet of leaping, wine-inflamed fauns and bacchantes. These occurred while Amoun and Cléopâtre were making love behind billowing veils wielded by slave girls, and—building toward a final swirling orgiastic dance—simulated the rhythms of what was supposed to be happening behind them. Critic André Levinson, not a great admirer of Fokine's work, saw the entire ballet as an "unbroken divertissement," with erotic visions providing the continuity.

Cléopâtre embodied ideas about the dancer that Fokine, according to his memoirs, had been pondering ever since his days as a student at the Imperial Ballet School in the 1890's. He disliked the *de rigueur* ballerina manner: ". . . a special proud stance: a unique habit of holding the head almost immobile, the stretching of the neck, and a peculiar gait." You could always "distinguish a ballerina from plain mortals," he complained, but "it was impossible to guess what role the ballerina was supposed to be interpreting, what image she was supposed to convey."

He also disliked, and presumably avoided in *Cléopâtre,* that predominantly frontal presentation of the dancers which suggests that the audience is royalty on whom no backs may be turned.

Undoubtedly the most thrilling moment was Cleopatra's entrance (invented, some say, by the ballet's brilliant designer, Léon Bakst, or, according to others, by Diaghilev's other great artist colleague, Alexandre Benois). The Queen, played by Bakst's discovery and Fokine's pupil, the wealthy and seductive dilettante Ida Rubinstein, arrived in a mummy case borne by slaves. Here is how Jean Cocteau remembers this provocative scene:

> The bearers set the casket down in the middle of the temple, opened its double lid, and from within lifted a kind of mummy, a bundle of veils, which they placed upright on its ivory pattens. Then four slaves began an astonishing maneuver. They unwound the first veil, which was red, with silver lotuses and crocodiles; then the second veil, which was green with the history of the dynasties in gold filigree; then the third, which was orange with prismatic stripes; and so on until the twelfth veil, a dark blue one, which, one divined, enclosed the body of a woman. Each veil was unwound in a different fashion: one called for a manège of intricately pat-

terned steps, another for the skill needed to shell a ripe nut, another for the casualness with which one plucks the petals of a rose; the eleventh veil, in what seemed the most difficult movement, was peeled off like the bark of a eucalyptus.

The twelfth veil, dark blue, Mme. Rubinstein released herself, letting it fall with a sweeping circular gesture . . . She was wearing a small blue wig, from which a short golden braid hung down on either side of her face. There she stood, unswathed, eyes vacant, cheeks pale, lips parted, shoulders hunched, and as she confronted the stunned audience, she was too beautiful, like a too potent Oriental fragrance.

Ida Rubinstein as Cléopâtre.
Photograph by Bert, Paris.

Another indelible image of Rubinstein has come down to us: she stands, languidly regal, with her hand on the head of her crouching slave, the lithe, pantherine Vaslav Nijinsky. Here, in particularly vibrant form, are two characters emblematic of late Romantic style, the Decadence personified: the Favorite Slave, a sensual, savage, yet strangely androgynous male, and the Fatal Woman, whose love brought death. Romanticism has been stood on its head. Instead of a hero like Petipa's Ta-Hor, who longs for a beautiful woman and performs heroic deeds for her, we have a hero ready to sacrifice his life and cast away his sweetheart for a single night of pleasure; instead of a virtuous heroine who prefers death to dishonor or a loveless marriage, we have a woman who uses her power to take a life—for a caprice, out of boredom.

The image of the dancer as an unequivocally deadly femme fatale, subject only to her own libido, with the hero as her prey, seemed innovative in ballet, but Diaghilev and Fokine neither originated this remorseless voluptuary nor held a patent on her. She stares at us from the misty and glittering Salome paintings of Gustave Moreau, from Gustav Klimt's imperturbable and gilded *Salome,* his *Judith* with the head of Holofernes. Wanton, she snakes at the core of Beardsley's undulating drawings of perverse desire. Singing, she sheds her veils in Richard Strauss's *Salome.* The Belle Dame Sans Merci with unspeakable appetites, encrusted with jewels and redolent of perfume, she commands the pages of Flaubert's *Salammbô* and *Herodias,* of Oscar Wilde's exquisitely decadent play *Salomé.* She appears nightly in Gautier's "Clarimonde" to drink the hero's blood. She is the devouring Sphinx Wilde wrote of, whose nails rake the flesh of her victims, even as they are luxuriating in her caresses. O. V. Milosz hymned her despairingly in his "Femmes et Fantômes": "With your eastern dance, wild as the flesh,/ And your mouth the color of murder,/ . . . O Salome of my Shame, Salome!"

Krafft-Ebing's *Psychopathia Sexualis* (1886) and Freud's *The Interpretation of Dreams* (1900) shed light on this image of woman and perhaps influenced new manifestations of it. She is not just a particularly coruscating emblem of the late Romantic fusion of love and death, she is also the castrating female who terrorizes young male psyches. Could this be why Salome was the most pervasive of these heroines? Is not beheading the ultimate castration? Or was it the combination of her seductive dancing with her necrophiliac lust for the head of John the Baptist that titillated spectators? Bram Dijkstra, in his recent fascinating study *Idols of Perversity,* has amassed convincing evidence of the overwhelmingly misogynistic and repressive images of females that nineteenth-century male scientists, artists, and writers fostered in order to counter (whether consciously or not) women's increasing power and clamor for rights.

It is interesting that Salome is one of the first roles that Ida Rubinstein chose for herself. Denied permission to mount Wilde's shocking play in St. Petersburg in 1908, she nevertheless commissioned the dance scene from Fokine, and apparently performed it. Wilde's play was also one of the expensive dance-dramas she produced after she parted company with Diaghilev. (Curiously, Diaghilev's production of *Salome* in 1913—a year after Rubinstein's—with choreography by Boris Romanov and Beardsleyesque decor by Serge Soudeikine, was not a success despite Tamara Karsavina's bewitching performance. Perhaps the public missed Rubinstein's almost hermaphroditic slimness and langourous savagery.) Salome executed her grisly dance in opera houses, on vaudeville stages, and in the movies (the little-known but critically acclaimed 1923 film starring Alla Nazimova). Loïe Fuller, toast of the Symbolists, essayed the role twice—in 1895 and in 1907—with the later version, *La Tragédie de Salomé,* a show of gorgeously turbulent lighting effects and a string of relatively abstract, if impassioned, solos: a dance with pearls, a dance in a peacock-feather costume, a dance in which she played with two six-foot snakes.

The Salome that caught the public imagination most intensely, was admired by Edward VII of England, played two hundred performances at the Palace in London in 1908, and spawned a horde of imitators was that of Maud Allan, a Canadian who had trained to be a musician and who, until 1905 and *The Vision of Salomé,* had been having a moderate success as an imitator of Isadora Duncan. Allan, dark and comely, barefoot, dressed in a filmy skirt and discreetly placed clusters of pearls, seems to have drawn her ideas principally from Wilde and Moreau, although her first performances, in Vienna, may have been inspired by the success of Richard Strauss's opera.

Her solo dance took the form of a flashback. Descending the darkened palace stairs, Salome relives the triumph of her dance before Herod, starts in fear at the sight of a dark object—the head—and begins to dance lasciviously with it. Some critics pretended to find nothing vulgar and much that was spiritual in her dancing, although contemporary descriptions make one wonder:

> Her naked feet, slender and arched, beat a sensual measure. The pink pearls slip amorously about her bosom and throat as she moves, while the long strand of jewels that floats from

Maud Allan in *The Vision of Salomé*. Photograph by Foulsham and Banfield.

> the belt about her waist floats langorously apart from her smooth hips. The desire that flames from her eyes and bursts in hot flames from her scarlet mouth infects the very air with the madness of passion. Swaying like a witch with yearning hands and arms that plead, Miss Allan is such a delicious embodiment of lust that she might win forgiveness with the sins of such wonderful flesh. As Herod catches the fire, so Salome dances even as a Bacchante, twisting her body like a silver snake eager for its prey, panting hot with passion, the fires of her eyes scorching like a living furnace . . .

Fokine and Diaghilev took this heroine—usually presented as a soloist onstage, or backed by extras—and embedded her in danced dramas. In doing so, they put ballet's traditional virtuosity to new uses and introduced a rapt public to a new style of Orientalism in ballet—to savage color combinations, to music that sounded barbaric to Parisian ears. The score for *Cléopâtre* was a pastiche, to be sure, but a pastiche of Rimsky-Korsakov, Tcherepnine, Mussorgsky, Taneyev, and others—composers who had listened well to the melodies of the Caucasus, where Russia meets the Orient. This was a ballet in one act, and plot was subordinated to sensation. Decor, figures, and music fused into a single vivid landscape.

Although Fokine is best known today for *Les Sylphides,* his frail, plotless evocation of moonlight and Romanticism and Chopin, and for *Petrouchka,* it was his Oriental ballets that captivated the early twentieth-century public: *Cléopâtre, Schéhérazade,* the "Polovetsian Dances" from *Prince Igor,* and the less well-remembered ones: *Les Orientales, Islamey, Le Dieu Bleu,* and *Thamar.* (One might, stretching the category slightly, include *The Firebird* among these, since there was more than a trace of the East in the ancient Russia it evoked and in the exoticism of the original Firebird, Tamara Karsavina, in her beaded and feathered draperies.) These ballets—all made between 1909 and 1913—which fit into and elaborated on an already swelling tradition of works for concert stage and vaudeville, brought *fin de siècle* Decadence onto the twentieth-century ballet stage in more morbid and sexually explicit images than nineteenth-century audiences would have tolerated.

More than Fokine's other ballets, they enabled him to transform ballet from an art that never deviated from academic canon into one that altered and expanded according to the dictates of its subject matter. Denizens of the mysterious East didn't wear pointe shoes, and their torsos were as mobile as could be wished; their sexual appetites made them as likely to stretch out on cushions as to stalk about on well-turned-out legs. Even a virtuous maiden, like Ta-Hor in *Cléopâtre,* displayed a sensuous plasticity. For Fokine, the scenarios he developed with his collaborators represented aesthetic worlds to be conquered.

He was almost of an age with Isadora Duncan—she was the elder by three years—and admired her greatly, but as revolutionary artists, their *modus operandi* differed. She would have nothing to do with existing dance traditions; he worked within the framework of ballet—adjusting here, loosening there, adding, subtracting. After all, he had passed through St. Petersburg's renowned Imperial Ballet School and Theater, made his debut in 1889 as a little slave with a ewer in Petipa's "Indian" extravaganza *The Talisman,* and been assigned to teach advanced students while still a young dancer in the company. He knew what he didn't like. In the manifesto that he claimed to have submitted to the director of the Imperial Theaters around the end of 1904, along with a proposed scenario for *Daphnis and Chloë,* he railed against formal mime, virtuosic steps displayed for their own sake, and the cumbersome ballet structures then in vogue. He was not just pleading for expressiveness, but for unity of concept—for music that would provide mood and atmosphere instead of simply danceable tunes; for costumes and decor appropriate to the subject and period being depicted; for steps that violated classical canons if necessary; for an end to peas-in-a-pod ensemble patterns. Although the naturalistic crowd hubbub he devised for his ballets reached back to the "realism" of Bournonville and Perrot, the undoubted catalysts were Isadora Duncan's spontaneity and Konstantin Stanislavski's reforms in acting, and he likened himself to contemporary painters, working in the *plein air* of human movement.

Still, it is doubtful if Fokine would have developed quite the way he did had he not fallen in with the immensely cultivated and worldly Serge Diaghilev and the circle of like-minded artists who had long been the latter's friends and colleagues. As I've already mentioned, it was Benois (or Bakst) who had the brainstorm that resulted in Cleopatra's stunning entrance. Benois was the one who wanted the lovemaking of Cleopatra and Amoun to occur onstage behind curtains. (It was Fokine, however, who found Bakst's idea for the end of *Schéhérazade* unaesthetic—the libidinous harem women stuffed into sacks and lugged away—and conceived the scene of carnage that made ballet history.)

The *Cléopâtre* that the public in Paris and London raved over differed from Fokine's 1908 *Une Nuit d'Egypte* in several revealing ways. It was not simply that Bakst's somber, monumental set and revealing costumes were an improvement over the makeshift decorations Fokine had had to devise in St. Petersburg. Fokine's original ballet utilized a scenario for a never-produced ballet with a score by Anton Arensky, which he found in the archives of the Imperial Theaters. There was a happy ending: a sympathetic high priest (the father of Amoun's faithful girlfriend) had substituted a sleeping potion for the poison. Many of the last dances were performed as a divertissement to entertain Cleopatra and Antony.

When Diaghilev decided that Fokine would be the principal choreographer for his 1909 Paris season of Russian music and ballet, *Une Nuit d'Egypte* underwent a metamorphosis. Diaghilev proposed jettisoning most of the Arensky score and substituting music that would be more "Oriental" in rhythms and harmonies. It was he who slashed the happy ending—perhaps already envisioning the moment in which Cleopatra draws the dying Amoun up by the chin to give him one last enigmatic glance before sweeping out with her servants, leaving the audience with a final tableau of Ta-Hor beating her breast over the body of her beloved. The violently beautiful fantasies of nineteenth-century Orientalist painting finally became flesh.

The exotic look and sound of the worlds that Fokine peopled was in tune with fashions in art and literature. Diaghilev, Bakst, and Benois—friends since their student days, balletomanes, collaborators on the remarkable magazine *The World of Art,* which Diaghilev edited and published between 1898 and 1904—were knowledgeable about trends in the European art world: Symbolist poetry and Art Nouveau found their way into the magazine's pages. They also valued Russian art of the past and present that was relatively unknown outside their homeland. They understood and shared the public's fascination with the Orient, and, even though parts of that Orient lay within Russia's domain or just beyond her borders, they were avid to reconstruct it as fable.

With the opening of the Suez Canal in 1869, the Near East began, in the view of many, to be "spoiled." European fantasies* turned inward, fed on themselves, even though there were new discoveries, new "rages"—like that for Persian miniatures. During the last part of the nineteenth century, the resumption of trade with Japan had rekindled interest in the Far East. Whistler painted his Japanese-inspired pictures, Pierre Loti wrote his *Madame Chrysanthème;* David Belasco and, in 1904, Puccini reworked it as

*For a fuller discussion of this, see the earlier chapter "Heroism in the Harem."

Madame Butterfly. In England and the United States, American impresario Augustin Daly staged the enormously successful *The Geisha* ("Whatever that may be—no doubt something musical," grumbled G. B. Shaw disingenuously). Visitors to the 1900 Paris Exposition (Diaghilev and Benois among them) ogled Cambodian dancers and raved over the Japanese dramas that actress Sada Yacco and her troupe were performing at Loïe Fuller's theater; Bakst and Fokine studied the Siamese troupe that came to Russia in 1901.

Aside from the "Siamese Dance" in *Les Orientales,* the Ballets Russes collaborators staged no Far Eastern ballets, but Japanese art had an impact on the *look* of the stage—not only via the influence of artists like Hiroshige and Hokusai, to whom a 1902 issue of *The World of Art* was devoted, but through the work of artists associated with Art Nouveau, such as Aubrey Beardsley, whom Diaghilev and Bakst admired.

The ballet designs, like these styles, emphasized asymmetry, the spiral line, a certain randomness in the placing of details. But the designs of Bakst, whether seen on the printed page or on the stage, have a feeling that is very different from the morbid clarity of Beardsley's drawings or the pastel tendrils of Art Nouveau; their lashing outlines enclose a mass that is as voluptuous and vividly colored and bold as the dances of the Russian Orient. André Levinson complained that at some point in every Fokine ballet the dancers held hands and snaked around the stage in a sort of het-up farandole, but there is no doubt that Fokine's surging, rippling formations also conveyed a vitality and a sense of lush growth, as well as a picture of an enclosed three-dimensional world.

The fantastic and barbaric effect of the ballets owed much to the music they were set to. The names of Rimsky-Korsakov *(Schéhérazade),* and Borodin *(Prince Igor)* were not familiar in Western Europe. Of those composers known as the "Mighty Five"—Balakirev, Mussorgsky, Borodin, Cui, and Rimsky-Korsakov—only Balakirev had trained abroad. They were all intensely nationalistic. In 1862 Balakirev had traveled in the Caucasus and in these mountains, which separate Russia from Turkey and Iran, had collected Persian and Georgian folk tunes. In his childhood, Rimsky-Korsakov had heard the melodies played by four Jewish musicians employed on his family's estate. These composers liked the freedom and expressive possibilities of the tone poem—which made their music ideal to accompany the color-drenched one-act ballets that Diaghilev was presenting. Their modal progressions, their strange and vivid key relationships, their irregular rhythms and obsessive repetition of rhythmic motifs seemed wildly exotic to Pari-

Costume design by Léon Bakst for Fokine's *Narcisse* (1912)

sians brought up on the work of Western conservatory-trained musicians. What Cocteau described as "spirals of sound so piercing . . . that one's nerves could hardly endure them," was an excerpt from Rimsky-Korsakov's opera *Mlada.*

Although works like *Cléopâtre* and *Schéhérazade* were acclaimed as authentic, they were no more "authentic" than Petipa's Oriental ballets. Petipa's allegiance was to the beauty of the classical vocabulary; Fokine's mission was the reform of ballet. Still, he and his collaborators concentrated on a consistency of vision, which created a semblance of accuracy. Speaking of *Schéhérazade,* Fokine admitted that he realized that "Orientals do not live or dance in such a manner." To be authentic, he would have had to procure real Oriental musicians: "The Orient, based on authentic Arabic, Persian, and Hindu movements, was still the Orient of the imagination. [Although] dancers with bare feet, performing mostly with their arms and torsos, constituted a concept far removed from the Oriental ballet of the time." The public and most critics raved over the exactitude of Bakst's backcloth for *Cléopâtre,* with its colonnade of pharaonic statues. Only a few may have noticed or cared that Rubinstein with her blue-dusted wig, filmy draperies, and jewel-studded halter looked less like an archaeologist's vision of the famous queen than like Albert Samain's evocation of her (". . . swollen with love like a ripe fruit . . .") or like another blue-haired siren of literature—the heroine of Flaubert's *Salammbô.* Since this princess of Carthage indulged in private rites with a large python and watched immobile while the barbarian leader who loved her was torn to pieces by a blood-lusting crowd, the spiritual kinship of the two women (the word "heroine" is hardly suitable) might well have created its own kind of "authenticity."

In any case, the public cared little about accuracy: it wanted—and got—sensual allure. In these ballets, depth is only the modulating of a surface. The characters lust and die in a surge of color, motion, sound, and pattern. *Schéhérazade* was the sensation of the 1910 Paris season of the Ballets Russes, inspiring fashion designers and interior decorators with its astonishing meld of intense blues and greens, vermilions and pinks, its turbans and harem pants. A Persian miniature of superbly erotic promise, it flaunted color combinations, harmonies, gestures that were all, in their own way, "forbidden." Those who know *Schéhérazade* only through recent revivals by the Dance Theatre of Harlem and the Houston Ballet or from the 1940's, when it figured in the repertories of de Basil's Ballet Russe and the Ballet Russe de Monte Carlo, might have trouble understanding its spectacular success. The eroticism and believability-in-fantasy are easily eroded by un-astute direction or shredded by one-night stands, nor could later audiences even begin to be as shocked and titillated as audiences before World War I. Nevertheless, seventy-odd years ago, *Schéhérazade,* in its concentrated voluptuousness, its single-minded depiction of lust and death, surpassed all rival Oriental fantasies as well as Fokine's own *Cléopâtre.*

Although the Rimsky-Korsakov suite of tone poems that Diaghilev appropriated as a score for *Schéhérazade* depicted several *One Thousand and One Nights* tales, the ballet is based only on the prologue to (and motivation for) Schéhérazade's storytelling marathon. As soon as the Shah and his suspicious brother have left on a hunting expedition, the Shah's wives and concubines, including his favorite, Zobeide, bribe the Chief Eunuch to open the doors to the slave quarters. While they are in the midst of an orgy, the Shah and his brother return—the hunt was a pretext for testing fidelity—and have everyone put to the sword. Amid the dead and dying, Zobeide stabs herself. The plot is minimal, which accorded with public taste: R. Brussel wrote of *Le Dieu Bleu* (1912), "To say that its story lacks depth is at once to praise its chief quality."

Zobeide isn't a major dance role. It was created around Ida Rubinstein's enigmatic sexuality and decorativeness. The swirls of dancing—no pointe shoes, little turnout, rippling arms, and a hint of swaying hips—are executed by court dancing women and assorted slaves, and finally by the cadres of wives and their paramours. The centerpiece is the Favorite Slave, with his animal leaps. He seizes Zobeide with a mixture of lasciviousness and humility; his embraces seem to emanate from a crouch. The sexual tang is not only the result of Nijinsky's epochal performance, it's a by-product of the choreography. Toward the end of the orgy, he is jumping repeatedly in the midst of the circling throng—coalescing the entire stage into an image of phallic thrust. And at the grisly climax, with guards stabbing and bodies falling—Delacroix's *Sardanapalus* done in living flesh—the Favorite Slave arches off the floor, feet quivering upward before he falls limp. It is death as the final orgasm.

Fokine's Oriental vein served him well, helped to assure him a place in history, and played itself out fairly quickly. *Islamey,* created in Russia in 1912 and not shown in the West, was a *Schéhérazade* clone, concocted because Rimsky-Korsakov's widow would not permit the use of her husband's score. *Thamar* (also 1912), based on a poem by Lermontov and set to a Balakirev score, went beyond *Cléopâtre* in its single-minded handling of lust and violence. In this ballet, the nymphomaniacal Georgian queen (a role created not by Rubinstein, but by Tamara Karsavina) orders that male

The Death of Sardanapalus by Eugène Delacroix: "The violently beautiful fantasies of nineteenth-century Orientalist painting finally became flesh."

strangers who pass through the gorge that her castle overlooks be brought to her for an hour of love—after which they are hurled out the window.

Fokine broke with Diaghilev in 1912, but returned in 1914 to stage Rimsky-Korsakov's opera-ballet *Le Coq d'Or*—yet another work featuring a beautiful and deadly Oriental female, the Queen of Shemakhan (Karsavina). And in 1934 he, Paul Valéry, and Arthur Honegger collaborated on what may have been the last stand of the *fin de siècle* Oriental femme fatale (outside of movies): Ida Rubinstein's *Semiramis*. As perhaps befits a vanishing icon, this cruel and omnipotent voluptuary mounted a tower and, sloughing off her passions, achieved a mysterious transfiguration.

* * *

The advent of the Ballets Russes inaugurated in Western ballet the kind of dance-drama that G. B. Shaw had cried out for at the turn of the century. Gone were the plots that served as coathooks for stunts. Dancers like Tamara Karsavina, Anna Pavlova, Lydia Lopukhova, Mikhail Fokine, Vaslav Nijinsky, Adolph Bolm made expressivity into another sort of virtuosity. The Oriental works in particular refurbished the shabby image of the male dancer that had dominated nineteenth-century ballet. Except in Denmark, it was female dancers, above all, who had fascinated the public. Diaghilev and his colleagues offered a sort of masculine dance the Western public had not seen. With the exception of Fokine and Mikhail Mordkin, most of the prominent male dancers—Adolph Bolm, who achieved such a success as the ferocious Chief Warrior in the Polovetsian Dances from *Prince Igor;* George Rosai; and those who came later like Stanislas Idzikowski, Leon Woizikowski, and Léonide Massine—were not *danseur noble* types. Although they had had scrupulous classical training, most of them were part of the Russian tradition of character dancing. Had they remained in Russia, they would have performed the gopaks and polonaises, the czardases and "Pas Espagnoles" that studded the divertissements of three- and four-act spectacle-ballets. They would have played slaves and jesters and barbarian warriors and supernatural beings—which is what they played for Diaghilev. The difference was that these were now leading roles.

Nijinsky, although he did dance Albrecht in *Giselle,* was one of these nonprincely dancers. The roles he played in the Ballets Russes must have been determined in part by his physiognomy, but also by Fokine's view of him as unmasculine, and by Nijinsky's intimate relationship with Diaghilev (which seems to have been more that of master and slave than one of two lovers equal in power). He often was cast as a slave, although he also played the charming Krishna-ex-machina in *Le Dieu Bleu;* the powerful but asexual Specter of the Rose; the poet in *Les Sylphides,* with his airy, sustained dancing and his chaste interest in the bevy of nymphs who lean on him; the pathetic Petrouchka; the impish Tyl Eulenspiegel (in his own 1916 ballet). None are conventional presentations of the male dancer as hero.

In the many photographs that were taken of him, he appears short, heavily muscled, with a remarkably long, strong neck and flat, high-cheekboned face. Cameras of the day weren't swift enough to record his amazing jump, but resting or poised for action, he projects an enigmatic combination of lethargy and latent force, sometimes of a savagery that is barely contained. It may have been his suppleness, the curvilinear look of his plastique that made people perceive him as androgynous, and beyond that, as feline.

What is also immediately clear is that his virtuosity (How many pirouettes? Ten? Twelve?) was rivaled by an uncanny ability to transform the look of his whole body from one role to another (which is, again, not the *danseur noble*'s task; as for Gary Cooper, one persona could suffice). In photographs of *Le Pavillon d'Armide*—another Gautier-inspired ballet with a voyage into the past—he is the innocent, dutiful, straight-backed eighteenth-century page, gazing at his mistress with adoring eyes. He appears innocent in *Schéhérazade* too, but it is the frankly sensual innocence of the

Vaslav Nijinsky as the Favorite Slave in *Schéhérazade*. Photograph by Baron Adolphe de Meyer.

primitive; his shoulders rise, his head is thrown back in sheer exuberance at finding himself the object of desire.

Along with the artistic freedom these roles engendered and legitimized in ballet, along with the power they exerted on the public imagination, they embodied an aspect of social change too. Over the century, the resolute, virtuous, and charming slave girl of Romantic ballet had turned into a woman who has taken charge of her destiny with a vengeance. Salome was an avatar of the Decadence—an image of woman-as-destroyer. Male choreographers and spectators projected their hidden desires onto a dream Orient—a created world whose exotic nature transformed and sanctioned those desires. But they projected as well the anxiety and terror aroused by the power of the female. To women in the audience, to performers like Ida Rubinstein or the exotic dancer who called herself "Mata Hari," and especially to Loïe Fuller and Maud Allan, who not only chose the role but created the choreography, Salome was perhaps an alter-image, through which it was proper to express their erotic longings, their will to power, and their suppressed fury toward men.

Ruth St. Denis in *Jephthah's Daughter* (a private performance). Although her skirt and veil are pulled back by wires, it is her artistry that creates the sense of motion. Photograph by Parkers, London, 1906–1908.

THE VEIL OF ISIS

Perhaps the most famous of all "Oriental" dancers, Ruth St. Denis, was one of the few *not* to undertake the role of Salome when the craze was at its peak. Although the critic of the *New York Globe* raved that in the Eastern dance-dramas she created for herself, she had no trouble "out Saloming all the Salomes," St. Denis had other goals than seducing an imaginary prophet and an excited public with her body.

The intelligent and theatrical young American dancer was in Germany in 1907, enchanting audiences with her program of what American writers of the day tended to call "Hindoo" dances, when a Salome was proposed. The art patron and dilettante Count Kessler introduced her to Max Reinhardt, who was planning to mount Wilde's play. St. Denis objected to the play—not enough emphasis on the dancing, for one thing—and Kessler persuaded the poet Hugo von Hofmannsthal, an ardent admirer of St. Denis's art, to write a new scenario for her.

The Salome never materialized, but the fragmentary letters and notes pertaining to the collaboration point up the difference between St. Denis and the hundreds of others who took on the role. Although Kessler told Hofmannsthal that he envisioned the dance blooming out of the poetry "as a flower of poison," his metaphor is one that St. Denis, much as she relished a meaty role, might have found distasteful—that is, were the role to express no more than this. The key to Salome's evil doings, she told Hofmannsthal, was "intense artistic egotism," and an enigmatic sentence of Hofmannsthal's in regard to the project reflects a moral twist that could easily have come from her: "In front of an idol under whose deadly eyes the gratified elation becomes torture . . ." For this performer, the head of

John the Baptist—after having been given its due as an erotic object, no doubt—was to enlighten Salome as to her own spiritual emptiness. In St. Denis's dances, evil was less a compelling force than a misguided desire to dominate nature, and it never triumphed. In later years, when she danced in Christian churches, Babylon herself was enlightened and rose from the ashes of her depravity to become the Bride of Christ (*Babylon,* 1937).

In 1931, around the time of her final rupture with her husband and erstwhile dancing partner, Ted Shawn, St. Denis finally got around to Salome—both acting in the Wilde play and performing the dance separately. Even from a clumsy film made of the dance in 1947, when she was sixty-five or sixty-six, you can get some idea of her approach. This Salome doesn't discard veils, she puts them on. With a typical and quite endearing alternation between seductive mysteriousness and down-to-earth behavior, the white-haired St. Denis, still slim and supple, grabs a piece of fabric from one of two solemn little girls sitting at the back of the draped performing area, and lets the dimensions, weight, and color of the cloth shape a brief dance. Then she tosses it aside and tries another. Herod, off to the side, leans ever more lustfully forward—barely able to contain his excitement when, at the end, she lies before him, exhausted and elated by her own dancing.

An ambiguous sexuality pervades the work, but the essence of the dance, as it was in so many St. Denis solos, is transformation. The various veils that she so skillfully manipulates subtly alter her image, and like *maya,* the veils of illusion that stand between the Buddhist soul and Nirvana, they must all be discarded before her true nature can be revealed.

St. Denis trafficked in images beloved by the Symbolist painters and poets—especially by those pleased to be labeled "Decadents"; but her bejeweled queens and goddesses of Eastern antiquity, her veils, pearls, lilies, incense, and peacocks caught the light differently. For the Symbolists, peacock motifs signaled lust and pride, what Philippe Jullian calls "soulless splendor." Loïe Fuller swathed herself in peacock feathers for one of the dances in her *La Tragédie de Salomé* to show how very bad she was. St. Denis chose instead to transform herself into a peacock to show the folly of vanity, using the exotic feathers to garb a Christian moral. In her 1914 solo *The Legend of the Peacock,* she played a rajah's vain and ambitious concubine, who, poisoned by his first wife, was doomed in the afterlife to stalk the deserted terraces beside her tomb, preening herself in solitude. She could admire Sarah Bernhardt in *The Sorcerer* and, fingers in ears, memorize the actress's superb poses without appropriating the Divine Sarah's sado-erotic repertory of heroines. In 1918 she *did* present herself as Theodora, but her view of this circus dancer turned empress didn't jibe with the role Sardou wrote for Bernhardt. In St. Denis's solo, ". . . stately, rich in color, filled with the atmosphere of court and greatness . . . ," the Empress recalled her former dances for guests at the Byzantine court and then returned to her throne. To St. Denis, dancing ability would not have been a skeleton in an empress's closet, or a sign of her present depravity, but part of her glory.

Her American optimism transformed images generated by European pessimism and surrounded them with an aura of spiritual health, which made her performing different from that of almost all other Oriental dancers who undulated on theater, concert, and vaudeville stages. American audiences felt better about savoring exoticism when it was tempered by moral uplift. Her 1910 *Egypta,* and the subsequent "Egyptian" dances that she composed or drew from it, were not heroic adventure tales bedecked with dancing, like Petipa's *Daughter of the Pharaoh,* nor were they sensuous and violent *tableaux vivants,* like Fokine's *Cléopâtre.* St. Denis's Egypt referred to *The Book of the Dead,* not to Théophile Gautier. In it, the river Nile was ceremoniously invoked; in it, Isis descended from her throne to search for Osiris. The grand subject of *Egypta* was the cycle of birth, death, and rebirth, often expressed in scrupulous, art-historical two-dimensionality—with the union of Isis and Osiris conceived neither as an occasion for erotica nor as a pretext for dazzling steps, but as a gorgeously ceremonious expression of the male/female principle in nature.

Like Isadora Duncan, St. Denis brought to dance a reformer's zeal and a taste for philosophy. Her glamorous, enigmatic presentation of herself depended not only on personal beauty, a limber body, an astute theatrical sense, and a familiarity with all the manifestations of fashionable Orientalism, but on a blend of these with ideas culled from a variety of interrelated sources: Christian Science, Buddhist texts, Vedanta, the writings of the American Transcendentalists, and the teachings of François Delsarte. For her, the Orient was not a roster of characters through whom she could express her wildest and most derelict fantasies. It was a landscape where the refinement of a human soul could be played out, without denying the human body. "Passage indeed O soul to primal thought . . . Passage to more than India" was Walt Whitman's hymn. It could well have been hers too. A spirit journey to India—to all regions of the East—was her means for furbishing an ideal image of herself onstage, through which she hoped to guide herself and her audiences to spiritual fulfillment, and to revitalize dancing by so doing. "I demand of the dance," she wrote, ". . . that it reveal the God in man." Like Isadora Duncan, she was a reformer; unlike Duncan, she usually went about her business in disguise.

"I am Kuan Yin on the Altar of Heaven," wrote St. Denis in a poem about her solo *White Jade* (1926), ". . . my body is the living temple of all Gods." The idea threads back to the days more than thirty years earlier, when young Ruth Dennis's mother (the "St." and the dropped "n" came later) gave her lessons in the Delsarte System of Expression. Mrs. Dennis, an intellectual and freethinking woman, could appreciate the extent to which Delsarte's theories might be taken as practical illustrations of notions about the correspondences between nature and the spirit which she had imbibed via Mary Baker Eddy and Emanuel Swedenborg. Romantic artists of the nineteenth century had been haunted by the duality of flesh and spirit; it would become St. Denis's mission to reconcile them. Delsartean theory not only helped to dispel Victorian constraints about the body (and women's bodies in particular), it dignified posture and gesture as mirrors of the soul's inclinations.

In her autobiography, *An Unfinished Life,* St. Denis speaks of the great impression made on her by a performance given in 1892 by Genevieve Stebbins, the American Delsartean theorist and teacher, and St. Denis's exemplary biographer, Suzanne Shelton, has pointed out (in *Divine Dancer*) that the themes and cyclic form of two of Stebbins's pantomimes, *The Dance of Day* and *The Myth of Isis,* surfaced years later in St. Denis's own *Egypta.* In the first, Stebbins enacted the passage of day from dawn to darkness; in the second, she suggested the evolution of a soul from nothingness or Chaos, through birth, life, death, and resurrection. These ideas of evolution, transformation, and the circular nature of existence permeated almost all of St. Denis's solo dances in one form or another.

Stebbins, who had studied yoga and Eastern philosophy, might have quickened the young St. Denis's interest in the East, but in an era where things Eastern were all the rage, her influence was only one among many. In turn-of-the-century America, "The East" might be encountered via imported art—say, in the Japanese masterpieces acquired by Asiaphile Ernest Fenellosa for the Boston Museum of Fine Arts—or as inexpensive prints; absorbed through the writings of the American Transcendentalists or the lush travel books of Pierre Loti or the quatrains of the *Rubaiyat of Omar Khayyam*; seen in a Broadway hit like *The Shogun;* imitated in a Delsarte exercise (Stebbins's "Eastern Temple Drill" is but one example), or entered into via amateur theatricals—at Olana, the estate of Hudson River School painter Frederick Church, guests were supplied with Turkish costumes for performances on a raised area at the base of the staircase, where handsome kilim carpets hung as curtains. When St. Denis was a girl, one of the boarders at her parents' New Jersey farm, theosophist James Lovell, gave her Mabel Collins Cook's poem *The Idyll of the White Lotus*. It was a book she continued to treasure, even when her first job in the theater—six shows a day as a semiacrobatic skirt dancer in Worth's Dime Museum in New York—allowed her little scope for lofty ideas about the mysterious East.

In the summer of 1900, after performing in London as a small-part actress and dancer in David Belasco's *Zaza,* St. Denis, like Serge Diaghilev, like Isadora Duncan, visited the Paris Exposition and thrilled to the performances given by the Japanese actress Sada Yacco. Loïe Fuller was sponsoring the troupe of Yacco and her husband, Otojiro Kawakami, in slightly Westernized dramas, like *The Geisha and the Knight* whose second act (an adaptation of the Kabuki play *Musume Dojoji*) gave Yacco—as a spurned maiden turned demon—a chance to display six solos designed to convince the monks guarding a temple that she is a dancing girl come to perform a rite.* Had St. Denis seen this—the most frequently performed play in the repertory—she would not only have responded to the exquisite images of Japanese prints set in motion, but perhaps also made note of how subtly one woman could suggest a variety of character traits and hold an audience without a single high kick, split, or backbend. It is unlikely that St. Denis's rehearsal schedule would have permitted her to see Yacco's New York performances earlier that spring before St. Denis sailed for Europe with Belasco's company; otherwise one might speculate that it was Yacco's artistry that inspired her to propose to Belasco that she race from the London theater where she was playing in his *Zaza* to dance a "Madame Butterfly" solo of her own devising in another theater, just before his one-act play of that title.

The solo was probably never performed, but in preparing for it St. Denis, who understood Belasco's penchant for consistency in theme and decorative detail, added readings in Buddhism to her agenda. It is possible that

*These solos (six out of the nine in the authentic *Musume Dojoji*) expressed various aspects of womanhood: a delicate geisha dancing, a young girl playing with a ball, a dance imitating a fish, a fan dance, an umbrella dance, and a fox-demon dance. Not only would these have appealed to St. Denis's nascent interest in transformation, but as St. Denis's biographer, Suzanne Shelton, has pointed out, the structure of performing variegated solos before an audience of monks may well have influenced St. Denis's solo *Radha* (see pages 131–133 of this chapter).

by this time she was already familiar with Indian Vedanta through the writings and lectures of the popular Swami Vivekananda, whose saffron robes, imposing presence, and command of English had stolen the show at the World Parliament of Religions at the 1893 Chicago Columbian Exhibition. Lessons on the bonds between health of body and spiritual health, such as could be found in his *Raja Yoga* (1897), would have slid comfortably into a mind primed by Delsartean theory and Christian Science. One state of *samadhi* as Vivekananda defined it would have appealed to a vibrant young dancer for whom acceptance of the Christian negation of the flesh would have meant the end of a career. In this state of enlightenment, one retains consciousness of soul, body, and the world, yet at the same time sees them all as permeated by Brahman. And his words of advice for the would-be teacher are ones that St. Denis (reading "artist" for "teacher") could take to heart: "What can he transmit, if he has not spiritual power in himself?" St. Denis, like Duncan, appropriated the electrical imagery of the day to describe her mission. In 1913 she told a reporter, "I send out what might be called vibrations which are felt by the audience." Transmitting spiritual messages was what she believed herself to be doing, and she could not convey what she had not first received. It was this sense of herself as a mystical dynamo that maintained the often precarious balance between sensuality and spirituality in St. Denis's dances, and, perhaps, in her life.

Every student of dance history knows the story: on a spring day in 1904, Ruth St. Denis, on tour with Belasco's *Madame Dubarry,* is sipping a soda in a Buffalo drugstore with a friend when she sees above the counter a vision that is to change her life. It is a poster advertising Egyptian Deities cigarettes. Flanked by pillars, foregrounded by growing lotuses, the bare-breasted goddess Isis sits in serene meditation. What matter if above the dark niche holding her throne are written the words "No better Turkish cigarette can be made"? For St. Denis, Isis became an icon of her imagined other self—the emblem and pattern on which to build a career. Here in chaste yet seductive and vividly theatrical form was the mysteriously potent image of woman as font of ancient spiritual wisdom and symbol of nature that had earlier been given literary identity by theosophist Helena Blavatsky in her two-volume study *Isis Unveiled* (1877). Rhapsodized St. Denis years later in her autobiography, "I knew that my destiny as a dancer had sprung alive in that moment. I would become a rhythmic and impersonal instrument of spiritual revelation, rather than a personal actress of comedy or tragedy."

The image crystallized first as Isis's Indian sister goddess, Radha, in a solo-with-extras that St. Denis began to work on in 1905—interestingly, around the time that she appeared as the goddess Sanumati in the Progressive Stage Society's production of Kalidasa's *Sakuntala*. St. Denis told a reporter about a less toney influence: the reopened Coney Island amusement park and its picturesque Durbar of Delhi—a simulated Indian court, with elephants, snake charmers, nautch dancers, and more dubiously authentic splendors.

Playing King Dushyanta in the single performance of *Sakuntala* on June 18, 1905, was Edmund Russell, the Delsartean high priest, whose readings from Sir Edwin Arnold's long Buddhist poem *The Light of Asia* were a popular salon event. It's possible that St. Denis discovered Radha in an early edition of *The Light of Asia* that also contains Arnold's genteel Victorian translation of the erotic Indian poem the *Gita Govinda* (which he titled *The Indian Song of Songs* in order to emphasize the soul-body aspect of the union of Radha and Krishna). In the first verses, Radha languishes while Krishna dances with five gopis, each of whom represents one of the senses. As she later admitted, St. Denis, intent on creating a lofty, satisfyingly exotic theatrical persona for herself, laid her Hindu dance-drama in a Jain temple and substituted Buddhist notions of renunciation for the erotic union of Radha and Krishna. She *knew* the East in her soul and didn't vex her nascent choreographic powers with questions of authenticity. "I did not go to India," she remarked grandly, "India came to me."

In *Radha,* as we know it today from photographs, scenarios, reviews, and a fairly dismal film of her 1941 revival (when she was sixty-two), a belly dancer's hip gyrations, a nautch dancer's spins, pseudo-*mudras,* and poses vaguely reminiscent of the flute-playing Krishna or the dancing Shiva—bent leg raised, torso twisted in opposition against it—mingle with the backbends and light waltz steps of the skirt dancer. The music, excerpts from Leo Delibes's *Lakmé,* contributed to the generalized exoticism of this dream India.

At the time that St. Denis created *Radha,* India's ancient temple dances had fallen into a state of decay and had become, by and large, the province of prostitutes. Had she wished to learn real Indian dancing, she would not have been able to. (She did, in preparation for her later *O-Mika,* take lessons from a former geisha in Los Angeles.) What is more important than quibbles over authenticity is the way in which this girl from New Jersey absorbed Krishna, Radha, gopis, and all into herself to create an expressive solo form, much as the Indian Bharata Natyam soloist does, albeit in a different way. To her, this was the way she "bridged the gap between the

St. Denis in *Radha.* Photograph by Otto Sarony, New York, 1908–1909.

god and the rite." As notable is the theatrical shrewdness with which she proclaimed her spiritual values, while giving the public a generous dose of seductive seminude femaleness.

As the dance begins, the idol Radha sits in the lotus position within a shrine, wearing what one headline described as ". . . a Little Jacket and a Skirt of Gauze and Maybe an Anklet or Two." After a procession of priests (extras recruited for the occasion) has filed in with offerings and knelt in homage, the goddess, awakened by their bells and incense, begins to breathe. Descending from her throne, she performs five brief dances; for each, the priests hand her props that shape the choreography and announce the "sense" being depicted. Supposedly visiting her priests in a dream to exhort them to free themselves from the bondage of the senses, the clever Radha demonstrates each one for their edification. Loops of pearls excite her eyes; the little bells that she shakes enchant her hearing; for "smell," she twines ropes of flowers around herself and arches back under the weight; a cup of wine intoxicates her. Building to the most dangerous sense, touch, she dons a long skirt and, reclining, kisses her fingertips, caresses her body. In the climactic "Delirium of the Senses," she spins in orgasmic forgetfulness of self, but, her sensational lesson ended, renounces these dangerous pleasures and returns to her shrine.

Audiences and critics who saw *Radha* early in 1906, when St. Denis was performing her solo twice a day at the New York Theatre on a variety bill that included Nena Blake and her Bronco Beauties, and Kosta, "The Man With The Revolving Head," were probably concerned neither with religious experience nor with authenticity. Many must not have known what to make of her. Even when she entertained guests at Mrs. Stuyvesant Fish's home or in the studio of sculptor Rowland Hinton Perry, some of the newspapers condescended (BAREFOOTED WRIGGLER, FROM FAR-OFF INDIA [IN N.J.], DELIGHTS SOCIETY). After her concert debut in an elegant invitational matinee at the Hudson Theatre, critic Alan Dale, whose article in the *Journal* was headlined . . . DANCES THAT HAVE AROMA, SETTINGS, AND NO STEPS, concluded that ". . . Miss St. Denis could really dance if she wouldn't be so awfully Hindoo."

Whether many of the wealthy and cultured spectators who found St. Denis graceful and artistic and tastefully exotic studied the program notes in which she explained the symbolism of her floor patterns or were led to explore Eastern religions by her dances is a moot point; she fit the decorative tastes of the day and gave them plastic life. There were no angles in her dances, she said proudly; with their curving poses, swirls of simple movement, delicate punctuations, and exotic textures, these first solos presented a surface as intricate and as flowing as Art Nouveau design.

Too, for many of her admirers, as for her, Eastern transformations were liberating. The hostess who wore a caftan, lounged on pillows, and filled her drawing room with incense could discuss poetry, painting, philosophy in a comfort that approached sensual indulgence while feeling that said comfort only enhanced understanding. It beat literary societies where you wore tight clothes and sat on hard chairs.

By the end of 1906, St. Denis rated being included with Isadora Duncan in an article in the *Literary Digest* entitled "New Artistic Expressions," and within four years of her debut, an increasing number of people were seeing her more as she wished to be seen—perhaps as a result of her successes abroad or of the sheer volume of high-minded words she emitted in interviews and the articles she wrote. Instead of being faulted for a lack of snappy steps, she could be associated with the India of Rudyard Kipling and Laurence Hope. It could be noted that *Radha* saw the light in the same year that Jules Massenet's opera *Le Roi de Lahore* was premiered. When she performed on a vaudeville program at the Grand Theatre in New York in 1909, a critic wrote approvingly that now working-class people could afford the uplift of her art, and another claimed to have experienced the idea of self merging in the universal spirit from watching St. Denis perform.

At that first concert matinee at the Hudson Theatre on March 22, 1906, *Radha* shared the program with two other solos that also could be considered prototypes of later St. Denis works. *The Incense* tells no story, but suggests a private ritual in terms both concrete and abstract. Because the solo is so simple, St. Denis was able to perform it almost until the end of her long life, and films of it made in the 1950's can still enthrall and move spectators. She appears framed in a doorway or opening in the rear curtain of the stage. Wearing a lavender sari and holding a small bowl, she is an anonymous woman come to a temple to "make *puja*." Her gestures, the gliding two-step that carries her from one to the other of two tall incense burners follow closely the rhythms and phrase structure of Harvey Worthington Loomis's vaguely "Oriental" piano music. Always one is conscious of her tender and tranquil face and of the curves that her every gesture and pose trace on the stage space. As with the Greek statues Delsarte admired and the *tribhanga* posture of the South Indian dancer, her body always finds repose in a serpentine line—one hip settling over the supporting foot, the rib cage counterbalancing this position and the head inclining toward the outcurved hip.

Her actions are few: she sets the bowl down, picks it up; she walks, swaying gently with each step; she folds her hands in salutation to the gods; she takes a pinch of something from her bowl and scatters it on the fire. But when she performs this last action, her arm *becomes* the smoke and rises in ripples and small upward spirals. Twice she stands in the center of the

St. Denis in *The Incense.* Photograph by Strauss-Peyton, 1916.

stage and, smiling beatifically, ripples both arms with uncanny suppleness, swaying slightly as if floating in some heavenly elixir. Once her arms rise in their ripples; once they descend—as if the heavenward waves of incense and prayer were bringing down an equally undulant benediction. Watching this solo on film, it is easy to understand how she must have fascinated audiences when she was young, and early still photographs of *Radha* indicate a similar expressive beauty and a kind of innocence.

The virtuosic arms also played a prominent part in *Cobras,* for in this more secular solo, with two green-stoned rings on each hand, she transformed them into snakes to be charmed. While a small group of extras created the atmosphere of a bazaar behind her, she made her apparently boneless limbs ". . . glide around her neck and body in endless soft turnings and twistings." She made them drink from a bowl and strike at the audience. It was a solo meant to be entertaining as well as mysterious and slightly frightening, and bears a family resemblance to such solos as her *Nautch Dance,* in which (at least in later versions that she perhaps juiced up for vaudeville tours) she played a street entertainer cajoling a crowd for baksheesh and muttering imprecations when her dancing fails to bring a storm of coins. The flip side of St. Denis's high-mindedness was an irreverent wit and more than a nodding acquaintance with earthiness. She always knew she was performing for an audience, but sometimes she let the audience know it—which may be why some critics, in her words, couldn't decide whether she was indeed ". . . an embodiment in living flesh and blood of a spiritual realization, or merely a resourceful dancer using picturesque names to label an ordinary aesthetic urge."

Unlike Duncan, St. Denis did not draw inspiration from music: "I'm not a music dancer; I'm an idea dancer," she said, interviewed in her old age. Dances came to her in the form of pictures, of landscapes; the music came later. These landscapes were not the vague pastoral ones in which Isadora danced, but very specific and, more often than not, indoor places. The sets she sometimes used, the sumptuous costumes and adroit lighting thrilled even those not easily seduced by effects—like H. T. Parker, music critic for the *Boston Evening Transcript,* who wrote of *The Legend of the Peacock* that it gave her the opportunity ". . . to wear the feathered panoply of a peacock's blues, greens and gold and to keep the expanding lines and the glinting colors in slowly rhythmed motion." As her veiled Isis stood to receive the homage of her priests, he marveled how ". . . the light falling upon her struck off glints now of black and gold and again of blue and silver." These luminous settings—a vine-draped temple in India, a Chinese shrine, a bazaar, a Japanese brothel, a banqueting hall in ancient Egypt—were keyed to a transformation of herself. Except in some lyrical, Duncan-inspired solos of the 1920's, St. Denis never danced as St. Denis; always she adopted an exotic persona to express her ideas about the rituals of existence or spiritual regeneration.

She could be peevish about this. Comparing herself to Duncan, whom she greatly admired: ". . . I had to be an Indian—a Japanese—a statue—a something or somebody else—before the public would give me what I craved." It had been, however, her choice, and, for her, impersonation was more than a common theatrical strategy—something any smart young performer would have learned; it was a necessary means to her goal of spiritualizing dance. She would achieve the Oriental outlook by identifying herself with a figure who already had it, a figure who could personify her ideal inner self. In the East, she knew, the gods danced. By casting herself as various goddesses, beyond human desire, she could *stand for* the enlightenment she sought. In later years, she scolded herself for the hubris of this, but kept right on doing it.

Her dance, like *Kuan Yin* (1919), might consist of a garland of sculptured poses. She might, as in *White Jade* (1926), descend from her pedestal, execute a few filigreed gestures, and return to decorative immobility. As *The Spirit of the Sea* (1915), draped in a stage-sized expanse of blue silk, she undulated from a prone position into the full-bodied waves of high tide, then sank back down. Yet even in such minimal solos, she tried to create an image of motion beyond the movements of the dance itself. The queen steps down from her throne, the statue from her pedestal, the goddess emerges from her shrine, and while she herself may not change, she affects the spectator's perception of her. As Isis begins to dance, she raises her veil, and in the enraptured words of Ted Shawn, ". . . her dance signifies to him of receptive heart that she will raise the veil of her mystery that he may see that she is benign and beautiful in her intentions toward all living creatures." Her dancing done, she draws her veil around her again, and the solo ends as it began. It is not circular, however, but spiral: we have seen her face.

In certain dances, the "other" motion was that of *becoming*. As with *Radha* or her spare 1908 *The Yogi,* she might act out the drama of enlightenment. She might dance to appease the gods and so change a human life. Transformation as theme could even mold an apparently secular dance like her popular *Dance of the Black and Gold Sari* (also known as *Dance of the Red and Gold Sari*): an Indian shop clerk demonstrates a sari to a prospective customer by skillfully draping it on herself in a number of fetching ways. Yet, as some recall St. Denis performing the solo originally, another meaning could emerge—the woman gradually becoming transfigured by the splendor of the garment.

St. Denis often used costume as metaphor—an idea she may have passed on to Denishawn student and performer Martha Graham. A change of clothes indicates a change in condition, a change of consciousness. *O-Mika* (1912), a dance-drama with spoken words based on a Lafcadio Hearn story, offered St. Denis as a sort of spiritualized Salome or a high-art performer of the quick-change numbers familiar to vaudeville audiences. As a prostitute entertaining her prospective customers with little dances (including a sword dance with her brother, who designed her decor and lights and filled in as a performer when necessary), she gradually divested herself of five layers of kimonos to stand revealed as the goddess Fugen Bosatsu to the young monk who sought her. (Charles Darnton of the *Evening World* delivered what must be considered an accolade to this exercise in theatrical Japonaiserie when he wrote that it made *The Yellow Jacket,* a long-running Broadway "Chinese" hit, "look like last year's raincoat.") The robing and disrobing recurred in numberless variations. Ishtar, descending through seven gates to seek her lover, Tammuz, in the underworld, was divested of some of her jewelry at each gate, as if she had to renounce her heavenly immunity to accomplish her task. (In the Denishawn Babylonian extravaganza *Ishtar of the Seven Gates* [1923], the goddess's jewelry—and her sublime isolation—are restored to her at the end.) When St. Denis took to performing in Christian churches in the 1930's, the Virgin Mary, too, dropped her veils and wrapped herself in gorgeous mantles of various symbolic hues.

A proto-abstractionist, St. Denis could become the symbol, the distillation of values both temporal and spiritual. When she, her husband, Ted Shawn, and their Denishawn company toured the Orient in 1925–1926, it was Shawn who bustled about learning Senegalese dances; acquiring the choreography and costumes for *Momiji-Gari* from the great Japanese actor of female roles Matsumoto Koshiro; and learning (in three days) from the equally celebrated Chinese actor Mei-Lan Fang, *General Wu says Farewell.* St. Denis—to judge from her memoirs, from those of Jane Sherman, one of her dancers, and from a home movie that "Brother" St. Denis shot—also took lessons, scrutinized local dancers, posed in kimonos and saris, but she internalized the flood of impressions. Struck by the contrast between India's past splendors—as she dreamed them—and the shabby, often sordid realities that faced her every day, she conceived a solo in which she would represent the Spirit of India. Clad in rags, old and sick, she would, with the help of a yogi, renew her strength and cast off her rags (the "transformation number" yet again) to dance once more in an embroidered sari with bells at her ankles. In this case, her physical beauty *stood for* spiritual power.

Her dancing could be earthy, but it was never muscular; its rhythms and forms all had to do with yielding. The search for interior stillness that motivated so much of her work also shaped it: like calm whirlpools, her rippling gestures led inexorably back to her own body—the lodging place of the god. As Suzanne Shelton wrote in the course of a remarkable comparison of Isadora Duncan and Ruth St. Denis, "They followed the polar paths of mysticism: one seeking the Self in the Universe; the other seeking the Universe in the Self." That spectators—or many of them—did sense the mysticism beneath the savvy theatricality and eye-catching stage settings is attested to by the scrapbooks of clipped reviews and agog feature articles that chart St. Denis's career. Here are H. T. Parker's words, these written in 1923:

> She is mistress still, as she was many years ago, of those wave-like motions of hand and arm and body that accord with her Hindu, her Buddhistic personages. She still works illusions of their glittering presence, tranquil posturings, rustlings and rufflings of the spirit within and the flesh without.

". . . the spirit within . . ." She *did* make it visible.

Shawn, whom St. Denis married in 1914, her junior by a dozen years, shared her notions about the spiritual values of dance. Had he not been a divinity student, turned at the last minute from the pulpit toward the stage? Hired, along with his charming partner, Hilda Beyer, to smarten up St. Denis's touring programs with ballroom numbers, he soon became *her* partner. Dispensing to the press the image of himself and Ruth as "great lovers" and perhaps spurred on by the Ballets Russes's first American tour in 1916, Shawn expanded the ideas expressed in St. Denis's solos into genre pieces, with dancers trained in the Denishawn schools taking part as populace or ensemble or secondary soloists, and himself as consort to the goddess. St. Denis's *Spirit of the Sea,* for instance, became a drama with an opening dance of playful "sand nymphs" and a Fisher Boy (Shawn), who throws himself into the sea rather than live without St. Denis, the incandescent sea goddess. He could add theatrical spice to her tales of renuncia-

Ruth St. Denis and Ted Shawn as the reformed Nautch Dancer and the Holy Man—an excerpt from the pageant *Life and Afterlife in Egypt, Greece, and India.* Photograph by Strauss-Peyton, 1916.

tion by interceding between the lost soul and the gods: the saucy nautch dancer tempts the meditating mendicant (with unaccountable muscles), is herself tempted, and sheds her veils and jewels to renounce temporal values in "a clinging gauze of white."

The packaging was shrewd. While watching Wagner's *Tristan* in Dresden, St. Denis had found that, as a spectator, she had difficulty sustaining an intense focus over great stretches of time. Therefore, she always built her more elaborate ballets out of brief, often very simple solos. By joining short dances together into suites, extracting popular numbers from dances too elaborate to tour (like the mammoth pageant, *Life and Afterlife in Egypt, Greece, and India,* created for the Greek Theater in Berkeley in 1916), the Denishawn company could fill a concert program or function as a vaudeville headliner amid the acrobats and comedians and talking dogs (Meredith and Snoozer, the "intellectual bulldog," shared a bill with Denishawners in 1916). Formulas, once arrived at, could be applied to other performers. When, during a period of marital and professional strife, St. Denis took fifteen women dancers on the road in a program of non-Oriental "music visualizations" of Brahms, Chopin, Tchaikovsky, Debussy, et al., Shawn created a fancy vaudeville dance-drama, *Julnar of the Sea* (1919), for Lillian Powell, and in 1920 made the young Martha Graham his co-star in *Xochitl,* "a succession of barbarically gorgeous pictures" in which she played a fiery Indian girl and he an Aztec emperor.

Shawn divined new possibilities in St. Denis's tall, slim figure with its resilient voluptuousness and almost rubbery flexibility, in her glamorous presence, and in her capacity for suggesting different characters through the tilt of her head or the shape of a gesture. Perhaps each cast the other as he or she visualized their real-life relationship. While she made him Tammuz to her Ishtar, Osiris to her Isis—in both cases, a lost lover sought and found by a sorrowing goddess—he made her the bewitching, worldly La Favorita whose favors he sought in his *Cuadro Flamenco,* in *The Garden of Kama,* a high-caste Indian lady to his god-in-disguise, the dancing girl he mastered in *Arabic Suite.* Shawn's description of this last—their first duet—lays out quite clearly how he envisioned their relationship:

> The curtain rose on a quiet desert scene with a Bedouin tent of striped camel's-hair silhouetted against a brilliant, wide sky. Ruth, in a soft and feminine costume of maize, coral, and turquoise, performed her interpretation of the dance of an almeh. At the finish of her solo, she sank down at the tent entrance to await the coming of her lord and master.

> In this role I entered, burnoose flying and white turban twisted high. Brandishing a curved sword, I executed a dashing vigorous dance full of leaps and turns accented by pounding feet. I was a veritable whirlwind of movement and, to the accompaniment of rolling drums and clashing cymbals, worked to a terrific climax of spiral leaps ending center front in a heroic pose with scimitar held high above my head.

To Shawn, St. Denis was both goddess and plunder—more worth the plundering because she *was* a goddess; the hero who could win her was a hero indeed.

Shawn capitulated wholeheartedly to St. Denis's Orientalist visions, but he was perhaps more dedicated to promulgating the image of the male dancer as A Man. If the Denishawn duet *The Garden of Kama* had not predated Rudolph Valentino's rise to stardom, one might think, from the photographic evidence, that Shawn had been out to make the celebrated Latin lover look to his laurels. In 1913, before he met St. Denis and was performing ballroom duets with Norma Gould, he told a reporter on the *San Diego Union,* "Dancing is a manly sport, more strenuous than golf or tennis, more exciting than boxing or wrestling and more beneficent than gymnastics." He created a few works for male ensemble during the Denishawn days, and the all-men company that he founded during the thirties, after separating from St. Denis, featured his dancers not just as savage, bare-breasted warriors, rough-and-ready American plainsmen, toiling slaves, and the like, but as athletes, both literally and in the abstract manner he acquired in a brush with German modern dance. In such works as the "Olympiad" section of his *O Libertad* (1936), he was careful to subordinate dancerly polish among members of his company to the simulation of real muscular effort and free-swinging vigor, and in *Kinetic Molpai* (1935) or *Labor Symphony* (1934), he arranged his men in patterns that were pristine and beefy at the same time.

Shawn's own image was more the product of charisma and skillful performing than physique. It was perhaps inevitable that he would often, during his early years with St. Denis, be referred to as the "American Mordkin": Mikhail Mordkin, touring the United States with Anna Pavlova in 1910, had startled audiences with his virile performing in such Fokine-style Orientalia as *The Legend of Aziyade,* as well as giving them a taste of strong classical dancing. But Mordkin was bold-faced and brawny, whereas Shawn, although tall and well formed, had small, round features, nested between plump cheeks, and a body whose bone structure and musculature was hid-

Shawn and St. Denis in one of a series of poses for *Physical Culture Magazine.* Photograph by White Studio, New York, 1917.

den by a layer of smooth flesh—although that flesh was rarely hidden by his costumes. (When he was an old man, a diet enforced after a heart attack pruned him down, and he was tremendously proud of his newly lean body and sent well-wishers snapshots of himself in bathing trunks.) He must have boiled when San Francisco critic Redfern Mason described him as "orchidaceous." But whether or not he ever came across as effeminate, he did often look like a little boy playing fierce and sexy. Which may have been one of the secrets of his appeal.

Shawn, also trained in Delsarte's system, believed with St. Denis that health of soul went along with health of body. The magazine *Physical Cul-*

ture printed an article by him entitled "Dancing for Powerful Muscles," and put a picture of him, looking bulky, on the cover. Before the Shawn-St. Denis liaison, she syndicated a series of articles on "How Dancing Develops a Beautiful Figure." Their determination to publicize the kind of dancing they did as healthful led to some curious offshoots. A book of health tips and exercises for women, entitled *The Art of the Body* (1931), is studded with photos of the author, Margaret Agniel, in decorative "Oriental" poses wearing nothing but a jeweled headdress. In her preface, she acknowledges her debt to Bernarr McFadden and Ruth St. Denis.

A newspaperman unwittingly articulated the difference between Denishawn Orientalism and the Diaghilev sort. "Simplicity," he said of the Shawns, and "goodness of character." And went on: "They express the wholesome and the sound to a wonderful degree."

The blend of exoticism and wholesomeness was supposed to pervade the company and school as well. Former Denishawner Jane Sherman recalls cultural evenings on tour, during which company members gathered in a hotel room while "Ted" and "Miss Ruth" read aloud from P. D. Ouspensky's *Tertium Organum* and led discussions about a possible fourth-dimensional reality. They also promulgated strict rules about dancers' behavior, liaisons, and dress, most of which were ignored.

A 1918 brochure advised the prospective Denishawn student that among his/her options were Delsarte, Oriental Dance, Egyptian Dance, Ballet, Greek Dance, Geisha, Creative, and Plastique. Yet Denishawners, paying their fees and sweating through classes, could be assured that both school and company were but "visual manifestations of divine mind expressed in dance's action and reaction." And here, too, there were lectures. The first Denishawn School in Los Angeles not only trained dancers such as Martha Graham, Doris Humphrey, and Charles Weidman, and movie stars like Carol Dempster, the Gish sisters, and Louise Brooks (actually a Denishawn dancer first and a cinema heroine later), but young women interested in learning grace and poise from the varied curriculum. A movie made to publicize the school shows students handing their wraps to waiting servants as they head for the swimming pool, and St. Denis and Shawn, in fashionable clothes, taking a little tea on the terrace.

In the studio-theater at 301 South Alvarado, which was draped with striped fabric to resemble the interior of a sheikh's tent, the Shawns, company, and students gave performances that showed Moroccan girls and Siamese temple dancers at all stages of development. In August of 1917, for instance, several dances were performed as part of an "East Indian Nautch."

While the performers were presenting a rajah's banquet with entertainment, the guests, among whom were composers Carrie Jacobs Bond and Charles Wakefield Cadman, were supposedly served "betelnut, spiced sherbet, and sweetmeat paste." A right-thinking journalist wrote enthusiastically of this cultural evening, "The barriers between performer and audience were thus broken down, without resorting to the cheap methods of the Winter Garden." (Perhaps St. Denis was recalling the days when, as Cleopatra, she was unrolled from a bundle of rugs to dance for costumed guests at Louis Tiffany's "Egyptian Fête.") It must be remembered, however, that Shawn was a man of entrepreneurial and managerial gifts. Under the aura of health and culture surrounding Denishawn was a lot of savvy salesmanship: any would-be performer with $50 to spend could learn a Denishawn dance and get sheet music and costume designs thrown in.

Although Denishawn's most famous pupils—Martha Graham, Doris Humphrey, and Charles Weidman—eventually tired of disguises, St. Denis and Shawn inspired many performers to picturesque and spiritualized exoticism. Former Denishawn dancers like Evan-Burrows Fontaine, Vanda Hoff (Saidee Van Hoff), Claire Niles, Florence O'Denishawn (Florence Andrews) purveyed the Denishawn brand of exotica to Broadway, variety, and concert audiences. Like proud parents, Miss Ruth and Ted clipped news of their successes and added them to their scrapbooks. Expert imitator Gertrude Hoffman added a St. Denis impersonation to her copy of Maud Allan's *Vision of Salomé* and her pirated Fokine ballets. Socialites taught by Denishawners draped themselves in veils and showed their guests what they had learned in the way of arm-rippling. Early issues of *The Dance Magazine* teem with pictures of dancers in Oriental regalia and yielding Oriental attitudes—some of them inheritors of the savage Ballets Russes tradition, others more saintly in the St. Denis manner, still others decorative and somewhat vapid climbers on the Oriental bandwagon. The names attest to their exoticness: Dorsha (Doris Bentley), Madam Laurka (who combined a dancing career with devising physical-therapy exercises and "medical gymnastics" for hospital patients—further linking Orientalism with health). The Texan who achieved world fame as La Meri combined scholarly research into ethnic dance with something like St. Denis's creative imagination and performing flair.

The Diaghilev and Denishawn strains of Orientalism—the pessimistic and the optimistic, the secular and the spiritual—mingled in curious ways among choreographers and dancers they may have influenced. Anna Pavlova's

Uday Shankar and Anna Pavlova in *Krishna and Radha,* 1923

principal choreographer, Ivan Clustine, supplied her with Egyptian ballets and Assyrian duets; she had a devouring-female number (*Orientale*) reminiscent of Fokine's. But when she toured the Orient with her company, she, like Shawn and St. Denis three years later, read Rabindranath Tagore's poetry, bought fabrics, learned from experts, and returned with a poeticized exotic repertory—*Ajanta's Frescoes, Japanese Dances,* and *Krishna and Radha* (in which the young and as yet unknown Uday Shankar partnered her at some performances). As Pavlova aged, exhausted by obsessive touring, the various Oriental and Spanish roles prolonged her performing life by offering some respite from the rigors of classical dancing.

One mysterious link between St. Denis and Pavlova is Roshanara, who died young in 1926. Born Olive Craddock in India, the daughter of a British army officer, she supposedly studied Indian dance there, yet pictures and reviews suggest that many of her dances were impressionistic rather than authentic. In 1912, having played Zobeide in *Schéhérazade* during the Diaghilev company's 1911 London season, Roshanara presented four of her own more virtuous dances on a program that Pavlova showed around England. Her repertory included an *Incense Dance,* a *Nautch Dance,* and a *Hindu Snake Dance* to selections from *Lakmé.* In 1916 she was touring America on the vaudeville circuit; the following year, she performed her Indian pieces for Adolf Bolm's Ballet Intime, interspersed with Bolm's own colorful ballets and original Far Eastern vignettes by the Dalcroze-trained Japanese dancer Michio Ito. Photos taken of Roshanara in 1911 in *Incense Dance* show poses strikingly similar to ones St. Denis assumed in her own 1906 *The Incense.* Whether Roshanara could have seen, or even studied with, St. Denis or whether St. Denis simply perceived her as a kindred spirit, news of her doings was regularly pasted into the Denishawn scrapbooks.

In Germany, not long after St. Denis's success there in 1906–1908, the soloist Sent M'Ahesa began to perform dances of a mystical and almost abstract Eastern nature. Accounts of her Egyptian *Bird of Death,* her Bedouin dancer and Siamese temple idol suggest something odder than St. Denis, something closer to emerging modernism, but the resemblance is there. ". . . One may well imagine," said Frank Thiess, "the interior of a temple, in which, all alone, the idol begins to carry out a fantastic dance in order to worship itself."

Perhaps the most potent image lingering from St. Denis's early-twentieth-century American Orientalism is that of this idealized self. But to the modernists who were her pupils, her legacy was not veils and exotic disguises. These they discarded. What they seized on was her conception of a dance as a vehicle for showing change, rather than for displaying the status quo, her idea that one didn't end quite as one began. They did not need the Orient as a pretext. In 1906, seeking a form of personal expression beyond the parts she was offered in plays and the traditional show-off roles then open to the dancer, St. Denis had found the East to be a storehouse of guises. These, in their mystery, beauty, theatricality, "otherness," facilitated the transmission of ideas that, as Ruth St. Denis, she would not have known how to convey onstage. Asia was her better self.

·4·

Modern Movers

Jane Dudley and Marjorie Bahouth, Hanya Holm students, improvising on a New York rooftop, ca. 1934. Photograph by Leo Hurwitz.

THE CREATED SELF

By the end of 1930, quite a few people in America knew what a "modern" dancer looked like, even if they had never seen one. The German duo of Harald Kreutzberg and Yvonne Georgi had performed in American cities during 1929 and 1930. The Denishawn-bred American dancers—Doris Humphrey and Charles Weidman, Martha Graham—had begun to tour with their own companies and their emphatically nonexotic works. And in December 1930, Mary Wigman, the leading exponent of the German *Ausdrucktanz,* the teacher of Kreutzberg and Georgi, crossed the Atlantic to begin her much publicized first solo tour of the United States.

Despite the charisma of Weidman and Kreutzberg, the image of the modern dancer that stuck in the public's mind tended to be female—a bony, barefoot woman whose long, severely cut dress made the high kicks of ballet infrequent and molded her body into a single fervent gesture, thrusting in dynamics, angular in design.

The stereotype was perhaps crystallized by the colorfully hostile journalism and much-bandied-about jokes that dogged modern dance through the thirties, like the remark that if Graham ever gave birth it would be to a cube. In the edition of *The Ziegfeld Follies* that opened in 1935, Fanny Brice, in dark leotard and wraparound skirt, regaled audiences with a fiercely satirical dance and a song by Billy Rose and Ira Gershwin, "Modernistic Moe," in which, according to one critic, she bewailed ". . . the difficulties of eating since her rhythms were modernized." The implication of the slurs seems to be that a career as a modern dancer interferes with the healthy, natural course of womanhood. Few people worried (as they might well have) about what being a chorus girl might do to one's life; the modern dancer's image was more threatening for a number of reasons.

Some members of the ballet world attacked it on the grounds of violating beauty and ignoring traditional schooling. Interviewed during the thirties, Mikhail Fokine aired this view to fledgling journalist Walter Terry: "Ugly girl makes ugly movements onstage while ugly mother tells ugly brother to make ugly sounds on drum." Beauty could make innovation palatable. Duncan had been independent and a freethinker, but she distilled and heightened a prevailing ideal of beauty.

The modern dancer-choreographer rarely went out of her way to look pretty or charming, and *never* looked helpless. Apparently uninterested in seducing an audience, this woman danced about deep matters. Whether dancing alone or with colleagues in clear, bold patterns, she moved to music that emphasized her powerful rhythms, music as lean and dissonant as she was. It was predictable that she would not instantly appeal to the general public. Furthermore, a woman who ran a dance company and created works that expressed a point of view on society was as disquieting to many as the "new woman" who sought a career and went knowledgeable to the polls. Even as dancers in the United States were allying themselves with such respected American values as independence, love of freedom, and relish of hard work, they were questioning other values. Not surprisingly, the modern dancer, who tended to view political conservatism and hidebound social conventions as unsympathetically as she viewed dance traditions, provoked among some people the derision hitherto reserved for the bluestocking and the suffragette.

Nevertheless, the dancer's austere image—forged in Germany during World War I and its edgy aftermath, acquiring new world identity as America slid into the Depression—was well suited to the age. So was the frequent darkness of her vision. If her dancing alarmed or disgusted many, was not spectacular or escapist enough for others, it did stir and excite some. Even as the thirties began, modern dance's predominantly youthful audience was fervent and growing in size—not at all put off by the fact that what Guillaume Apollinaire had said of vanguard artists of 1913 was certainly true of vangard dancers in 1930: "Generally speaking, modern art repudiates most of the techniques of pleasing devised by the great artists of the past."

In a review of Kreutzberg and Georgi, H. T. Parker of the *Boston Evening Transcript* astutely appraised the German version of the new aesthetic:

> By common consent, this post-war era, in which the modernist dance has come to be, in which Mr. Kreutzberg and Miss

Georgi have grown up, is a machinal age—scientific, mechanized, mathematical, metallic, hard-surfaced and, as the comprehensive vernacular has it, hard-boiled. The modernist beauty is a mathematical beauty.

Hence in the dancing of this German pair short, sharp phrases; broken, contrasted, interwoven rhythms; quick, keen thrusts; snapped lines; angles rather than curves; geometrical, rather than sensuous patterns. Like them or mislike them; find them ugliness or pleasure, they do express the time spirit.

In America, as in Germany, modern dancers quickly made an impression on colleagues in music, theater, and the art world who were glad to see dancing that complemented innovations in their own fields. In 1930, when Leopold Stokowski presented the first American performance of Igor Stravinsky's *Le Sacre du Printemps* and Arnold Schoenberg's *Die Glückliche Hand,* he asked Martha Graham to dance the Chosen Maiden in Léonide Massine's choreography for *Sacre* and Humphrey and Weidman to mime the leading roles in the Schoenberg work. Norman Bel Geddes's production of *Lysistrata* was enlivened by Humphrey's and Weidman's choreography, and members of their company studded its cast. (The following year, Graham would design the chorus's movement for Blanche Yurka's *Electra* and dance several solos in it herself.) When in January 1930, Graham, Humphrey, Weidman, and (Helen) Tamiris shared a tactfully balanced week of concerts at Maxine Elliot's Theatre in New York, under the rubric of Dance Repertory Theatre, John Mason Brown, the eminent drama critic, was asked to deliver a speech at the party preceding the season. "The Dance in Modern Theatre" was its subject, and it put an imprimatur of sorts on the new forms.

On the opening program of the Dance Repertory season, Graham and Humphrey each presented a notable dance. Neither was brand-new: Graham's *Heretic* and Humphrey's *Life of the Bee* had premiered the previous spring. Unalike as these works were, they had more in common with each other than with any dance creations of the past. They encapsulated in pristine form some of the values that were to dominate American modern dance of the thirties and represented the "new" dancer in all her stubborn glory.

Heretic was one of the first dances Graham made for the company she formed in 1929. Photographs and descriptions of it, as well as a 1986 revival, make you imagine that for Graham it constituted a kind of primer,

Martha Graham and group in *Heretic.* Photograph by Soichi Sunami.

or perhaps the framing members of a structure to which she would later add a roof and siding—it was that brief, plain, and sturdy.

Heretic was set to a ten-bar Breton folk song, played seven times on the piano in an arrangement by Louis Horst, Graham's musical director, accompanist, mentor, and lover. Twelve women—wearing long dark dresses, narrow and almost translucent, with nets confining their hair—are arranged in a semicircle to form a living wall that blocks, that confronts the soloist (Graham in a white dress, her hair hanging free). The women's arms are crossed, their feet planted wide apart. As Elizabeth Selden, a brilliant chronicler of modern dance, wrote of all the figures in Graham's work during this period, they seem ". . . riveted to the floor, yet straining against it." Each time the spirited, marchlike melody begins, the solitary figure makes her "statement," which questions, disputes, attempts perhaps to plead with the rigidly conformist group that opposes her. When she has finished, the women rise slowly onto their toes and suddenly, stiff-legged, thump their heels back down to the floor. After this rebuttal, in silence, the group shifts, each new position struck with the explosive force of a blow, then held. Some of the dancers lunge at the woman, some look as if they might be spitting at her, some turn their backs. There is another melody for the soloist, poignantly sweet, on which she expresses with a simple move or two her growing helplessness, her firm resolve; then the group reassembles, prepared to repeat its obdurate stamping. Bessie Schönberg, who performed in *Heretic,* recalls Graham as being very lyric, but for the group, ". . . it had to be harder and harder. She had almost eliminated arms. They were either thrown straight up to heaven or all held down. She trimmed the group into wooden figures in a way, but, of course, alive."

The words to the song, which Graham did not use, frame a dialogue between medieval church and heretical sect: the more the upholders of the faith threaten to take from the heretics, the more resolute their spirits become. In Graham's dance, however, on each repetition of the music, the group slams out an uncomprehending rebuttal to the soloist's gestures; and at the end, the visionary outcast sinks down, her voice obliterated. It's as if Graham took as a theme the trial of Joan of Arc, pared it down to the bone, and presented it as a tragic conflict between dissimilar and unequal forces, between the spirit of innovation and the obstinacy of tradition.

Humphrey had been working at group composition longer than Graham, and her *Life of the Bee* is a more complicated work, more fluent in space as befits its subject. To the sound of distant buzzing (people backstage humming on paper-wrapped combs),* the dancers skitter sideways across the stage in a curious stance that evokes an insect's physiognomy—chin down, but eyes up; knees angled stiffly and turned out; weight on the toes; arms straight out to the sides, fanning with increasing scope and vigor; fingers spread. They wear short tunics of black and gold brocade, with slightly longer curved "tails," and black cloches; lines are traced on their faces. Their agitated flying coalesces into formations that refer to clusters of hexagonal honey cells and the pyramidal shape of the hive, while they rhythmically fan into mobility an infant queen. Her initial clumsiness is neither sentimentalized nor prettified. Almost immediately she must engage in a battle to the death with an intruder queen. The hive will tolerate only one leader; the well-being of the group necessitates the sacrifice of the individual. While the group waits, the two queens circle each other, pelvises jerking back and forth, hands jabbing viciously. There is no emotionalizing by the antagonists; they concentrate on the fight. The vanquished queen† is unceremoniously rolled offstage. After a primitive dance of triumph, the "bees" follow the victor off in a long, flying line.

---

*In a later reconstruction, for the Juilliard Dance Theatre, Humphrey set the dance to Paul Hindemith's *Kammermusik, No. 1.*

†Apparently, the intruder (Humphrey) was the victor in early performances; later, it was the young queen who won the fight.

Members of the Humphrey-Weidman company in *Life of the Bee,* with Cleo Atheneos as the Young Queen. Photograph by Soichi Sunami.

These two dances—considered remarkable by those who saw them and performed in them—asserted the modernists' preoccupation with serious subjects, with elemental drives and emotions. Economy of gesture—both have that—and a biting clarity of design. Both explore, in dissimilar ways but with a similar degree of abstraction, the tension between individual destiny and the will of society. In both, the dancers—soloists as well as ensemble members—were presented as bold, intense, lacking conventional virtuosity. The roles of the women in the group demanded of them human vitality and personal heat, but also pruned them into impersonal forces as meticulously meshed as the gears of a machine. Their bodies incarnated dual facets of modern design: the strong jagged lines of angle and zigzag, and an overall streamlining for action, with no decorative details to catch the wind.

"There were no 'pretty' bodies," Jane Dudley, once in the Graham company, remembered of the thirties. "We looked like a collection of modern sculpture."

This, of course, is only part of the picture, just as the image of the modern dancer as grim, plain, and angular is a generalized one. Sometimes the costumes were silk, the skirts full. Often legs flew high in the air. There were certainly curves. Humphrey—red-haired, delicately built, ravishing—was a profoundly lyrical dancer by nature. Tamiris was vivacious, possessed of what one of her chroniclers later called a "heroic and maenadic femaleness." There were wit and lightheartedness. Weidman, with his lean, clever body and mobile face, was a masterful comedian as well as the modern equivalent of the *danseur noble.* Small, fair Hanya Holm—who came to New York from Germany in 1931 to head a Wigman school and stayed to become an important figure on the American scene—had an impish side. And, although most of the time Graham was viewed as Marc Blitzstein saw her in 1931, with a "macabre and mystical face" and a "spirit which is deep and narrow," she could be hilarious, especially when mocking prudish manners or dance conventions. All could display an artless sensuality, and, compared to ballet dancers, they looked impulsive, capable of unexpected vagaries.

Nevertheless, despite the individual variations in style, despite the theatricality of the works,* and despite the fact that around 1934 the leading modernists began to modify their starkness, they still showed more severity and more seriousness onstage than the general public expected from dancers, and less concern with traditional glamour than almost any dancer before them had dared show.

The idea of a stage dancer whose squared-off strength and vigor reflected not only modernity but a kind of primitivism cropped up first in Europe, long before any system for producing such a dancer existed. Vaslav Nijinsky, considered by many to have been one of dance's first modernists, went far beyond any previous ballet choreographer in his demands on the dancer. Members of Diaghilev's Ballets Russes, who first performed Nijinsky's *L'Après-Midi d'un Faune* (1912) and *Jeux* (1913), had to accommodate, not without protest, to a compressed, flattened-out style. Their bodies felt cramped, abused by the steps. Years later, Tamara Karsavina wrote of *Jeux,* "I had to keep my head screwed on one side, both hands curled in as in

---

*Two of Humphrey's "serious" dances, *Water Study* and *Life of the Bee,* were included in the Broadway revue *Americana 1932,* along with Weidman's more certifiably entertaining *Ringside.*

one maimed from birth." Those chosen to be in Nijinsky's epochal *Le Sacre du Printemps* struggled to toe in, to distort their bodies into positions that struck them as ungraceful, to move with primitive force and suddenness, to count the rushing, barbaric rhythms of Stravinsky's music. (When Massine staged his *Sacre* in America, he not only had Martha Graham but, among the celebrants, cadres of Humphrey-Weidman and Graham dancers. Who knew all about being primitive and sudden and counting irregular meters.)

Yet even as Nijinsky was attempting to remodel his recalcitrant classically trained dancers, an entirely different kind of dancer was taking shape. Around the time Nijinsky was learning to count Stravinsky's music with the aid of Marie Rambert, a pupil of innovative and influential composer/theorist/teacher Émile Jaques-Dalcroze, the young Mary Wigman was deciding she didn't wish to become a teacher of Dalcroze Eurhythmics.* Nor did she wish to become a ballet dancer. Gracefulness and delicacy had no significance to her, no relevance to the times as she perceived them. Furthermore, when she began to think about becoming a performer, she was a strong-boned, strong-willed woman of twenty-seven, hardly a candidate for a corps de ballet—had Germany had a potent ballet establishment, which it didn't. And she rejected the cliché of the desirable *fräulein* as firmly as she rejected that of the ballet princess. Years later, in 1926, a clever fictitious "interview" concocted for *Die Weltbühne* had Wigman saying that the dancers who worked with her best understood the ecstasy of dancing when they were hungry, and she hoped that they had the look of women no man would want to marry.

Unlike Nijinsky and his classically trained associates, Wigman had almost nothing to unlearn in order to make her body express contemporary ideas. With Jaques-Dalcroze, she had learned rhythmic exercises and plastique studies, Duncanesque in feeling, that correlated musical phrasing with motion in space in clear, orderly ways. In 1913 a friend, the painter Emil Nolde, thinking Eurhythmics too confining for such a bold person, steered her toward Rudolf von Laban's school.

In Zurich or during summers and war years in lively Ascona—where artists, students, theosophists, and health faddists argued, labored, and cel-

---

*The Swiss Émile Jaques-Dalcroze (1865–1950) developed a system that came to be known as Eurhythmics. Its initial aim was to help music students develop a more acute rhythmic sense through dancelike studies that not only corresponded to the rhythmic aspects of music, but related movement to pitch and timbre. His exercises and his alphabet of gesture also promoted grace, coordination, and spatial awareness; and his experiments with large groups and with the innovative stage designs of Adolphe Appia were influential both on professional choreography and on amateur pageants. All his work was connected to Utopian notions about the perfectability of life.

ebrated—Laban and his pupils worked on a variety of projects. There were the systems of notation and movement analysis that he was developing; there were his "dance plays" and early experiments with movement choirs for huge numbers of laymen and professionals. But although many of the philosophical and pedagogical notions that were to underlie German modern dance came from Laban, the "new dancer" he said he was seeking wasn't built on his own body. Above all, he taught Wigman and his other pupils to tap personal sources.

She was free, for example, to shape herself more in accord with art ideas than with existing dance styles—and was struck by Eastern art in particular. (When Graham's dancers, trained in severity, saw her perform in 1931, they noted with some disapproval an Oriental sensuousness and intricacy

Mary Wigman in "Todesruf" from *Opfer,* ca. 1931. Photograph by Charlotte Rudolph, Dresden.

of gesture, the seductive sway of her hips, a way she had of floating, when it suited her to do so.) She sympathized, too, with the work of Expressionist painters she knew, like Nolde and Oskar Kokoschka. The first sketch of her *Witch Dance* was created toward the end of 1913—the year in which Nijinsky's *Sacre* had its scandalous premiere—and with it she committed herself to venting the demons within.

The grimness of the war and the postwar years in Germany affected her, as it did many of the artists who were her contemporaries. Often, in her solos, a force seemed to be threatening to take over her body. In peaceful dances, the force was benign, a delight to yield to; in darker ones, she struggled against it. When in 1920 she established a school in Dresden and trained a company to perform with her, conflict and duality molded the group work too.

Rejecting the various roles that ballet provided was as easy for the American choreographers as it had been for Wigman. At the time they began to choreograph, they had little or no experience of fine ballet, let alone innovative ballet. Had they, for instance, seen *Les Noces,* Bronislava Nijinska's maverick masterpiece of 1923, they would surely have felt a kinship with her presentation of the dancer as simple, forceful, almost crude, arranged in patterns as architectural as they were ritualistic.* The ballet dancers they did see—always excepting the ravishing Anna Pavlova—were too concerned with display for their taste, too seasoned with artifice and aristocracy. In works that tended to be deplorably second-rate, these dancers were showcased as princes, sultans, demigods, nymphs, princesses, harem girls—governed by dreams and betrayed by love. What right-thinking contemporary American with ambition would aspire to such roles?

But they had more than ballet to rebel against. Unlike Wigman, who had begun to dance and to fix her own way in choreography simultaneously, Graham, Humphrey, and Weidman (generally acknowledged as "pioneers" of American modern dance, along with the expatriate Holm and, sometimes, Tamiris) had been professional dancers for years before they began to seek out individual styles and modes of expression. Touring with Ruth St. Denis and Ted Shawn, all three had become adept at transforming themselves into geishas and samurai, Cambodian royalty, fiery Spaniards, and the ruffle-skirted, frock-coated crowd who populated Shawn's Ameri-

**Les Noces* was first seen in the United States when it was revived for the Ballet Russe de Monte Carlo's 1936 tour.

can genre pieces. So there was the Denishawn tradition to disavow and "the weakling exoticism of a transplanted orientalism," as Graham unkindly put it in the heat of modernism. There was watered-down Duncan dancing to avoid, the look of the languid, garlanded amateur nymph, and with it, eventually, the notion of "art dancing"—hoop and scarf numbers (at which Humphrey excelled), pretty plastiques, and dramatic solo vignettes evoking a character in history, a figure in a painting.*

Denishawn, however, had given these dancers a good grounding in theater, and several valuable ideas. They could use the Delsarte training Denishawn had provided, not as a "system of expression," but as an aid to understanding and analyzing how the body responds to emotion.† The ideas of Jaques-Dalcroze were helpful too. As a girl, Doris Humphrey had learned some Dalcrozian theory and exercises at Mary Wood Hinman's dancing school in Chicago, and she continued those studies with a Dalcroze teacher at Denishawn. By the time she began to aid St. Denis in the creation of "music visualizations," she probably already understood more about the sensitive handling of mass and expressive keying of design to music than the less analytical St. Denis. The form had limitations. Augustus Bridle made them quite clear in the *Toronto Star* in 1924, when he reviewed the St. Denis–Humphrey interpretation of Beethoven's *Sonata Pathétique*:

> The dancers become the notes of a chord; every now and then one breaking away for a melody or a quick run, now and then two climbing as two melodies; always the melodic figure conspicuous as coming out of and flowing back into the harmonic structure of the ensemble which keeps moving, covers the stage with varying color, uses all bodily motions, accents every downbeat, punctuates every pause, tableaus every rest, carries the sustained notes by becoming sculpturesque poses, and so leaves nothing to the imagination . . .

Nevertheless, the music visualizations were plotless, as close to abstraction as Denishawn came, and a strong basis for the musically subtle and morally sinewy group works that Humphrey later made.

---

*Madonnas must have been a popular subject, because in 1928 Tamiris did a "takeoff on the Madonna interpretations," and in 1929 John Martin, *The New York Times*'s dance critic, directed a parody called *Madonnas at Play* for the company run by Dalcroze-trained Elsa Findlay.

†Humphrey's later classification of movements as occurring in opposition, succession, or parallel to each other seems to be rooted in Delsarte theory.

Denishawners rehearsing a music visualization at Mariarden in New Hampshire, summer of 1922; among the dancers are Charles Weidman, Doris Humphrey, and Robert Gorham (center), Martha Graham's sister Georgia (center of group at left), and Louise Brooks (center of group at right). Photograph by Cutter Studio.

Doris Humphrey with Cleo Atheneos, Ernestine Henoch, and Francia Reed in *Dionysiaques* (1932). Photograph by Edward Moeller.

The American choreographers began to make dances in the Denishawn mode, and they didn't become "modern" overnight either. Most of the eighteen short pieces in Martha Graham's first New York concert in 1926 featured romantic images (*Intermezzo, Clair de Lune*) or exotic ones (*Three Gopi Maidens, A Study in Lacquer*). She had her "madonna" (*From a XIIth-Century Tapestry*). All the works were praised primarily for their pictorial beauty. Among them, perhaps a few presaged another approach: *Danse Languide,* for instance, to Scriabin's piano music of the same name. It was undeniably pretty. But it was also, reputedly, cool in tone, spare in design; a photo shows three young women—Graham's students at the Eastman School in Rochester, New York—whose bodies and monochromatic draperies form long, complementary curves. Art Nouveau metamorphosing into Art Deco. Doris Humphrey's *Air for the G-String* (1928) also blended spareness with grace and curving lines. As shown in an early film, five women, wearing robes that sweep the floor behind them in long arcs, walk with measured, softly surging steps to the Bach music. The patterns that the robes leave on the floor, the spiralling of the bodies within the silk trains reflect the music's meticulous organization as well as its long melodic lines. The women lean back slightly, pelvises forward like the female figures in medieval paintings, but these are angels that might have been designed to adorn the seductive geometry of the Chrysler Building.*

By 1927 Graham was studding her programs with harsher material, like the solo *Revolt* (originally called *Dance,* after the Honegger music it was performed to). It had, some recall, the force of a manifesto—the dancer, body braced, elbows jutting, fighting some invisible force. In 1928 she revealed a social conscience in *Immigrant* and the antiwar *Poems of 1917.* It was in the spring of that same year that Humphrey made *Air for the G-String,* but she also veiled her prettiness in "moldy green" in service of the forlorn eeriness of Henry Cowell's *The Banshee.* By 1929 these two women seemed well on the way to defining dance and their roles in it. In addition to creating group works that year, both performed solos notable for their minimalism and force: *Speed* (Humphrey) and *Dance* (Graham). The titles were ironic: Humphrey did not move fast—rather she evoked the illusion of motionlessness that high speed can create; and Graham, who tagged her work with a quote from Nietzsche, "strong, free, joyous action," kept her feet planted on a small platform and moved primarily between her shoulders and her knees.

Yet, perhaps mindful of their audience, the choreographers phased out

*Completed in 1930

prettiness and exoticism gradually. In 1930 Doris Humphrey was still captivating people with her 1924 Denishawn hit "Hoop Dance" (*Scherzo Waltz*); Charles Weidman was performing his *Japanese Actor, XII Century* and *Singhalese Drum Dance*; and Graham's *Danse Languide* and *Tanagra* still appeared on programs.

For these American choreographers, as for the German ones, the task of redefining dance and dancer involved both a subjective and an objective approach. Rather than plundering an extant vocabulary and imbuing it with originality and expressivity via sequence or brilliant performing, they tried to invent their own vocabularies. Their method was first to intensify the connection between emotion and form as had the Expressionist artists (". . . the form is the outer expression of the inner content," Wassily Kandinsky said in 1912; "Out of emotion comes form," echoed Martha Graham in 1927). In August of 1927, Doris Humphrey wrote to her mother:

> I'm putting some of my ideas into practice—although they are not unique, I should rather say *the* idea that everybody is putting into use. Miss H'Doubler and the Germans, and Ronny and all the Bentleys* and so forth are all using the same principle, which is that of moving from the inside out; so I don't feel that I'm stealing anybody's stuff.

The procedure fit the times: Freudian theory encouraged self-exploration. And the deliberate search for an emotional center produced a correlative physical principle: "moving from the inside out" also means initiating movement in the center of the body and letting the limbs follow (as the lash follows the impulse given by the whip handle, Graham observed to her students; a much more vigorous image than Duncan's pulsing waves). Graham likened herself, at a primitive stage in her development, to the dancer in a primitive culture. An atavism both humble and grandiose. She would start from scratch, concentrating on the pulse in the body, on the rhythmic action of foot against ground, and articulate the head, arms, and hands as her work evolved. According to John Martin, she announced, too,

---

*Margaret H'Doubler: pioneering dance educator, long associated with the University of Wisconsin. Ronny Johansson: a Swedish dancer then teaching and performing in the United States. One "Bentley" is surely Alys, who wrote several music books and taught modern dance in Studio 61 in Carnegie Hall. I can't ascertain what other Bentleys Humphrey was referring to. It is possible that she meant teachers of Alys Bentley's ilk.

that the hands were the last part of the human body to develop. In *Frenetic Rhythms* (1933), he noted triumphantly, she had come to the hands. The story is the stuff dance legends are made of, but certainly it was with deliberateness that Graham incorporated into her choreography aerial work, turnout, legato movement, speed, and ground-covering patterns. By any standards, the solo *Dithyrambic*—two years away from *Dance*—with its unprecedented length of fifteen minutes and its "barbaric crescendo of falls"—was a virtuosic piece.

To the American choreographers, pure self-expression connoted self-indulgence. They tried, by distilling and magnifying their own responses, to make themselves stand for humanity. Not simply "a woman mourning," but all the bereaved of the world, a symbol of grief; not just "a foolish woman," but the essence of frivolity. They discovered that the basic processes that ruled their bodies could, by analogy, express the aspirations and struggles of mankind.

Nietzsche was their lodestar, as he was Isadora Duncan's and Mary Wigman's—Wigman "danced Nietzsche" during World War I, for the Dadaists at the Café des Banques. His writing fanned their interest in ritual, in the chorus of Greek drama. In *Also Spracht Zarathustra,* Graham found enthralling ideas about the perfection of the dancer into an ardent supercreature, all affectation burned away in a dedication of energy so white-hot that it fused performer with dance. But it was from *The Birth of Tragedy from the Spirit of Music,* Isadora's "bible," that the choreographers derived principles that supported their art.

The crux of it was the Apollonian/Dionysian tension present in the art of antiquity and in human lives—between the desire for stability and calm on the one hand and, on the other, the passions that throw you off-balance, the call to danger. This duality, recast in physical terms, formed Humphrey's "fall and recovery," the basis for a style and a technique to which Weidman also contributed. For Humphrey, the "drama of motion" lay in the struggle to resist the pull of gravity, in the ecstasy or the tragedy of complying, in the suspended moment before equilibrium was regained. The theory gave rise to the many swings, falls, rebounds, and tilts that identified the work of the Humphrey-Weidman company, to the arcs that a slow, falling gesture traced on the air and the oppositional pull of limbs with which a person thrown off-balance tried to avoid collapse. In *The Shakers* (1931), the almost constantly unbalanced positions of the dancers, their jagged, thrusting motions articulate the battle between principle and sexual desire, between order and religious frenzy. In the solo *Two Ecstatic Themes* (also 1931), Humphrey presented in very pure form the idea of gravity as a force both actual and symbolic. At first, in "Circular Descent," the woman, feet apart, sways, bending deeply backward from the knees. She seems to be trying to evade a force that would press her to the floor, but at the same time she wishes to yield to it, to embrace it. Once on the floor, she immediately begins to rise ("Pointed Ascent"), probing the space above her with small, angular gestures of elbow and hand—directing oblique spurts of force that bring her to a calm and victorious stand.

Breath played a major role in the dynamic of fall and recovery and other aspects of Humphrey's style. It was the source of many vital, nonmetrical rhythms, of fluid successional movements. To Graham, breath was also crucial—not so much everyday even-rhythmed breathing, but the gasp, the sob, the slow sigh of relief, and the ways in which these—heightened and abstracted—could affect the dancer's muscles and skeleton. Her theory of "contraction and release" was built on the act of inhaling and exhaling. The dancer, whether sitting, standing, kneeling or lying down, caves in as if suddenly hit with a blow to the center of her body. But "caves in" is the wrong term if that implies any relaxation of tension. This impulse may be a small one or a series of small ones that affect the body only slightly and momentarily, but it may also be huge, causing her arms to swing sharply forward, her head to bow over, completing the curve in her back. Or perhaps her head lifts, her arms stretch out, her hands cup in miniature contractions of their own. As Graham developed her technique, a contraction might hit the dancer sideways, make her twist, spiral, or be spun to the floor. It might attack percussively, then deepen slowly, resonating throughout her body. But always, no matter how drastic the fall, there is a release, a rise, an advance, an inhalation. The dancer waits—poised, charged—for the next crisis. Graham's approach to the Nietzschean duality was fundamentally more intense and constricted than Humphrey's, less likely to create flow or generous shapes in space. It accorded well with other notions Graham acquired—such as the Eastern one that a gesture, like a boomerang, must return to the sphere of the one who performs it.

Ideas like these created a pervasive dialectic of struggle in modern dance, which immediately and unmistakably set it apart from other forms of Western dance. Engaging parts of the body in adversary relationship to each other, accentuating the power of gravity, could express inner battles or the dancer's fight against some imagined force. Charles Weidman's *Studies in Conflict* (1932) must have been a veritable fever chart of opposition. The Dane Rudyar music was played three times. First Weidman struggled with himself, then with an opponent (José Limón), then with a small group of men—certain of the movements recurring in each section.

Space, too, could be an adversary. During her first years in New York, Hanya Holm wrote that the American choreographers were less aware than the Europeans ". . . of the dramatic implications in the vision of the individual pitted against the universe." A program from one of the lecture-demonstrations that the company she formed in 1936 gave around America announced that Bernice van Gelder would perform a study in "Attraction toward depth," while Elizabeth Waters would essay "Conflict between the attractions of two opposing focal points."

What ballet dancers sought to conceal—effort—modern dancers revealed. The floor wasn't the grass of an imagined glade on which to gambol; it was a drumhead to strike and rebound from. It might be a trap. A leap didn't express the soul's transcendence or earthly passions; it represented a triumph of strength and daring, a hard-won soaring of the human spirit. Any distortion, angularity, or imbalance resulting from emotional verity or physical struggle was not only acknowledged, but affirmed.

We know that the major choreographers of the 1930's had distinctive styles, cherished their individuality, weren't even the best of friends. Yet as we study the photographs, the dark, imperfect home movies in libraries and film archives, we say—sometimes even before we recognize the choreographer—"thirties modern dance."

I can, of course, air the view that time blurs distinctions. I can attribute the similarities to fashions in practice wear. But it's evident that, concerned with private response and original work as these artists were, they were not blinkered to everything but their own inventiveness: they looked hard at the world around them and inquisitively at trends in the other arts and within their own discipline. The state of society, the themes and credos they shared with artists in other fields gave their works, their bodies, and the roles they chose for themselves a certain family resemblance that is as striking as are the individual differences.

It might be supposed that the German dancers, who preceded the Americans into modernism, influenced American modern dance at the outset, yet it remains a moot point how directly and to what extent they may have done so. Certainly during the late twenties, there was a transatlantic traffic of performers, students, and ideas. Louis Horst, who played such an important role in stimulating Graham's development, spent part of 1925 in Vienna. Harald Kreutzberg first came to the States in 1928 with Tilly Losch as his partner. Louis Horst accompanied them on the piano, as he did so many of the New York concert dancers. When Kreutzberg and Georgi arrived, they used Graham's studio to rehearse in. Laban visited America in 1926, although he didn't concertize. In 1926 and 1927, the nineteen-year-old Eugene von Grona gave recitals in New York. Hans Wiener, a.k.a. Jan Veen, appeared around the same time; it was he who, in 1929, gave New Yorkers their first look at a "movement choir"—albeit one humbler in scale than Laban's—sixteen girls moving rhythmically in formations. Ted Shawn brought Wigman pupil Margarethe Wallmann to teach at Denishawn. All the Americans seem to have known and liked Ronny Johansson, who, although Swedish, had been exposed to ideas originating in Germany.

During the twenties, *The Dance Magazine* whetted American curiosity with articles about the German scene—one on Laban in 1926, with photos of his "thoroughly modern dance creations," one on Wigman in 1927, "The Dancer of Dresden." Some publications speculated whether the Americans were, so to speak, "going German." The writer of an unsigned article that appeared in *The New York Times* early in 1928 confidently stated that Graham's February concert might be considered a farewell to her old style, "before she goes over to the new German technique." As critics became more knowledgeable, they discerned differences between the dark-souled Germans, with their attraction to the grotesque, and the more open, straightforward dancing of the Americans. Yet quite a few early reviews presuppose some influence from Germany on the major American modernists, if only as a catalyst.

Later, at least through 1936, young American dancers went to Germany to study and returned to teach what they had learned. While Hanya Holm adjusted her ideas to suit American temperament, American bodies, and the pace of American life, some native-born artists used the word "German" to lend cachet to their publicity, as shown in an item released by Dorsha* around 1931 to explain her "progressively modern" choreography: having studied Italian and Russian ballet; dances of Cambodia, India, and Java; Russian and Spanish folk dances; Isadora Duncan's dance; and "the recent German methods," she has "out of this background, formed her own style and evolved her own forms when none suited her requirement."

It is more obvious that the stripped-down look of American modern

---

*Dorsha (Hayes)—who also performed, at various times, as Doris Bentley, Dorina, and Joan Rhys—director of the small Theatre of the Dance on West Sixty-fifth Street in New York.

dance was in keeping with austerity of the Depression years. Its distortions and conflicts were expressive of the social and political unrest, its emphasis on strength and action symbolically relevant to the activism of the day—the movements and programs to combat hunger, unemployment, injustice. You feel something of the same impulses molding the blunt but shapely prose of Ernest Hemingway, the polished primitivism of e. e. cummings's poetry, the bleak dissonances in the music of Henry Cowell or Wallingford Riegger.

Social dancers had acquired a modern accent earlier, when the chiseled abruptness of the tango began to upstage the fluid, sensuous, three-dimensional patterns of the waltz. The tango rhythm's impudent pounce, lilt, and stalk was picked up by concert dancers. The later Charleston had some of the distortions and angularities the dancers' work displayed, the still later foxtrot its severity and rhythmic interest.

Modern dancers were also translating into physical terms some of the same formal issues that concerned many twentieth-century visual artists. In intensifying their gestures and pruning them of what Kandinsky called "obtrusive beauty," the choreographers tacitly allied themselves with the German Expressionists as well as with the American Precisionist painters who were their contemporaries. What the painters and sculptors embodied in shape and color, the dancers expressed primarily through action. They didn't plunge too deeply into abstraction—committed as they were to articulating human feelings, relishing the concreteness of the human body. Instead, like the Precisionists, they presented recognizable subjects reduced to their essential forms, often inspired by the boldness and blockiness of primitive art. This simplification facilitated the aggrandizement of the dancing figure, spoken of earlier, helped the audience to identify one woman with humanity.

If you look at a couple of extant films of Graham performing her great solo of 1930, *Lamentation,* she may at first make you think of a peasant woman, rocking from side to side, listening for the step of someone who doesn't come, giving herself over to keening the death of a loved one. There she sits on a small bench, feet set wide apart, body encased in a tube of lavender jersey, with only her face, hands, feet showing. The figure bears a striking resemblance to Ernst Barlach's seated bronze *Russian Beggar-Woman* (1907), and, indeed, Graham is known to have been inspired around this time by Expressionist artists Barlach and Käthe Kollwitz. But the dance isn't about weeping, it's more like a question put to destiny. Sometimes the woman tilts on her bench, as if she were trying to wrench herself out

Martha Graham in *Lamentation.* Photograph by Soichi Sunami.

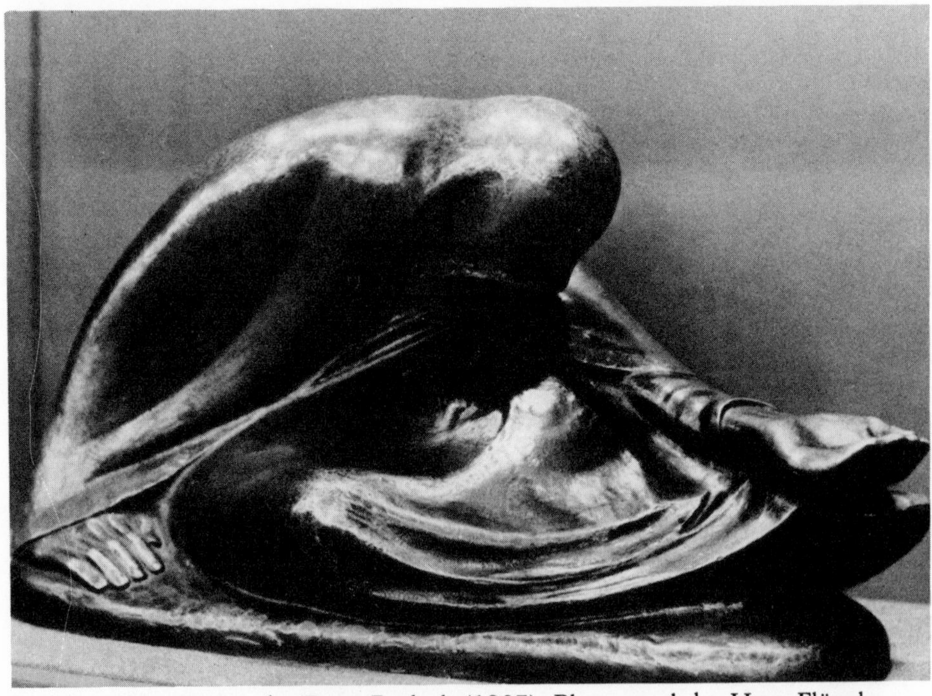

Russian Beggar Woman by Ernst Barlach (1907). Photograph by Hans Flügel.

of the earth. The pull of one part of her body against another, away from her heart, creates diagonal folds of tension in the fabric, until it seems the embodiment of her grief, and that grief becomes a palpable thing that constricts her movements and against which she must fight. It is this presentation of mourning in formal terms—as active struggle and shifting design, rather than as the self-preoccupied emoting of a single woman—that gives *Lamentation* its essential and universal power.*

Around this time, the dancers learned—possibly from the Cubist painters, possibly from movies—that several aspects of the same event could be presented simultaneously. I don't know how Helen Tamiris construed herself when she composed, in 1931, the "Crucifixion" in her solo suite of *Negro Spirituals*. In the film she made of it late in life, I see her as witness, or would-have-been witness ("Were you there when they crucified my Lord?"). But her narrow stance, feet pointing sideways, and the wide, flat twist of her body against those feet suggest the immobile cross itself. The falling back of her head, the hiking of one stiff shoulder, the final spread of her arms make us feel the pain of Christ, but her own bunched-together fingers stiffly hitting the air beside her flanks are the "nails" of the persecutors. And her restraint, her almost impassive face give her the look of a piece of folk art, akin in its powerful simplicity to the folk song it illumines.

Architectural ideas seem to have been particularly stimulating to the dancers, maybe because dance and architecture both mold space with three-dimensional forms. Perhaps, too, the dancers sympathized with the prescriptive idealism of certain modern architects, like Le Corbusier, whose *Toward a New Architecture* appeared in English in 1927. If, as he believed, finely designed mass-produced housing could induce serenity in the inhabitants and so contribute to a more harmonious society, thoughtfully designed dances could surely enlighten and uplift an audience.

Dancers compared their own aversion to decorative movement to progressive architectural principles, and, indeed, the ideas, and the language used to express them, are strikingly similar. No doubt realizing that "angular" was often an epithet when used to describe dancers, Graham told reporters—and allowed her manager, Frances Hawkins, to print the words on a flyer—that her angular gestures could be considered as segments of a curve; they *implied* a curve, which the spectator's eye could complete. Con-

*The way *Lamentation* is currently coached by Graham and performed by members of her company indicates that her tastes have changed. The considerable amount of emoting in terms of dynamics and facial expression, in my opinion, tends to rob the solo of its impersonal force and trivialize the emotion.

versely, when Philip Johnson and Henry Russell Hitchcock proclaimed, in their *The International Style: Architecture Since 1922,* that surfaces should be ". . . unbroken in effect, like a skin tightly stretched over the supporting skeleton," they might have been describing Martha Graham.

The credos of the modernist designers, as explained by Paul Frankl in his *Form and Re-form* (1930), also have the ring of statements issued by the choreographers of the same time:

> To create work in the tempo of the new spirit means to break with the old conventions and to establish new aesthetic values. New materials demand new forms and treatment consistent with their inherent nature. To distort, mutilate, or disguise the innate essence of a medium in a vain effort to make it conform to the conventions of the art, to superimpose upon it some preconceived "Style"—this is the very denial of any sound comprehension of the meaning of Style.

Although the dancers didn't have any "new materials," akin to molded plastic or tubular steel, they developed technologies that did redesign the body in terms of the Machine Age, with the help of rationales similar to those employed by the architects and designers.

Affirming the material as it stood was part of the process. Modern designers initially had trouble producing curves in steel, hence the much-discussed angularity of their work. Although angularity in dance may have been influenced by art ideas, it, too, was an unavoidable feature of the material: functional movements of the human body produce angles more often than curves.

"The things we make," said industrial designer Walter Dorwin Teague, have to be "candidly expressive of the materials and methods used." They also had to be, he felt, "candidly expressive . . . of the purposes for which they were intended." In early houses designed by Le Corbusier and like-minded peers, the steam-heat radiator often sits undisguised—indeed, is featured—against one wall. They found it beautiful. However, since not every utilitarian object is de facto beautiful, the thing manufactured had to express not just its function, but the ideal of its function. As painter Fernand Léger pointed out in a sage little article he wrote in 1924 called "The Aesthetic of the Machine," as the automobile became less boxy and closer to the ground in the interests of swiftness—less like a tottery carriage bereft of a horse—the more perfectly intent and appearance would coincide and the more handsome it would be.

A dining room in Dessau designed by Marcel Breuer in the twenties

So, in addition to discovering new ways of expressing human feeling and dealing frankly with bodies, the dancer-choreographers had to build those bodies into instruments that would express a contemporary ideal of superb function, grounded in nature. Walks, runs, skips, leaps, falls, swings were natural human actions. So, within reason, were kicks and turns. It was natural for a person to walk with feet slightly turned out and to step onto half-toe occasionally. It was not natural to live on tiptoe and turn out 180 degrees.

Pasted inside the cover of a Humphrey-Weidman scrapbook is a picture of Stella Walsh straining at a 45-degree angle into the 220-yard run that set a record in women's indoor track; the choreographers admired the look of a body in peak performance, rather than one attractively arranged. You can see this in old dance films. In these raw, black-and-white fragments, people are dancing on the lawn or in the little theater on top of the old Commons building at Bennington College in Vermont, where almost every summer, from the beginning of the Bennington Summer School and Festival in 1934 through 1938, Holm, Humphrey, Weidman, and Graham all taught and performed. They're dancing across the concrete stage of the Greek Theater at Mills College in California, where the festival was held in 1939. Watching them, you get the impression that impetus mattered more than picture-perfect line: as dancers burst into the air their feet aren't

Harriette Ann Gray of the Humphrey-Weidman company. Photograph by Fritz Kaeser, ca. 1938.

always pointed. But they stride as if intent on covering as much ground as possible. Their torsos are mobile and powerful: when they bend back, they don't arch above the waist; the body is one long line, hinged at the knee, as if bending back might be the prelude to catapulting forward. Graham inculcated in herself and in her dancers the finely tuned awareness of an animal. Her women were to imagine they had fox ears, pricked for every sound; they were to imagine, when they lifted their faces to the sky, that they had a third eye on their chins.

The look, for male and female dancers alike, was a virile one, yet it seems to have been less allied to feminist politics than to iconoclastic views of art. As it happened, many of the idols to be pushed aside embodied qualities traditionally considered "feminine"—ornament, fluidity, prettiness, gracefulness, lightness. Louis Horst told Graham's dancers to put those ideas out of their heads.

Like many artists in other fields, the choreographers were concerned with articulating America—rural and urban, past and present. Dealing with their own here-and-now city life meant dealing with harshness, a rapid pace, and machinery.

By the time World War I ended, the artists had finally come to terms with the machine—ending a long period of head-in-the-sand behavior. Erik Satie's score for *Parade* included the sounds of a typewriter, and George Antheil wrote a *Ballet Méchanique* (1924). Photographers dwelt as fondly on oil wells and storage tanks as they did on daisy fields. Fritz Lang's 1927 film *Metropolis,* along with such plays as Karel Capek's *R.U.R.* (1920), Elmer Rice's *The Adding Machine* (1923), and Eugene O'Neill's *Dynamo* (1929), investigated the perilous relationship between humans and technology. The Precisionists painted images of urban and industrial scenes, often with a sardonic edge. Charles Sheeler called his elegant 1931 painting of railroad tracks and a factory with smokestack fuming *Classic Landscape;* Charles Demuth's grain elevators were *My Egypt* (1927). Beating them all in irony was Morton Schamberg's meticulous 1918 painting of a section of plumbing pipe with inverted U-joint: *God.* Theater designers like Russell Wright and Norman Bel Geddes turned to industrial design, and as the thirties advanced, machines shaped a new aesthetic that dominated almost everything made in America.

In 1917 the Italian Futurist Filippo Tommaso Marinetti had written in his manifesto on Futurist Dance: "We must imitate with gestures the movement of motors, pay assiduous court to steering gear, wheels, pistons, prepare the fusion of man and the machine." Giacomo Balla had already made such a dance in 1914; his *Macchina Tipografica* was composed of twelve human pistons, and a six-person wheel, which they drove. In 1923 one of Nikolai Foregger's famous "machine dances" had the performers representing a transmission. The young von Grona performed mechanical dance pieces in New York in 1927 to percussion accompaniment and, wrote critic-dancer Lillian Ray, with a "machinelike accuracy of gesture." In the thirties, Ted Shawn posed the dancers in his all-male company beside a huge dynamo at General Electric's Schenectady plant, to advertise his *Labor Symphony* and, in the process, beef up the company's reputation for power and modernity. Unlike Chaplin in *Modern Times,* they were not to be intimidated.

But although some choreographers, both in Europe and America, used machine images quite literally in their dances, the major American choreographers I've been writing about couldn't have indulged in this contemporary brand of fantasy without violating their concept of dancing as expressive of human living. In 1934 Martha Graham firmly disassociated herself from all "machine dances": "The dance today does not express a machine. How can man be a machine or imitate a machine? There has been a change of tempo brought about by the machine. We can only express this tempo."

So a percussive attack was called for, a power, a pace, a certain sleekness. During the early thirties, *Rocky Mountain News* often featured a woodcut by industrial designer John Vassos: the jagged pictures were not of dances, but of modern cityscapes; perhaps they were meant to spur the dancers on. An American dancer who studied in Germany observed that after Hanya Holm settled in America, her style developed speed and sharper angles. The dancers emphasized rhythm—not a mechanical rhythm, but its vital analogue (Mary Wigman thought that rhythm was what machines and dancers had in common). In 1936 an up-to-date critic on the *Rocky Mountain News* praised the "streamlined dancing" of Holm's newly formed company by comparing it to the latest mechanical marvel. This dancing, he wrote admiringly, "moves across the stage with the zip of a modern train slipping thru Eastern Colorado into Denver."

Even though the modern dancers' dynamic softened and modulated over the thirties, the percussive accents must have remained noticeable enough for Edwin Denby to have written, relieved, in 1942: "Five years ago, they were concerned chiefly with the emphatic aspects of movement. They socked the active phases of gesture, stamp, jerk, thrust, or step. They gave slow motion a knife edge or contracted with paroxysmal violence. A dance seemed like a series of outcries." Any dancer of the eighties who's ever taken a class in which early modern-dance exercises were taught knows what he means. The dancers of the thirties had a phenomenal ability to move the entire body with a combined force and swiftness that makes dancers of today wince.

But articulating America also entailed dealing with enduring values and traits. Tamiris printed a "Manifest" in the program of her second New York concert (1928): "Art is international, but the artist is the product of a nationality and his principal duty to himself is to express the spirit of his race . . ." Being an American dancer meant conveying a sense of immense frontiers and the pioneer hardihood that conquered them. The vigor and sparseness of Shaker and American Indian design attracted the choreographers, and the narrow, stamping dancing that emphasized the polarity of earth and sky. So, for different reasons, did the syncopated rhythms and mobile hips of black American styles. It was not just that these sources provided specific national imagery—as in Graham's *American Document,* Humphrey's *The Shakers,* Tamiris's *Negro Spirituals,* and many other works—they produced an expansiveness of gesture, a sharpness of outline; figures onstage appeared clearly set apart from each other, as if more virtual space than actual space separated them.

The American past also provided metaphors for the individual creator. The woman in Graham's marvelous solo *Frontier* (1935) sits resolutely on *her* fence, gravely surveys and marks out *her* turf, salutes the space around her, and kicks up her heels at the pleasure of it all. The pioneer woman's attitudes toward solitude, independence, and territory are also those of the mature artist coming into her self-discovered kingdom. The conquest of the land is analogous to the forging of her body, to the mastering of her art.

GROUP SPIRITS

"Now individualism must give place . . . The universal is being made manifest through the employment of mass." This is not a Utopian Socialist making a speech; it's Martha Graham writing on the state of the art in 1929—around the time she enlarged her trio into a company of twelve. That same year, Doris Humphrey, who had been working with group forms for some time, fervently likened the ensemble in dance to the symphony orchestra in music, and wrote to her parents that ". . . the group can express more subtleties, more power, more variety than one single dancer ever could, no matter how intelligent or talented he might be."

During the twenties, many artists and intellectuals had thought of the self as something to be cultivated apart from society. Freudian theory and its popular party-chat ramifications validated this idea. Not surprisingly, solo dance concerts were prevalent. The success of an artist like mime-dancer Agna Enters may have been due in part to the vividness of psychological insights her vignettes could offer.

In the thirties, however, many creative people saw themselves as contributing members of society. The stock market crash of 1929 and the ensuing Depression forced an awareness of the impact that impersonal forces of government and finance could have on the individual, and vice versa. Attractive theories, such as those Harold Lasswell propounded in *Psychopathology and Politics,* smoothed the transition from Freudian self-preoccupation to social consciousness: personal drives, even inadequacies, could be transformed into political and social action.

The issue was how the artist could construe his responsibility to a society that had—via radio, book clubs, phonograph records, and the new talking

Doris Humphrey with (from left) Cleo Atheneos, Helen Strumlauf, Ada Korvin, Katherine Manning, Ernestine Henoch, Eleanor King, and Letitia Ide in *Dionysiaques* (1932). Photograph by Edward Moeller.

movies—more understanding of "culture" than ever before. Some painters and sculptors, newly concerned with the people and needful of money, became interested in enriching public buildings with statuary and murals—like the recently rediscovered aerodynamic abstraction painted by Arshile Gorky on a lobby wall at Newark Airport, one of many murals created under the WPA by painters on monthly stipends.

In 1936, when the Spanish Civil War had screwed every liberal conscience one turn tighter, Charles Weidman addressed himself to the artist's problem in a many-scened parable called *Quest*. The treatment was stylized: Weidman played the artist, Humphrey his alter ego; other dancers were cast as "Ideas." The difficulties were real: one scene depicted the hero trying to work in a repressive dictatorship that professed to honor artists. It was unavoidable that the modern choreographers, radical in their art, would also be radical in their politics, or, at the very least, Democrats; and the last part of Paul Love's program note for *Quest* conveys rather grandiloquently the left-wing ethic of the day: "The artist goes forth and calls others. He knows finally that the artist cannot live in an ivory tower, aloof from the masses of mankind. He is an integral part of the masses and must make that known."

How? Although many choreographers, particularly the younger ones, turned to agit-prop work during the thirties, the concern of choreographers like Weidman was neither to propagandize overtly, nor to make dancing popular; it was to articulate the impact of the individual on society and vice versa. The choreographers, of course, continued to create solos for themselves, Humphrey and Weidman performed duets, and occasionally a small-scale group work appeared, like Weidman's *Studies in Conflict* (1932) or Graham's *Course* (1935) with its duets and trios for seven women. But to mirror the world of factions and nations in ways that left no doubt as to the choreographer's moral stand—that was of prime importance. And to do so, companies of dancers were needed.

Establishing a company required establishing a system for training dancers in a new and idiosyncratic style, and, unlike Isadora Duncan, the German and American choreographers who came after her had no fear of tainting inspiration by organization. Duncan had made use of gymnastic exercises, but considered them purely as a physical warm-up; real dancing was something one had to be inspired to do. The modern dancers, on the other hand, tried to intertwine their ideas about expression with daily practice in dance technique. The classwork evolved directly from the kinds of dances they wanted to make and, in turn, fed back into the stagework. Classes could be a laboratory for choreography and an incubator for company dancers—spiritually exhilarating and intellectually stimulating, as well as physically demanding.

The modern choreographers were not alone in placing almost as high a value on form and craft as they did on expression. At the Bauhaus School, artists, architects, and designers were taught handicrafts and machine work. Nor were they alone in thinking that it made sense to do what ballet had been doing for centuries: to develop the bodies and sensibilities they needed by sound principles, rather than by scattershot direction. In Russia, Constructivist Vsevolod Meyerhold developed "biomechanical exercises" to make his actors more efficient in the use of their muscles; Nikolai Foregger's syllabus of exercises, eccentric moves, and gestures called "tafiatrenage" (taffy-pulling) promoted the skill and control necessary to perform his "machine dances" and defined a style at the same time.

By the mid-twenties, the extravagantly anarchic heydey of Futurism and Dada was over. The emphasis was no longer on tearing down, but on building new forms. It is tempting to consider the interest in technique, the preoccupation with order in relation to the postwar frame of mind: mightn't a stressing of reasoned and harmonious relationships in art and life avert another catastrophic war? The viewpoint is not much more utopian than the assumption that if nations signed the Kellogg-Briand Pact of 1928, swearing to renounce war, peace would prevail.

Once the choreographers had established companies, group structure and day-to-day life gave ideology a leg up, and vice versa. A charismatic dancer-choreographer surrounded by a group of performers he/she had trained: the Individual and Society. The societies were close-knit, trimly scaled, twelve to fifteen strong at the most. Holm's company numbered only seven women in addition to herself. Immense movement choirs like those Laban built in Germany never caught on in America, despite the allure of Laban's visionary notion that participating collectively in an art experience would raise the individual's awareness of his place in some universal harmony. Perhaps the Americans saw, as Laban eventually did in 1938, when he decamped from Nazi Germany, that choirs of thousands moving together could be disconcertingly symbolic of the manipulable mob. In any case, mass dancing wasn't terribly practical on a professional basis during the Depression, and working with amateurs didn't interest the major American choreographers very much. The Bennington four were each given a chance—Graham in 1935, Humphrey and Weidman each in 1936, Holm in 1937, all of them in 1938—to create a work for expanded group; normally their companies were closer in size and nature to the Greek choruses of antiquity.

Hanya Holm's plan for *Trend,* her highly praised 1937 work for expanded company, shows just how focused a choreographer could be on the significance of group-soloist relations. It was in those terms that she outlined for *Magazine of the Arts* the scenario of the dance—a study of a society gradually sinking into decadence and then experiencing "Resurgence." At no other period in dance history can I imagine anyone speaking of a dance as John Martin did of the ending of *Trend:* ". . . the breakup of mass into its essential units of force and their reassembling into more intelligent voluntary associations for joint effort." Earlier in the thirties, students at the Wigman School that Holm ran in New York were improvising on themes of leadership and how it may be transferred, and on how a group reacts to a leader.

The role of leader or instigator was popular with the choreographer-dancers, not just because it reflected the status quo of company life, but because it could so aptly express political verities. The individual could, for better or worse, alter the will of the group. The group could introduce new ideas: in Marcia B. Siegel's analysis of Humphrey's 1938 *Passacaglia,* "All the major dance themes are presented first in germinal form by members of the group, then later enlarged and given back to the group by the leaders." Such ideas might have been a way of embodying—facilitating—the transition from the individualist sensibility of the twenties to group awareness.

The soloist might play the villain: Humphrey's *With My Red Fires* (1936, her "big" piece of that Bennington summer) was enhanced by her terrifying performance as the destructive Matriarch, inciting a crowd to violence. Holm stalked through her anti-Nazi *They Too Are Exiles* (1939) as The Possessor. More often, in these microcosmic republics, the choreographers chose benevolent roles for themselves, as Humphrey and Weidman did in the Utopian *New Dance:* wise leaders, they molded the group by example and then, democracy achieved, stepped back into the ranks.

Primitive Mysteries (1931), one of the very few group dances of Graham's from this period to be revived in recent years, is a masterpiece for many reasons, not the least of which is the tender, subtle, and varied ways it expresses the interdependence of group and soloist.

Graham had begun work on the dance after she and Horst returned from a trip through the American Southwest in the summer of 1930. The barren landscapes she had seen, the sharp light excited her, as they had painter Georgia O'Keeffe. Her vision was fueled by the ceremonial dances of the Pueblo, Hopi, and other tribes, by the blocky *santos* and dressed-doll Virgins in Mexican Catholic churches, by the plangent accommodations Catholicism and paganism made to each other among christianized Indians. Perhaps her own recent experience dancing the Chosen Maiden in *The Rite of Spring* had some influence. Dancers who appeared with Graham in early performances of *Primitive Mysteries* say they thought of her, in her almost diaphanous white dress, as one of them, a young woman chosen from among them to play the Virgin in this ritual. For each of the three "mysteries," she enters and leaves the stage among the group of twelve women in their long, deep blue dresses; she senses with them in the silence the common pulse of their abrupt, weighted, thrust-forward stride. But each of the dances is climaxed by a circle that affirms her as its center.

Throughout this austerely beautiful dance, group and soloist complete each other's actions and give them significance. In "Hymn to the Virgin," the young woman begins by running with pattering steps to join now one, now another group of four women. (A third group sits watching in a line

Martha Graham in "Hymn to the Virgin" from *Primitive Mysteries,* surrounded by (from left) Anna Sokolow, Lily Mehlman, unidentified, Freema Nadler. Photograph by Edward Moeller.

upstage.) On each of her fourteen trips, a picture is created. Here she brings a group into a stiffly angelic circling dance. Here their hands ray around her to form a halo. Twice she leans on them as if they were both votaries and a living pedestal to support her. Four women bend forward, each extending an angled arm; while she stands in front of them on one leg and swings gently, their four arms, touching, rock to her rhythm like one large cradle. In all the images of annunciation, nativity, blessing, comforting, rejoicing, praying, it's often hard to be sure who is initiating, who responding, so deep is the understanding between the women.

After the stark "Crucifixus," in the final "Hosanna," an acolyte, a successor maybe, emerges from the group to create with the soloist a series of grave, iconic tableaux that conjure up images as Christian as the deposition from the cross, as pagan as a four-armed goddess. And all the while, the group forms wreaths and avenues of robust yet contained jubilation around them. The profoundness of the dance is rooted in its fineness and significance of form, but how poignantly Graham affirmed the solo figure's need of the group and the ambiguities of her role as its leader.

In 1929 Doris Humphrey wrote a letter to Letitia Ide, whom she was inviting to join her company. Conscientiously she laid out what dancers could expect of her, what she expected of them. Then, quite carried away by her image of the group: "I want to visualize with it the visions and dreams that make up the entire impetus and desire of my life. The group is my medium, just as marble is the sculptor's material." A few lines later, she writes (I'm tempted to say "adds hastily") that she realizes that the group is made up of individuals who will be noticed by the audience and on whose individual virtues the strength of the group depends. Humphrey, perhaps alone among her peers, was aware of the ironies in her position and confessed as much to her parents in a letter she wrote in 1929:

> Then there is the group. With one hand I try to encourage them to be individuals—to move and think regardless of me or anyone else—and in rehearsals it is necessary to contradict all that and make them acutely aware of each other, so that they may move in a common rhythm. I'm probably crazy to try to do exactly opposite things with the same people.

The blend of autocracy and democracy shaped not only the day-to-day processes of company life, but the dances themselves.

The first dancers Humphrey worked with on her new ideas were students at the Denishawn School in New York, where she and Weidman taught, and they called her "Miss Doris," as they called St. Denis "Miss

Ruth"; before long, plain first names were the custom in modern-dance circles. Graham was by temperament somewhat aloof, but Holm—if Alwin Nikolais's memories of the forties can characterize the late thirties as well—was more *gemütlich:* they'd have morning class, Nikolais says, repair to Holm's apartment upstairs to cook their lunches, hurry back down to the studio for more work.

Since there was little money coming in or going out, no large administrative staffs stood between the choreographers and their dancers. This is not to imply that the major companies were collectives, as some organizations were, like the New Dance Group (founded 1932) in its red-hot heyday. In 1935 there was some dissatisfaction among Humphrey-Weidman dancers with the way things were going. According to one of them, Eleanor King, they called a meeting, which Doris and Charles attended graciously but unenthusiastically. The directors liked one plan: that the dancers tax

Doris Humphrey and Charles Weidman leading a lecture demonstration with members of their company (José Limón is the first man on the right). Photograph by Edward Moeller.

themselves on moneys earned through the company, in order to establish a fund for medical emergencies. They firmly said no, however, to a proposal that the dancers have a voice in determining repertory or engagements.

Strong belief systems held companies together when there was little else to sustain the members. Concert dancers—even those in prestigious companies—could expect to be poor and often hungry. To augment the small fees they earned from their dancing ($10 or $15 per week, perhaps, when they were performing, nothing for rehearsal), they taught others to dance, posed for artists, worked in shops, waited on tables, made out however they could. When Humphrey and Weidman worked on Broadway—he with relish, she with tolerance—they got their dancers jobs too. Companies accommodated to work schedules by holding classes and rehearsals late in the afternoon, at night, or on weekends. Graham and Horst taught at the Neighborhood Playhouse and were able to get scholarships there for many of her dancers, ensuring them training in speech, acting, music, and dance, as well as a small weekly stipend.

Unlike the adolescents who filled the ballet schools and corps de ballet, these dancers couldn't aspire to opera-house fame or glittering technique. They were adults, some of them with college educations behind them. Or strong-minded youngsters, like Anna Sokolow whose teenage ambition was to be a Graham dancer. There were mavericks like José Limón, who'd planned to be a painter until he saw Harald Kreutzberg dance, and then turned up at the Humphrey-Weidman studio on Fifty-seventh Street—all eagerness, intelligence, and so little control that, at first, those in class with him kept well out of his way. These dancers had ideas about how they wanted to move, about what aesthetic ideologies they could subscribe to.

The carefully instilled docility of the ballet dancer was not something the choreographers felt they *ought* to want. Too many dancers loved to be mastered, Humphrey complained. Obedience was replaced by devotion. Lacking largesse, those who directed companies had to keep that fervor alive. One way to do this, of course, was to make their own ideas seem magnificent. Another was to convince the dancers they were participating creatively in the making and performing of ground-breaking dances, even at those moments when their actual input was tiny. I think, too, that for the choreographers to prize their dancers, they *had* to be able to think of them as creative. If the dancers had any gumption, they'd want to make their own dances, thought company leaders—inconvenient though it would be if they quit the fold to pursue independent careers. Even those without great talent for choreography were urged to try it—not a bad idea: since most choreographers were their own leading dancers, you practically had to be a choreographer in order to get a starring role.

The leaders advised reading, or tacked lists of books to the dressing-room wall. Eleanor King recalls a Humphrey list that included Nietzsche, Havelock Ellis, Gordon Craig, Adolphe Appia, and Virginia Woolf. And they did not teach technique alone. Holm, like Wigman, built technique, composition, improvisation, percussion accompaniment, and pedagogy into a complete education for the dancer. Humphrey and Weidman taught composition and, even in technique classes, often encouraged the dancers to approach phrases in their own way. Graham wasn't interested in developing choreographers herself, but her dancers studied composition with Louis Horst at the Neighborhood Playhouse School of the Theatre on West Fifty-sixth Street and/or at Graham's studio at 66 Fifth Avenue, where the best studies were presented in informal concerts. Some of Graham's dancers—Lily Mehlman, Anna Sokolow, Jane Dudley, Sophie Maslow, Lilian Shapero, and others—showed their own work in concerts around town while still in the Graham company, just as Eleanor King, Ernestine Henoch, José Limón, and Letitia Ide had their "Little Group" while still dancing with Humphrey and Weidman.

Pride of group fostered clannishness, although the companies did occasionally appear together, as in the two seasons of the Dance Repertory Theatre, the odd concert—Graham and Weidman shared one at the New School in 1932—or a benefit—International Labor Defense, Dance for Spain. The friendly Tamiris was a leader in many projects to bring choreographers together or improve conditions for dancers. But backstage, and in the summers at Bennington (where Tamiris was not invited to teach or perform), lines were drawn. One of the Bennington legends has it that Grahamites congregated under one tree during time-outs, the Humphrey-Weidman disciples under another. Color proclaimed the company in residence for the whole of a particular summer. According to Helen Priest Rogers, female students in 1935 wore tan cotton leotards. The Graham company dressed in blue—a two-piece leotard that exposed the midriff, made of wool or silk jersey (and usually made by the dancer), a matching wrap-around skirt, and a bolero for after class. Rogers remembers how proud she was that summer because she was one of those chosen to augment Graham's group in *Panorama,* and got to wear the blue outfit that told everybody so.

Audiences couldn't *see* the creative contributions made by members of the group. Mary Rivoire did have a small significant role in the last part of

Helen Tamiris and members of her company in *How Long Brethren* (1937)

Primitive Mysteries, but no one could have guessed that she was, at that time, particularly inspiring to Graham, or that as a Catholic she could confirm in rehearsal the meaning of this or that gesture. Spectators certainly saw the group-as-creative-medium vision Humphrey had spoken of in her letter. Reviews reflect this—approvingly or disapprovingly—in their language; "it," rather than "they," is often the subject of sentences about the group. Lincoln Kirstein, writing in 1937 of the fascination and aversion Graham's earliest group works had generated in him, referred to ". . . her Spartan band of girls seeming to me to press themselves into replicas of the steel woman she was." Elizabeth Selden applied the phrase "artistic militarism" to the rows and ranks of uniformly clad dancers moving as one to attack the space. Long after the thirties were over, Walter Terry said of Holm's seven women, "I still remember that they had more personality than the dancers in any of the other ensembles"; although pictures of Holm's work and her own words reveal that she, too, used them to represent one or more factions, unified in style and intent. Tamiris was reputed not to impose a style on her group, yet a photograph from her greatest success, *How Long Brethren,** provides in pretty raw form the vision of the individual as vibrant instigator and the group as a unit about to be swayed, or not swayed, by her eloquence.

*Under WPA auspices, it ran for forty-two performances in 1937, on a double bill with an edited version of Weidman's full-length *Candide.*

However, the homogeneity of the modern-dance group was not the homogeneity of a formal ballet ensemble. The modern techniques weren't academic ones, objectified by centuries of contributions and alterations. They were personal discoveries by a handful of charismatic individuals. They were on the boil and changing every day. This dancer's way of holding herself, that one's splendidly erratic leap might at any time be incorporated into the style. Yet the performers, although in on the great adventure of developing new ways to dance, had to—wanted to—model themselves upon their leaders to some degree. In one sense, the group was an extension of the choreographer-soloist, reflecting and magnifying her visions, just as hoops and scarves once enlarged her movements in space. A quite innocent and thoroughly practical ad placed in *The New York Times* subtly reinforces this notion: "MARTHA GRAHAM, America's Greatest Dancer, Is Now Available For 1935–36, in Solo Concerts or with Her Famous Group of 12 Dancers." The wording makes one envision the famous twelve (like the twelve disciples of a charismatic leader centuries earlier) primarily as devices for extending the scope and radiance of the solo performer.

The members of a ballet ensemble danced in unison, orderly counterpoint, or antiphony. The Radio City Rockettes formed a gaudy image of a huge machine made up of delectable girls. The people in a modern-dance group, too, dressed alike and were engineered into patterns, but they provided a more self-willed image of unity. Not only were they never onstage simply to entertain the audience or to provide background for the principal dancers, they *never* were without intent. In Humphrey's *Piano Concerto in A Minor* (1927–Grieg), she allied herself with the solo instrument and the group with the orchestra, but her sensibility was dramatic as well as musical: the soloist was trying to influence the resisting group to join her, that is, to take on her movements.

And however much the choreographer's predilection for clarity could make it appear that intent, like style, could be applied by master plan, the dancers all felt it differently. They experienced the style individually too. They hadn't been chosen for a uniform body type or height. The photograph of Humphrey's *Dionysiaques* (1932) on page 179 shows four women shaping the same pose in subtly divergent ways. A picture of four Holm dancers striking out into space captures four different versions of the step. In strenuous movements, like leaps, the prodigal energy that the choreographers prized blurred the edges and allowed considerable individual spirit to shine through.

A scrap of film from Graham's 1938 *American Document* demonstrates

(From left) Harriet Roeder, Henrietta Greenhood (Eve Gentry), Louise Kloepper, and Miriam Kagan in Hanya Holm's *Dance of Work and Play* (1938)

the combination of formality and impetuousness that I find so enthralling in dances of the period. The dance looks quite free and gay compared with works of the early thirties, but at a certain point, trios of women begin crossing the stage—pioneer sorts, you imagine. Within each trio, one woman strides along, one arm angled overhead; a second travels sideways, splitting her legs wide in exuberant jumps; a third sprints in profile, slanting her body forward in a perilous angle with the ground every time she leaves it, rocking back as she lands. The bold conquest of the frontier, the work that accomplished it, the exhilaration it produced are compressed into a single vivid design.

Was there a paradox, as Humphrey feared there was, between instilling creativity in dancers and then asking them to "breathe as one," according to the dictates of the choreographic scheme? Only if you consider freedom of expression as the impulsive personal liberty it was for Isadora, or became for some of the anarchic vanguard dancemakers of the sixties. In the thirties, concerted action, well organized, was a fine thing; voluntarily submerging a part of your identity in order to contribute to a group you believed in was a noble way of expressing freedom of choice.

The masterful "Variations and Conclusion" from Humphrey's *New Dance* spells it out in idealistic terms. As the dancers pinwheel around the stage in orderly lines, they take turns bounding out in vibrant little solos, duets, trios to assert their individuality. After this, the group—a tight phalanx on a pyramid of boxes—whirls from side to side in a four-count pattern, and again individuals burst out; but all are allotted a phrase seven plus seven plus ten counts in length in which to reiterate their "ideas" in condensed form, at the end of which time (five of the ensemble's four-count measures) they must resume their places in the turning mass of people.

Despite the fact that Humphrey's ideas about the group were anchored to their time, they were potent enough to travel fifty years into the future, so that dance critic Burt Supree could write, in 1983, of a revival of *The Shakers,* "The people, so much in unison, thrust by the same rhythmic, emotional tide, are individuals, not a chorus. It makes all the difference in the world."

Gender played a role in shaping the look of the modern-dance ensemble. Most of the dancers were women. From the beginning, Laban and Wigman had had excellent male students—Kurt Jooss and Harald Kreutzberg, for instance—although when Wigman toured America with a company, no men performed with her. In America it was difficult to come by male dancers with a degree of skill. Dancing was not considered a fit profession for a man, and concert dance, paying as poorly as it did, was the least acceptable form. Holm had only her seven women. Graham's company was all-female until 1938, when she took on Erick Hawkins. Tamiris didn't use men in her concert work until 1939 and *Adelante,* a WPA-sponsored project, for which she commandeered a corps of men and, for the lead, choreographer and ex–Humphrey-Weidman dancer William Matons.

Maybe these choreographers hadn't been able to find men with fervor to match their women's. Maybe they didn't want them. They understood women and women's bodies because, presumably, they understood themselves. They didn't need men either, since courtship was a theme the moderns thought had been worked to death by ballet choreographers. The strong women stood for Everyman.

Even the Humphrey-Weidman company performed primarily as two groups until 1932, although, of course, on the same program. Humphrey

Doris Humphrey and Charles Weidman in *Rudepoema* (1934). Photograph by Helen Hewitt.

choreographed for her women, Weidman for his men, whom he brought up to scratch, the story goes, by charging only for classes they did *not* take. Often when men and women came together, as they did in *The Shakers* (1931), they moved alike. Weidman worked at developing a "masculine" style in his classes, although the women were so strong it was hard to top them. In the three pieces that comprised Humphrey's great *New Dance Trilogy,** men and women danced different themes at times, since they were assumed to have differing attributes, differing roles and functions in society. But care was taken to present them as equals (i.e., able to execute the same steps), and it was rare that a man lifted a woman.

The stressing of order and formal balance in composing the group's patterns onstage could indict an overritualized society or hymn a desired harmony, but political and personal ideologies weren't the only things that made an uncluttered look desirable in a modern-dance ensemble. Theories about form, about abstraction, about what constituted a contemporary style affected the handling of the group just as it influenced the style of the individual dancer-choreographer.

An article Doris Humphrey wrote in 1932 nicely sums up the rationale behind her meticulous patterning:

> Four abstract themes, all moving equally and harmoniously together like a fugue would convey the idea of democracy far better than would one woman dressed in red, white and blue, with stars in her hair.

Modern dance dispensed with locale. The jagged thrusts of dancing, the rush and swirl of the group, the tensions that shot between forces created a landscape shaped entirely by movement and the intent behind it. No modern dancer lolled in front of painted scenery intended to represent a particular small town in Iowa, or carried jugs of make-believe wine around while a local princess cut capers in the foreground. The casual traditions of ballet impersonation played no part (the program says they're Naiads, but it's clear to everyone that they're onstage principally to show beautiful dancing). Bold, astutely chosen gestures linked the modern dancers with large impulses like vindictiveness, desire, celebration, protectiveness; and what characterization there was emerged through the shapes and rhythms these impulses imparted to individual and group.

In many photographs—those of Humphrey's *Dances of Women,* for instance, or Holm's *Trend*—people are spaced out like components of an architectural structure. The look isn't static, though; even frozen by the camera, the stage has a spatial dynamism of the sort that preoccupied many sculptors and architects. You can see the influence of visionary theater de-

---

**Theatre Piece* (1935), *With My Red Fires* (1936), *New Dance* (1935), set to commissioned scores by Wallingford Riegger.

Hanya Holm's *Trend* (1937)

signers like Gordon Craig and Adolphe Appia, who transformed the picture-box stage with its flat, *trompe l'oeil* backdrop into a three-dimensional arena. The set that the immensely gifted Arch Lauterer designed for *Trend,* the flexible set of gray steps and cubes that the Humphrey-Weidman company used in many works extended the design of the dance into vertical space and heightened the sense of volume. Lighting designers—Lauterer and Pauline Lawrence prominent among them—added to the architectural effect by using light to dramatize space and emphasize the modeling of the body. The look is very different from one you often see in photos of German or German-derivative dances. In these the dancers crowd together, each one expressing the same intent in different ways—a gesticulating mountain of humanity.

As choreographers, like many artists in other fields, sought the essence of their discipline, they often arrived at basic geometrical forms for deploying a group in space. These were not simply attractive shapes to be used; decorativeness, remember, was anathema. They were the resonant elemental structures that underlay organic life and the ratios that expressed its laws. In 1923 Wassily Kandinsky, who had been influenced by Rudolf Steiner and his theories about the mystical properties of shape and color, lectured at the Bauhaus on Mary Wigman's "Circle" and "Triangle," from her group work *The Seven Dances of Life.* In 1927 Doris Humphrey was toying with a geometric idea, a solo (Egyptian in flavor, she thought) to "express" the triangle, two girls "stating" the curve, a dance for three establishing parallel lines. All, of course, to be correlated with human experience.

At the time, Humphrey was reading one of Jay Hambidge's books on "dynamic symmetry" (probably *The Elements of Dynamic Symmetry*). It was, she confessed in a letter to her parents, a little complicated in its mathematical ratios, but she found much to fascinate her in Hambidge's explanation of the "lost" principle that had shaped Greek and Egyptian art. Dynamic symmetry, found, for example, in the spiral of a snail shell, suggested growth even in stasis, while static symmetry, like that in the snowflake, implied completion. Hambidge's talk of "form rhythms," of the figure he so piquantly referred to as the "rectangle of the whirling squares," was the sort of thing to fire a woman like Humphrey, as can be judged by a photo of the culminating moments of her 1928 *Color Harmony.* All the

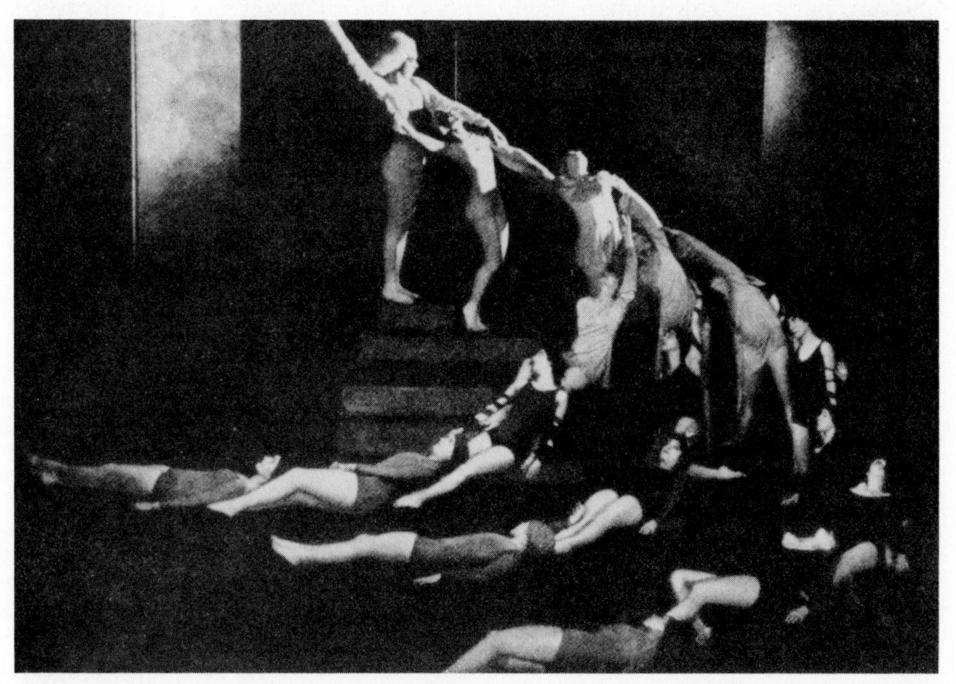

The final pose of Humphrey's *Color Harmony.* Photograph by Vandamm Studios.

"colors," each associated with a mood or type of human temperament, spiral upward toward the single silver-white light (Weidman) who comprehends them all.

We have more of Humphrey's thoughts on paper than we do those of her contemporaries, and certainly she let planning monitor intuition more than many of them ("It was here that I used symmetry for the first time as the best way to express cohesion and completion" is a typical remark). But all the modern choreographers seemed alert to the idea that larger forms could be as expressive as individual gestures. Graham's use of the circle in *Primitive Mysteries,* her "wall" in *Heretic* are cases in point. So is an amazing backward walk that the women execute in the "Steps in the Streets" section of her *Chronicle* (1936); their scalloping paths and averted gazes heighten the illusion of a restless, fearful society, edging circuitously backward because it can't move forward.

In dance, forms and patterns dissolve and evolve in ways that they can't on the painter's canvas. Skillful design can create the illusion of a motion, perhaps of a truth, larger than anything the individual dancers are doing, more significant than the mere massing of people. Humphrey's 1928 *Water Study,* performed in silence by a group of women, beautifully expresses the analogy between the human being and universal processes. Here are its opening moments. The dancers are spread across the stage, kneeling in profile, curled over. Slowly they unfold their bodies, then curl down again, as if the swelling and subsiding of breath were causing them to grow and shrink. The gesture travels across the stage from right to left, several women beginning to rise as those behind them are completing the action. Each time they repeat the unfolding, it expands in size. On the third swell, the women circle their right arms over their heads and thrust them forward, curving; the last few dancers slide out onto their stomachs. So each dancer—rising, arching, striking, subsiding—contains a tide in her body, while their cumulative movements seem to flow across the stage, as if one huge wave were breaking.

Some years later, in *With My Red Fires,* Humphrey built a more drastic pattern. The Matriarch is inciting the crowd to violence against two transgressors. These two are, ostensibly, her daughter and the girl's lover, but the Matriarch is more than an aggrieved mother; she's a ruler demented by power. In lines, raying out around her like spokes from the hub of a wheel, the "citizens" crouch and scurry, bodies thrust forward, heads straining, fists clenched. The music stops; everyone freezes. Then, to a ratchety sound in the Riegger score, the Matriarch gestures stridently, and the people run forward again in wheeling lines. You see them as craven, blindly obedient bloodhounds on the track, but as the pattern repeats, you begin also to see a mechanical wheel of destruction that this leader is inexorably winding up, one chillingly analogous to that which Adolf Hitler was setting in motion.

The thirties was a period of ferment—in ballet as well as in modern dance. In spite of economic difficulties, a lot of works got made and performed, particularly in New York. There were talented choreographers who were not, and had never been, associated with the major modern companies. Like Miriam Blecher, a leader of the New Dance Group, who turned down a chance to join the Graham company and whose highly political *Van der Lubbe's Head* won first prize at the first annual Dance Spartakiad in New York (1933)—a rather Germanic event that was held only a few times.

However, it was the so-called pioneers, the select few I've been talking about, who forged the new image of the dancer: a man or woman whose ideals soar, but whose feet are planted firmly on the common soil; one who displays a physical daring that is related to human struggle, rather than to superhuman virtuosity; one capable of a lyricism that is neither feeble, nor decorative, nor sentimental. This vision of what a dancer should be was accepted by followers of the major figures and by those who considered themselves more independent; they then subjected it to stylistic transformations and adapted it for their own uses.

The concept of the dancer as a member of a forceful, unified ensemble suited the political dances that proliferated during the thirties. The ideal of plainness meant that dance could be modified for the nonprofessional and still have an effect—important in that the antielitist aspect of modern dance required some commitment to the adult amateur. In the heat of social art-making, militant groups of amateurs or paraprofessionals, like the Needle Trades Workers' Dance Company or Edith Segal's Red Dancers and her Nature Friends Dance Group, could make potent statements on topics ranging from the general—the menace of fascism, the corruption in capitalism—to the specific—tales of fallen heroes in the labor movement, like Tom Mooney, or the plight of the Scottsboro Boys. One critic of the day mocked the Little Duncan Dancers on the grounds that their style was unapt for the protest dances they were bent on creating.

By the mid-forties, Holm and Tamiris were expending most of their creative energies on Broadway musicals. Humphrey stopped dancing in 1946, as a result of a crippling hip injury, and Weidman's career lost much of its impetus. Although before her death in 1958 Humphrey was to create several superb dances for the José Limón Company (*Day on Earth,* for example) and the Juilliard Dance Theater, she never surpassed her great work of the thirties. Graham, on the other hand, went on in the forties to invent a new form of dance-drama and to compose the works on which her reputation rests: *Letter to the World, Cave of the Heart, Dark Meadow*—the list is a long one. She has decried some of her own early works as childish.

It's hard to assess properly the dances of the late twenties and thirties: most of them were discarded when the choreographers moved on to new ideas. Still, considering what these innovators accomplished then, the work seems far from simplistic—more a matter of willful primitivism, a salutory, if stubborn, return to basics. In pruning away from dancer and dance everything that did not matter supremely to them, they forced a wealth of new flowerings.

·5·

The Heroines Within

Martha Graham in her solo *Frontier* (1935). Photograph by Barbara Morgan.

"If people would only stop looking for a literary meaning in my dancing," Martha Graham complained to dance critic Margaret Lloyd in March of 1935. Anyone who has sat in a theater watching a dance like *Letter to the World* (1940) or *Night Journey* (1947), thunderstruck by Graham's approach to narrative and to famous figures from literature, might well be mystified by her words.

But she was speaking on the brink of a gradual change of focus. The year 1934 had been the inaugural summer of the Bennington School and Festival. For the first time, Martha Graham, Hanya Holm, Doris Humphrey, Charles Weidman—the acknowledged leaders of American modern dance—were teaching and performing for the same eager students. If the atmosphere at pastoral Bennington was relaxed compared to the scrounging life that the dancers and choreographers led in New York, it was also stimulatingly competitive. Hints of what "Doris" was working on, or of "Martha's" latest interest, leaked from behind closed studio doors and were whispered among student dancers. Partisanship ran high. At a time when the choreographers were moving out of the first belligerent stage of modernism, Bennington's climate encouraged a broadening of perspective, a redefining and elaborating of the spartan principles of dance that these pioneers had constructed in defiance of previous traditions.

Graham's shift toward a less austere form of drama in her group works and toward making specific references to history or literature could be discerned almost immediately. By the fall of 1934, John Martin, comparing Graham's new *American Provincials* to her *Heretic* of 1929 noted some changes. The themes were similar. In the first part of *American Provincials,*

"Act of Piety," Graham cast herself as a fervent Puritan; in the second, "Act of Judgment," she became the victim of the mob's religious frenzy. Martin noted that the older work had been more abstract and admired its "stark impersonality," while the new one was ". . . almost as much of the theatre as of the dance." He ended with what turned out to be a sound prophecy:

> Though it is in every sense an American study, it is also something larger. It is, indeed, that concept without time or place which the Greeks called Medea. And what a Medea Miss Graham could play if she put her mind to it!

By the time Graham put her mind to playing Medea—in 1946, with *Cave of the Heart*—she had perfected a stage image of herself as one who was no stranger to dark passions, and who was self-aware as few previous dancers had been. In the process, she had transformed the stage into a habitation where time and place became flexible concepts vibrating together.

"The theory of relativity . . . is only the scientific expression of 'the new landscape' of the twentieth century," wrote the art historian Wylie Sypher, "a landscape revealed for the first time in cubist painting and the cinema." He might have added, "and later in the dance theater of Martha Graham."

But although the landscape that she created and the dancer-images with which she peopled it were not only up-to-date but startlingly new in Western dance, they were also rooted in antiquity. One of the remarkable things about this contemporary artist, this "modern" dancer, is the way she drew on ancient myths, Eastern theater traditions, cinema techniques, and twentieth-century psychological theories to filter her own responses to the world and render them more potent and essential as art.

It is not difficult to understand what gradually drew Graham toward dance-theater. Her style and the technique that she taught to others had, from the outset, been based on how the human body responds to emotional and physical drives; her own performing was always dramatic, no matter how abstract her presentation of character and feeling. She had the gift for looking spontaneous, as if she were compelled to make these gestures and no others. With her small, taut body, sturdy legs, carved face, she could suggest immense restraint and frenzy and every nuance between these extremes—including flippancy and coquettishness. Early in her career, two critics who weren't frequent Graham-watchers (or even, I suspect, frequent dancegoers) intuited qualities that were intrinsic to her personal style. "She

Martha Graham in a Denishawn Javanese dance, photographed in 1922 by Nikolas Muray

has a quality of aloofness that is almost eerie in its fascination," wrote a critic in Omaha, upon seeing her perform with Ted Shawn in 1921. And in 1931 George Beal of the *Boston Post* described her this way: "Dark, swiftly moving, tense, she seems more a creature freed from restraint than one trained to perform a routine of movements set to music."

Theater had always been a part of her professional life, beginning with the exotic impersonations that the Denishawn company required of her. Since 1928 she and her artistic mentor and musical director, Louis Horst, had been associated with the Neighborhood Playhouse School of the Theatre, where Graham taught memorable classes to actors as well as to dancers. She had staged the movement for Katharine Cornell's *The Rape of Lucrece,* for John Houseman's production of Archibald MacLeish's *Panic,* and for Blanche Yurka's *Electra* (in which she also performed three solos). It was perhaps inevitable that she would have noticed that the public for theater—even innovative theater—was larger than that for the bold, severe dances she was creating. But she could not make even subtle alterations in her approach until she had figured out a way to handle narrative that would have nothing in common either with Denishawn genre pieces, with their reliance on disguise, or traditional story ballets, with their alternation of pantomime and diversionary dancing. Whatever literary themes she was going to choose, she needed to ensure that the dancing itself would not become "literary," that each work would be what she later called ". . . a kind of fever chart, a graph of the heart."

Dance is action, not reaction, she said once. By the end of the thirties, she was ready to find ways of presenting reaction as action, and herself as the protagonist of a dance-play.

With *American Document* (1938), themes that had not hiterto appeared in Graham's work began to surface: the coming of men into her company—Erick Hawkins in 1938 and Merce Cunningham in 1939, others soon after—inevitably signaled a change in her choreographic concerns. With the men, role-playing and narrative entered the work; sexual love became not only a viable theme, but an alluring one. Between 1938 and 1944, Graham explored—illumined—a variety of forms drawn from American ceremonies, theater, literature, and popular entertainment: minstrel show (*American Document*), circus (*Every Soul Is a Circus,* 1939), poetry (*Letter to the World.* 1940), passion play (*El Penitente,* 1940), puppet show (*Punch and the Judy,* 1941), wedding (*Appalachian Spring,* 1944). During the latter half of the 1940's, she turned to making dances—*Cave of the Heart* (1945), *Dark Meadow* (1946), *Errand into the Maze, Night Journey* (both 1947)—that drew their themes from history, myth, and Greek drama, crystallizing a form that was to occupy her for the rest of her creative life.

The business of roles was tricky. It had always been one of the tenets of Graham's generation of modern dancers that one danced as oneself, didn't pretend to be someone else. What mattered most to her, I think, was that

Erick Hawkins and Martha Graham at Bennington College in the summer of 1939, during an outdoor rehearsal of the "Puritan Love Duet" from *American Document.* Photograph by Barbara Morgan.

she be able to make her own emotional life the subject of her art without banality or self-indulgence. In the 1930 solo *Lamentation,* she had been able to abstract elements of an emotional situation so that she could embody the archetypal "woman grieving." Simple self-expression hadn't interested her then; ten years later, self-impersonation didn't either. Nor did she want to transform herself into great heroines. However, by identifying the issues of her life with those of other women in myth or history or literature, she could avoid confessional art, achieve a kind of distance, and foster a stage image both personal and aggrandized. When she wrote in her notebooks,* "The Woman-soul. The immortal woman in woman," she did not intend, as Ruth St. Denis had, to transcend womanhood; she was trying to make her womanhood transcendent.

"The interest in myth was in the air," remarked New York painter Adolf Gottlieb, recalling the 1940's in an interview. The artists had discovered Carl Jung. Whether they read his books or learned his ideas in studio and barroom conversations or, like Jackson Pollock, in the course of psychoanalysis, they found both liberation and discipline through his theory of the collective unconscious. Mark Rothko gave his paintings of this period such titles as *Antigone, Sacrifice of Iphigenia, The Omen of the Eagle*—names that have the substance and the ring of Graham's dance titles. These myths, he said, ". . . are the eternal symbols upon which we must fall back to express basic psychological ideas. They are the symbols of man's primitive fears and motivations, no matter in what land or what time, changing only in detail, never in substance."

Rothko was echoing Jung's own stirring words: "The man who speaks with primordial images speaks with a thousand tongues . . . He transmutes personal destiny into the destiny of mankind . . ." No wonder Graham found such ideas inspiring; they taught her how to objectify her emotional life and forge a dancing persona. Although it was not until 1950, at a crisis point in her own life, that she sought the help of the pioneering Jungian analyst Frances Wickes, Jungian thought had provided her with an approach to personae long before this time. Jottings in her *Notebooks* show that copious reading drew her not only to Jung, but to such authors as Frances Cornford, Joseph Campbell, Robert Graves, Jessie Weston, and

*Graham has always jotted down sketches for dances, steps, passages from literature in almost a stream-of-consciousness style. Dancers and other colleagues have spoken of these notebooks. A few were published in 1973 by Harcourt Brace Jovanovich as *The Notebooks of Martha Graham.*

Maud Bodkin—writers whose analyses of religion and literature were shaped by Jung's ideas or kin to them. Campbell she knew: he was married to Jean Erdman, one of her dancers. In rare free time at Bennington, a colleague recalls, she also socialized with Erich Fromm and his wife—perhaps eager for the insights into dreams and fairy tales that he could give her.

Jung's idea of the collective unconscious and the power of the archetypal images that dwelt there stimulated her as they did the painters, but, obviously, in a different way. "The more clearly man can discern the essential meaning of the archetype in relation to the time and place where he stands, the more the individual can be said to choose his own way . . . ," wrote Wickes. This became one of Graham's missions: to define the archetypal images of woman in herself and in relation to her world, and to set them dancing.

The pages of her *Notebooks* are peppered with references to such archetypes—virgin, maenad, prophetess—and to the three aspects of the White Goddess that Robert Graves relates to ancient fertility cults and calendrical lore. Sometimes Graham separated out several aspects of woman or a woman and played them all herself. Her 1950 solo *The Triumph of St. Joan* showed her as Maid, Warrior, and Martyr. (Later, in the 1955 work *Seraphic Dialogue,* she used three different dancers to express these aspects plus a central Joan who conjures them up and meditates upon the stages in her life when each was dominant.) In *El Penitente* (1940), a trio as clear and bold in style as a Diego Rivera mural, one of the peasants who marches in to perform a Mexican-Indian passion play is the Penitent who assumes the burdens of Christ; another man functions as a distant vision of Christ—an animistic figure in a mask like a cow's skull, who stays in the background, offering mild, stiff gestures of benediction and rebuke. The woman, however, changes roles according to the Penitent's need. She dances as his vision of a peasant madonna, sturdy and sweet, tilting from side to side within the arch of blue cloth that she carries like a portable niche. Then she becomes a tempting Magdalen (a later addition of an apple as a prop links her also to Eve)*; the dark, anonymous burden he must drag in the "death cart"; and finally the Mater Dolorosa or the saintly Veronica—wiping the sweat from the sufferer's face, holding the huge wooden cross to give him a moment's respite.

In a Graham dance, a role might be split and parceled out among other

* Arch Lauterer designed the original props for *El Penitente.* The apple may have appeared along with the mask when Isamu Noguchi redesigned the production in 1944.

Merce Cunningham, Erick Hawkins, and Martha Graham in *El Penitente.* Photograph by Barbara Morgan.

dancers too. When, toward the end of her long performing career, age limited her activity onstage, another could dance as "the young Clytemnestra," "the young Judith," "the young Heloise." In early works, two or more performers might suggest subtler aspects of a single character. In the beautiful *Letter to the World,* in which the dancing provides an emotional subtext for selected poems by Emily Dickinson, the central role of the poetess is shared by two performers; one dances, the other speaks the poems and moves in and out of the action in a dancelike manner. On the program at the Bennington premiere in 1940, Graham defined the difference between them in terms that showed she was no stranger to psychoanalytic thinking: "The one in red is the impulsive hidden self of the one in white."

If identifying herself with archetypal females objectified her own feelings, it also imbued the characters she played with enormous intensity and specificity. However many roles Graham assumed and whether she listed herself on a program as Jocasta or—more obliquely—as "one like Medea," their dilemmas were hers. *Deaths and Entrances* (1943) was supposedly inspired by the lives of the three Brontë sisters, but it is worth remarking that Graham herself was one of three sisters. At the time that *El Penitente* was made (1940), Graham and Hawkins, who played the Penitent, were lovers. Perhaps one did not have to be as perceptive as Edwin Denby to understand that ". . . it was not the relation of man to the Divine but the relation of a man and a woman that seemed the true subject." Or that "it was as though Miss Graham had used the Spanish-Indian farmers' expression of religious faith as a metaphor for her own faith in the strangeness love can have."

With *Herodiade* (1944), a duet for herself and a woman attendant, she began to venture deeper into Jungian territory and to equate herself with various monumental heroines of myth poised at some turning point of their lives, at what Graham has referred to as "the meadow of choice, the passage to another area of life." The situation is akin to that in the "crossroads dream" of Jungian theory, to that "moment of choice" that Frances Wickes has described, "when the ego must decide to step across the threshold into the perils of the unknown that lead to greater self-knowledge or to retreat into the safety of the known." Here, too, the choices were Graham's choices. Whatever prospect the heroine of *Herodiade* is attempting to confront, it could have been shadowed by Graham's own aging. She was fifty when she made this dance. Speaking of the three sculptures built of flat, white, interlocking shapes that Isamu Noguchi designed as a set, she interpreted the one known as "the mirror" in this way: "When a woman looks into a mirror . . . and if she's approaching the time when she's no longer young, she sees her skeleton, she sees her bones." No wonder she found it hard to relinquish her roles to others as her physical powers waned: they adhered to her body and to her heart.

Graham's dilemmas were not just those of a woman, but of an artist. The dances that she made after 1943 acknowledged that the equilibrium between impulsiveness and restraint affects artistic creation as well as emotional life. In a lecture that she gave to Juilliard students in 1952, she identified the terror that permeated *Herodiade* with "the fear of the artist,

Martha Graham and May O'Donnell in *Herodiade.* Photograph by Arnold Eagle.

of a blank white page when writing a composition, the fear of the empty studio when starting a dance . . ." Elsewhere, she noted that in the epic 1958 *Clytemnestra,* the heroine's rage at Agamemnon over the death of Iphigenia has something of the artist's rage over a work destroyed, a creative impulse subverted.

In her *Notebooks,* musing on Orpheus, on Persephone (". . . Where does a parallel come in the life of the artist—Who are the more immediate ancestors?"), Graham equated both the role she might play and the woman/artist she was with the questing hero of myth, one who cannot refuse the "call to adventure," who enters the labyrinth, the dark wood, hell itself to achieve what must be achieved. Here too Jung was implicitly her guide. This kind of artist/seeker, with what Joseph Campbell has referred to as "willed introversion," doesn't journey through a mythical geography, but into himself, the most perilous region of all, searching not for treasure or lost love or the supreme talisman, but for his primordial totemic self. The demons to be vanquished aren't fire-breathing dragons or bullheaded monsters—even though such a figure appears onstage in Graham's *Errand into the Maze*—or human enemies—although it is ostensibly Holofernes that the heroine of the solo *Judith* (1951) seduces and slays. The demons are one's own fears and doubts, the dark places of one's psyche. The rewards of this perilous descent can be great, for, in Wickes's words, ". . . to know the Self that dwells within one's self . . . is to arouse a hidden life of unforeseeable, unknowable creative possibilities." Beyond self-knowledge and purification, perhaps even beyond happiness, the goal is more potent artistry.

Memory was the path to the heart of the maze, as it is in psychoanalysis—memory and intimations of the future. Graham's task became to find the theatrical symbols through which she could externalize this inner search and make it visible and comprehensible onstage.

Deaths and Entrances was the first dance in which memory played a crucial role. It is almost always through the imagination of the most important of the three sisters that emotions and actions are revealed. She stands at some terrible crisis of her life, which may be madness. In one climactic solo, she seems to be shuddering herself to pieces. As the dance begins, she is about to move a chesspiece; at the end, as her sisters stare amazed, she triumphantly places a goblet on the board. Since the goblet has been established as a resonant symbolic object, it's as if she's laying her soul on the board to triumph or be checkmated. In between these moments, her imagination roves through her past, as in a cinema flashback; the journey is interior, fantastic, transformed by recollection. The other sisters dance, revealing their own temperaments, or their temperaments as she perceives them—sly for one, almost vicious for the other. Three girlchildren foreshadow the women the sisters will become, and pass through what seems to be a large, gloomy house carrying various objects—the goblet, a shell, a vase, two smaller vases—that trigger the heroine's memories. She summons up a party in which young couples dance; the party becomes a funeral. Her funeral. The men she loves, The Dark Beloved and The Poetic Beloved, enter the stage not of their own volition, but when her memory wills them to come.

Edwin Denby, in his great essay on *Appalachian Spring,* distinguished between Graham dances that he thought of as happening outdoors and those, like *Deaths and Entrances* and *Herodiade,* that were taking place indoors. "In these pieces," he said, "when Miss Graham suggests in her gesture a great space around her it is, so to speak, the intellectual horizon of the character she depicts." In the dances she made after 1944, the entire stage space gradually came to stand for this landscape of the mind, and the other characters to exist only in her memory or imagination. Oedipus's tribulations come to theatrical life through Jocasta's perspective. The events of the Trojan War are summoned up by Clytemnestra as she paces the boundaries of the Underworld trying to achieve self-knowledge, expiation, and eternal rest; the interplay between Hector, Achilles, and the others, envisaged from Mycenae and recalled from beyond the tomb, become formalized, preordained. Seen from her heroines's viewpoint, Jason, Oedipus, Agamemnon are handsome, childish giants posturing stiffly—men who are always sure of themselves and their women.

None of this may sound as if it would be a sure-fire hit with audiences, but, although many people continued to be baffled or outraged by Graham's dances, the "psychological" element was exciting to others. In the 1940's, when she created some of her most original and vivid theater pieces, her fierce images of men and women would have seemed not only contemporary in their concerns, but novel. To the intellectuals and artists from all disciplines who had always been part of her audience, as well as to all those with the same awareness of psychoanalytic theory that she had, and with a similar literary heritage (Dickinson, the Brontës, Eliot, Crane, Hawthorne, among others), her work must have looked like dance for grown-ups. Except for Antony Tudor and Graham's peers in modern dance—like Doris Humphrey or, later, her protégé José Limón—choreographers were not wont to traffic in psychological depth or complexity. Contrasted with the

Deaths and Entrances: Martha Graham and the "three remembered children": (from left) Pearl Lang, Nina Fonaroff, and Ethel Butler. Photograph by Arnold Eagle.

dilemmas that the heroines of most story ballets faced, hers were vast and intellectually demanding. And they had the additional allure of being passionate—couched in a vibrant physical language that, despite its formality, was far more erotic than that of ballet.*

Powerful advocates of modern dance, like John Martin of *The New York Times* and Mary Watkins of the *New York Herald Tribune,* had long pushed the idea that watching dancing of the sort that Graham made was meant to be stimulating and enlightening, rather than simply entertaining. That there was a certain intellectual cachet to being among Graham's audiences is evidenced by a remark that a Los Angeles critic claimed to have overheard in the course of a Graham performance in 1946. Said soothingly by one woman to her mystified companion: "The thing to keep in mind is that everything is a point of departure."

It was natural that Graham should find her archetypes in Western myth, natural that her interest in myth and ritual should draw her to Greek drama. (She had imbibed Nietzsche, remember, and Erick Hawkins was a classics scholar.) Yet subject matter and prophetic dancing choruses aside, her "Greek" dances have almost more in common with Japanese Noh drama than they do with Sophocles.

Denishawn's exotic repertory may have kindled her interest in authentic Eastern forms. References to Noh crop up in her *Notebooks,* and in the 1970's, a suggested reading list posted for students at her school included Noh plays. In Taos in the summer of 1937, she was reportedly eager to talk to Ramiel McGehee on the subject of Eastern theater.† Certainly she could have seen the various Asian performers who visited New York. In 1930 Tokujiro Tsutui appeared in Kabuki dances. That same year, the famous Chinese actor Mei-Lan Fang gave members of Graham's company tickets to his New York performances. Born in the same year as Graham, Fang spoke, sang, acted, and danced a variety of female roles—some of which depicted extreme emotional states with great elegance. Uday Shankar came in 1933, Ram Gopal in 1938.

---

*In 1963, long after one would have thought no one could be shocked by a Graham dance, two United States representatives, Peter Frelinghuysen of New Jersey and Edna Kelly of New York, saw *Phaedra* in Europe and found it so sexually explicit that they protested before a House Foreign Affairs subcommittee against the State Department sending Graham's company abroad as an example of American culture.

†Graham's biographer, Don McDonagh, reports this encounter. McGehee had lived in Japan and, for a time, given English lessons to the emperor's son.

Graham also knew the Japanese dancer-choreographer Michio Ito, and had performed with him, first on Broadway in the Greenwich Village Follies (1923–1925), later in one of the Neighborhood Playhouse's symphonic evenings. Both taught at the playhouse and had taught earlier at the Anderson-Milton School. Although Ito combined a Japanese sensibility with the more contemporary formality he had learned from Émile Jaques-Dalcroze in Switzerland, he had been trained in Kabuki as a boy, and possibly in Noh as well. He had staged W. B. Yeats's *At the Hawk's Well* in modified Noh style in London in 1916. The year he and Graham opened in the Follies, he staged a translation of the Kabuki play *Bushido* in New York.

Graham was not the only artist of her day to appropriate Eastern theater techniques and transform them to suit contemporary Western art forms and personal expression. Playwrights like Yeats and Thornton Wilder found in Japanese drama ways to disentangle theater from nineteenth-century traditions and give plays some of the flexibility of the new twentieth-century art, cinema.* Describing his own early one-act plays, Wilder wrote:

> In *The Happy Journey to Trenton and Camden,* four kitchen chairs represent an automobile and a family travels seventy miles in twenty minutes. Ninety years go by in *The Long Christmas Dinner* . . . In Chinese drama, a character, by straddling a stick, conveys to us that he is on horseback. In almost every No play of the Japanese an actor makes a tour of the stage and we know that he is making a long journey.

Similarly, what Graham borrowed from Eastern theater, and so brilliantly recast, were devices to negotiate a necessary elasticity in time, space, and states of mind. With these, she could accomplish her interior quests, and formalize the emotional experiences that preoccupied her.

Consider *Night Journey,* first performed in 1947, with Graham as Jocasta, Hawkins as Oedipus, Mark Ryder as the seer Tiresias, and a chorus of six women.

As the curtain rises, Jocasta stands swaying, holding a length of white rope. Tiresias enters with huge, purposeful steps, swinging his staff ahead of him to announce his presence, and knocks the rope from her hands.

---

*Structural analogues to flashback, montage, cross-cutting, and slow motion in Graham's theater pieces relate her concerns to those of moviemakers. An extraordinary outline in her *Notebooks* for a never-made film of *The Scarlet Letter* indicates the way her mind and eye worked.

With this act, he forces her to think back on her incestuous relationship with her son, that she may not die without understanding. In her solo—shuddering, twisting this way and that, covering her eyes, throwing her arms up, sliding to the floor—she expresses a desire to avoid the anguish her memories will bring. The music, by William Schuman, was conceived less to provide a rhythmic floor for dancing than to suggest a turbulent emotional climate. Jocasta, more than any other Graham heroine, is a victim, but even she must search her soul to see if she could have, should have, prevented the liaison. Although she is Jocasta, she is also any woman who, in taking a younger man for a lover, must wonder how much of her love is true, how much a search for youth, how much thwarted maternity.

Finally, she lies on the ambiguous structure that Isamu Noguchi designed. It is the incestuous marriage bed, but it also resembles a gridiron—a torture rack—and, if one looks closely, it is composed of a primitive male and female figure entwined. The smaller objects that form a stairway up to it look like dolmens, a miniature Stonehenge. A narrow swag of fabric—half column, half beam of light—twists down from a point high above the stage.

While Jocasta lies dreaming on her tilted rack, the young Oedipus strides onto the stage and into her memory, just as he once strode victorious into Thebes to claim her, the dead king's widow, as his booty. Members of a female chorus bow stiffly before him with olive branches. He walks up the sculpted stepping-stones and down the side of her bed, looming unseeing above her, as if for him, at this moment, this was the ramp leading down to the throne room and not her bedchamber. At his bidding, Jocasta sits docilely on an hourglass-shaped stool and watches this handsome, sensual bully dance for her. Slapping his hips, striding boldly on straight legs, he begins to spread and fold his odd Chinese puzzle of a mantle—stepping into its openings, crossing it, repeatedly thrusting his fist through it in such a way that he ends with it wrapped thickly around his arm.* Jocasta's little shudders may indicate only arousal, but they may also portend the horror that is to come/has been. When the two dance together, Graham as choreographer (and, once, as performer) brilliantly keeps the focus shifting. He stands on the stool behind her, possessively wraps his leg around her and presses his heel against her midriff; she watches his foot descend, as if, seeing beyond the ceremonious act of being claimed, she half knows it for her own flesh and blood. They stride formally together, wrapped in his

*This is not the original solo, but one that Graham worked out with Bertram Ross, who inherited the role after Hawkins left the company in 1950.

Bertram Ross and Martha Graham in *Night Journey*. Photograph by Arnold Eagle.

cloak, their bodies twisted away from their profiled heads and feet like figures on a Greek vase. Within a single spiral of movement he falls on her waiting body, and she rises to rock him across her lap like a child. As he stands, legs bent, she kneels on his thighs in lotus position, clinging to him, but this image of copulation tilts fluidly, and the next moment she is cradling his head. What is (or seems to be) merges with what was, because both are the creations of memory.

The chorus of women dance their horror at the sin they are powerless to avert. The archaic, carved-looking shapes their bodies take on in no way lessen the fury of their movement; standing on one leg, they fiercely pulse their rib cages, gradually sinking to the floor as they do so. Meanwhile, behind them, Jocasta and Oedipus, within the frame of the bed, slowly, quite formally enmesh themselves in cat's cradles with the white rope. The fatal umbilical cord also gives symbolic shape to their erotic entangling, and they strain against it even as they yield to it.

Now Tiresias enters, a macabre and implacable figure, using his staff like a pogo stick. There is no lengthy exposé. He touches the rope that wraps mother and son: it drops at his touch, and they fall apart. A few more succinct images, and it is over. Oedipus plunges straight as a tree into the waiting arms of the chorus and is slid out onto the floor, a rebirth into terrible knowledge. While Jocasta lies on her bed, he leans over her, plucks the large ornate pin from her dress, jams it over his eyes, and, holding it like a mask, quietly gropes his way offstage, already fading from her thoughts. She rises, repeats some of her first dance, stands for a moment as if to let all this become imprinted on her mind; then, with resignation shrugs slightly and the dark drape on her costume falls to the floor, leaving her in a long, pale shift. She again wraps the rope around her neck, and as the sound of Tiresias's stick recedes, she pulls it taut and falls slowly backward.

Memory was as crucial to Noh as it was to Greek plays and to Jungian analysis. As the Eastern theater scholar Faubion Bowers writes:

> The primary point to be remembered in the analysis of a *Noh* play is that action is generally recollected and that the plot hinges on an event that has already taken place in the past. This means that the dramatic situation is not necessarily acted realistically before one's eyes. Rather it is poetically recalled and discussed by the characters and chorus, and their movements become dreamlike glosses to the idea carried by the words.

In a Graham dance like *Night Journey,* the reenactment, passionate though it is, also has the feeling of being a "dreamlike gloss"—not on chanted words, but on a well-known story. Narrative is only the scaffolding for feelings.

Noh heroes, unlike Graham, reenact their stories alone. In the poignant *Utou,* for instance, the protagonist returns as a ghost to bid farewell to his wife and child. The actors who play these two roles are little more than an onstage audience for the climactic dance through which the hero reveals the eternal torments that are his punishment for being a birdcatcher and a destroyer of life, dancing too his detestation of his former life, his sorrow and remorse. Yet, although Graham peopled her memories and reentered her own story, the important thing for her also is how the process reveals or transforms her.

The ability of a character to move freely from one performance mode to another—narrating, meditating, acting out—is integral to Graham's structures. The Greeks and Trojans in her *Clytemnestra* introduce themselves formally before they enter the drama. Striding one by one in a ceremonious procession, each steps behind a tall, flat Noguchi sculpture with a vertical slit in it, like an abstraction of a mask. One of two singers speaks for them, "I, Helen . . ."; "I, Paris . . ." Graham first suggests the struggle between Mycenae and Troy in a procession of couples engaged in formalized combat before she limns particular moments in the tragic war. In a similar manner, a Noh actor may introduce himself and explains his mission before entering into dialogue with another. In the Indian Kathakali drama *The Sacrifice of King Daksha,* the audience sees Sati reviled and humiliated by her own father, Daksha; the scene is presented again from another perspective when Sati "tells" her husband, Shiva, about her experience, reenacting both roles. While it would be inaccurate to propose a direct connection between such Eastern theater works and Graham's dances, they are using similar techniques to get at the same thing: a shift in perspective. The flexible vision of space and time that links Graham's dances, Cubist painting, and cinema with Einstein's theory of relativity—as mentioned earlier in this chapter—also links them with those ancient Eastern philosophical concepts that underlie Asian theater and so surprisingly reinforce the findings of contemporary physicists.*

Characters often enter a Graham dance as ceremoniously as a Kabuki

---

*For a fuller discussion of this, see the chapter on Merce Cunningham, "Illusion of Choice—Acceptance of Chance."

Appalachian Spring (1944), with Martha Graham and Erick Hawkins, May O'Donnell as the Pioneer Woman, and Yuriko and Nina Fonaroff visible in the background. Photograph by Arnold Eagle.

actor proceeds down the traditional walkway (*hanamichi*) to the main stage. I can imagine no Western choreographer before Graham who would have staged a wedding celebration like the one she created in *Appalachian Spring,* with the four principal characters entering in grave single file with perhaps six feet of space between them. But then, this dance, with its tenderly illustrative Aaron Copland score, has little byplay and almost no local color; it is about the impact of untamed space and prospective solitude on people's vision of life. In the danced soliloquies for the Bride and the Husbandman, we see projections of the future, images of the past, as well as the vows they take in the present. The man's delving, shooting, plowing motifs, his satisfied gaze into the frontier distances tell us that his thoughts are on what he will make of this land. The fussy little skipping steps of the Bride in her first solo, her hands fluttering at her shoulders, suggest that she is a city-bred girl; her brushing gestures at sky and ground bespeak her present joy and timorousness before the vastness of the wilderness. At other times she envisions in her dances days yet to come: what if he were away and the baby came? What if . . . ? During these moments of premonition, everyone else onstage is still, not seeing her, as if the passage of real time has become suspended, while thoughts that would in reality have flashed through a person's mind in a second expand into a meditation on the future.

Time in a Graham dance is determined principally by the amount of thought the heroine devotes to something. Some events in *Appalachian Spring,* like a happy little country dance that serves as a wedding duet, operate in what seems like real time, while other events are compressed or elongated—again a feature of many Eastern styles. Proceeding across the Noh stage, the actor playing the priest Rensei in *Atsumori* confides to the audience, "I have come so fast, I am already at Ichi-no Tani." It is the tradition Thornton Wilder mentioned. The journey could be condensed because nothing important occurred during it. More crucial events might be prolonged: in *Sodegaki Saimon,* drawn from Act III of the Kabuki play *Oshu Adachigahara,* a poignant ritual suicide scene occupies perhaps twenty minutes of an hour-long drama. Likewise, in *Night Journey,* Tiresias's revelation, Oedipus's blinding, Jocasta's actual strangling happen even more swiftly than we imagine they might have, yet the moment between Jocasta's holding the rope and the pulling of it taut around her neck expands into a twenty-minute dance, of which the duet—the key to the revelation—occupies the most time.

Bertram Ross's description of Graham's entrance in her solo *The Triumph of St. Joan* makes one imagine a formality, a controlled passion, an interior voyage even more potent than that of the Noh actor:

> She used the opening music [the score was commissioned from Norman Dello Joio] to do nothing but walk very slowly across the stage, from one side of the stage to the other, toward a pinpoint spotlight. Hanging downstage center was an enormous shield-like banner by Frederick Kiesler with a fleur-de-lis painted on it. The top of it disappeared behind the proscenium arch, and the bottom edge reached the floor. It was wide too. Martha just kept walking very slowly until she disappeared behind it, and you waited and waited, and finally she appeared at the other side. She never quickened that walk. She was wearing a voluminous cape, and you couldn't even see her feet move; it was just as if she was on some sort of

> slow treadmill. A wimple covered her hair and throat, so all you could see was her face—getting brighter as she got closer to the light. She always felt that the light was not ***on*** her, but was coming out of her chest, that she was making her own path of light. It looked that way too. Absolutely! It was a remarkable moment in the theater. I thought that it was as if this figure had been walking for centuries.

Distance or effort may also be distorted in a Graham dance, as is often the case in Eastern forms. A character may behave as if he sees another far away, or doesn't see him at all, when the other actor is in fact standing a few feet away from him. In, say, the battles of Kathakali drama, two enraged combatants may look as if they're performing a well-planned little couple dance. The blinding in *Night Journey* has this ceremonious downplaying of effort. So does the climax of *Errand into the Maze:* the woman who has entered the labyrinth to slay the monster finally conquers him simply by facing him for the first time, grasping the yoke that he bears across his shoulders, and stepping up onto his thighs. You see strain in her body, resistance in his, but that is all; slowly she forces him to the floor, and he is dead. The "Corinth" where Medea lurks in *Cave of the Heart* is barely ten feet away from the Thebes of her imagining, where Jason sports with little Princess Glauce.

The woman in *Errand into the Maze* journeys into a space of her own creation, and it changes shape as it changes for her (in the same way that we experience the inside and the outside of Jocasta's palace simultaneously). She steps on tiptoe—resolute, yet terrified—along a snaking white tape laid on the floor, and we imagine that she is threading the maze by a narrow path. As she passes through a white crotchlike structure, she hauls the tape in and lashes the opening shut. She could be shutting the monster out, or shutting herself in with him, and once she is "inside," the whole stage expands to become her battleground.

The heroine controls sequence as well as space and duration. As can be seen from the descriptions of *Night Journey* and *Appalachian Spring,* performers not involved in the action of the moment do not have to leave the stage; instead they wait immobile, or all but immobile, as if in another place or suspended in another dimension of time. They dance, not when it is appropriate to a plot, but when it becomes necessary for them to present a viewpoint, or when the heroine summons them into her consciousness. In *Seraphic Dialogue,* three cloaked "aspects" of Joan of Arc—Maid, Warrior, Martyr—enter in a procession and sit along a Noguchi bench, like living decor waiting to be summoned by the central Joan-who-remembers. (This stillness of characters not involved in the action is also an element in Japanese and Indian theater forms.) In the case of Graham's dances, the cryptic archetypal structures that Isamu Noguchi set about the stage often provided home bases for various characters, places to which they could retreat when out of the action, as if they still lurked at the edges of the heroine's mind.

Noguchi's enigmatic sets and props are ideal furniture for the dreamscapes Graham's people inhabit; he once said, admiringly, that she used them as extensions of her anatomy. Many of them combined the look of antique monoliths with the flexibility of traditional appurtenances of Japanese theater—like the strip of blue cloth that is laid down to signify water or the fan that can be a sword. A character in a Graham dance often alters the set—taking on the function of a Japanese prop man without surrendering his/her identity. She of the Ground in the amazing *Dark Meadow,* Graham's only overtly Jungian work, turns Noguchi's phallic dolmens to indicate changes in the seasons. In the 1960 *Alcestis,* a massive inverted L could be tipped to become a bed for the heroine's unquiet slumbers. Since the environment, too, is a creation of her memories, she and her deputies control it.

Objects change their nature as smoothly as they do in dreams. The rope with which Jocasta strangles herself is also the umbilical cord. The curious cloak that Oedipus wears may—as Bertram Ross thought it did—represent the fatal crossroads where Oedipus killed his father unknowingly; it also engenders some powerfully phallic images. The crimson velvet carpet spread for Agamemnon to strut home on, when lifted by the curved wands attached to its ends, becomes a snare, a pincer-shaped pair of curtains; these are opened again and again by attendants to reveal Clytemnestra stabbing Agamemnon in his bath with a ritualistic slowness that makes it all the more grisly. The woman in *Dark Meadow* known as One who Seeks treads her somber journey along a long strip of gray material—simultaneously stepping on it and pulling it over her so that she devours her own path; the cloth is her grief and her destiny, and perhaps the gray pall that the mourning Demeter spread over the earth. The three performers in *El Penitente* use various props, masks, or additions to their costumes to signal the phases of the rite and vision. Costumes can link a character with the landscape or express something about interior dilemma. Jocasta's shedding of her dress doesn't facilitate her suicide; it tells us that she is finally ready to shuck off her life.

It was perhaps in *Cave of the Heart* that Graham made the most remark-

Martha Graham in *Cave of the Heart* holding Isamu Noguchi's "dress"

able use of props to translate inner states into dance action. Scrabbling along on her knees, at a pitch of frenzy and rage, she draws from her bosom a little snake of red cloth and feeds it into her mouth, devouring this serpent of jealousy that has awakened in her heart.* At the end of the dance, she steps into Noguchi's shimmering, trembling wire structure under which she has lurked to ponder Jason's faithlessness. It becomes a garment that reaffirms her supernatural power, a symbol perhaps of the golden fleece she once guarded, but as she spins, she seems to fuse with the consuming jealousy that led her to murder, and, as she stands encased in its rays, you can see it also as a cage.

In the great Greek plays of antiquity, drastic events occur offstage. The chorus describes them in language and movement; the protagonists relieve them in agonized poetry. You do see such events in Graham's dances: her adaptations of Oriental theater devices enabled her to show horrifying climaxes, like Oedipus's self-blinding or Jocasta's suicide, with a formality that owed much to brevity. However, such events were of secondary importance. As Graham said when she was making her Joan of Arc solo, she wasn't concerned with deeds, but with ". . . the possible condition of ecstasy which led her to do what she did . . . These are interior landscapes and not the episodes of her life." In a Graham work, as has been noted, more time, more dancing are given over to meditating on courses of action, agonizing over the immediate consequences of action, recalling past actions than to committing those actions. When two people dance together, they are most often expressing sexual love or combat, although they may also engage in ceremonies together, or in brief, enigmatic interchanges that suggest conversation. Numbers of dancers may, like the Greek chorus, take on the emotions of the protagonists; they may also dance with unequivocal passions of their own, as do the Furies with such hair-raising beauty in *Clytemnestra,* and the five couples in *Dark Meadow* with such potent and ritualized eroticism.

Emotion, then, became the true subject of the dancing; Graham's vocabulary modulated and developed to express its nuances. And, however much she refined this vocabulary as a technique suitable for teaching in class, it remained unmistakably something built on her own spirit and body—as

---

*In Japanese folk tales and legends, excessive passion—jealousy in particular—often causes a woman to be transformed into a serpent or be reborn as a serpent. Graham's adaptation of the serpent motify as a symbol of mental and emotional transformation not only made psychological sense, it provided electrifying drama. *The Serpent Heart* was the original title for the dance.

such, inherently dramatic. The contraction in the center of the body with the expelling of breath, its expansion and release with the new inhalation—together these could be shaped to suggest panting, sobbing, quivering with fear. The falls that she had devised as a dialogue with gravity could express states ranging from delight to despair to loss of consciousness. Over the years, she intensified the spiraling and twisting of the torso and subjected these to dynamic shading; they might convey either small doubts and indecisions or the pull of enormous alternatives. When a man lifted a woman, she might perch on his thighs, cantilevered out from him like a figurehead; rarely did she cuddle up to him. Yielding, whether to gravity or emotion, if it occurred at all, was never pictured in a compliant back arch. In some of the back falls, long an important part of Graham's technique, the dancer hinged backward at the knees, her arms reaching out, her hands cupping, yet as she fell, she pressed her knees together and twisted them across her body, while her head fell back—as if she were denying what she was simultaneously craving.

In the "Greek" dances, Orientalisms and archaicisms that had resided in Graham's muscles from Denishawn days were transformed and put to use: hieratic, two-dimensional postures, like those in an Egyptian frieze; arms rippling into oppositely curved crescent moons; carefully presented flexed feet, "Java crawls," "Bali turns" (these from her notes); poses reminiscent of an Indian dancer drawing a bow. Yet none of these was ever decorative; violent emotion pressed and twisted them and canted them off-balance; the pull of one part of the body against another—always a part of Graham's style—now bespoke indecision, inner turmoil. And the rhythms seemed most often to proceed from emotion—outcries and expostulations that rode through the music, as if these scores commissioned from contemporary composers were there to provide an aural climate rather than to support the rhythms of the dancing. The dancers brought the passionate flurries of activity to their own climaxes, freezing momentarily the way a Kabuki performer does, to fix an emotional stance in the viewer's mind.

For the Graham-created dancer, there was (is) no such thing as a casual moment, no relaxation of tension. "Rest comes through change, not collapse," she is reported to have said. "There is no facility in Martha's technique," said Jane Dudley, who knew. And Robert Cohan, who danced for her for twenty years, remarked about her rehearsal process: "She loved you to want her to kill you." The high pitches of feeling she was trying to convey could only be expressed by a body working at an extreme physical pitch, protected from the maudlin by the dancers' discipline and the structural finesse of the gestures.

The costumes for the dances I've been talking about, most of them designed by Graham, reveal the body as an instrument of passion, and the impetus of the dancing the way she wanted audiences to see it. The long bias-cut jersey dresses she so often used cling to the women's torsos, making visible the high, ecstatic lift of the rib cage and its reverse, the contraction. Graham's long torso was one of her most expressive parts; she herself rarely showed her legs (the "Warrior" solo in *The Triumph of St. Joan* was an exception). Whether the garments are conventionally skirted, or ingeniously sewn so that what looks like a skirt from the back is trousers from the front, they emphasize the high kicks of the legs as gestures flaring out from the center of the body. The fabric swirls in their wake, its afterflow sometimes offering a counterstatement to the movement. The men, often bare-chested, wearing only trunks or trousers, tend to look more totemic, less complicated.

The many superb dancers that Graham has trained have been crucial to her success. Yet, given that the impetus for her work came from her inner life, it is not surprising that an astute observer, poet Robert Horan, should have noted in 1947 that in her dances,

> . . . there is really no hero. There is a parade of figures that interrupt or influence her destiny . . . there are combatants and lovers, and until recently these are almost identical in Graham's work; there are phantoms or conspirators. But there is never a figure outlined with the same *human scale,* and portraying anything like the complex range of psychological motivation which she reserves for herself.

John Martin put it more baldly. In the spring of 1944, he wrote a Sunday piece in *The New York Times* entitled "Graham in Retrospect." In it he noted that:

> Her compositions are intensely personal, a projection into spectator consciousness of her inner processes . . . Every work, therefore, is essentially a solo, even though there are twenty people on stage.

It is interesting that Martin noticed this in 1944. (How much truer it became later.) No matter how gifted her dancers, how expressive as performers, in any work in which she appeared, they—as Noguchi had ob-

Erick Hawkins and Merce Cunningham battling during *Deaths and Entrances*

served about his sets—were bound to seem, if not extensions of her anatomy, then anatomical extensions of her mind, alternative propositions. If she made no solos for herself after 1943 except *The Gospel of Eve* (1950) and the Louisville Symphony commissions, *Judith* (1950) and *The Triumph of St. Joan* (1951),* it is because she didn't need to. This is not to say that her dancers weren't vital to her. She used them astutely, and their bodies and skills influenced her creation. In the group works made between 1938 and 1944, she poised herself between two antithetical males: Hawkins and Cunningham. Hawkins was called "The Dark Beloved" in the somber, seething *Deaths and Entrances,* and he played this role—sexually alluring,

*The last-named was made just after Hawkins left her company, when she was without a suitable male partner, and it's possible that growing strife between them prompted the other two solos. The Louisville commissions would not have resulted in company works, but a duet might have been possible.

masterful, potentially dangerous—in more than one dance. Cunningham was "The Poetic Beloved," a slightly mystical, even androgynous figure: he was the blithe acrobat to Hawkins's whip-wielding ringmaster (*Every Soul Is a Circus*), the winged Pegasus to his swaggering husband (*Punch and the Judy*), the gentle Christ figure in *El Penitente,* the fanatic Revivalist in *Appalachian Spring.* After Merce Cunningham left the company, Graham made no more dances that presented this double image of man. Perhaps the male roles also embodied a duality within herself: sensuality and idealism; or the taskmaster/perfectionist and the undisciplined, irrational visionary. She found other ways of articulating the tension between these.

Another crucial character in many Graham dances might represent an archetypal prophetess or sibyl. The mysterious Pioneer Woman in *Appalachian Spring,* the Chorus in *Cave of the Heart,* She of the Ground in *Dark Meadow,* the Attendant in *Herodiade* are all such figures. Their gestures can be soothing, yet they seem to stand at a little distance from the action, as if they know the whole story and await its unfolding. Less ambivalent than the heroine with her hobbled steps and twisting torso, this character dances with breadth and weightedness—handing the heroine a garment, moving part of the decor to signal a change in the climate. The inexorable Ancestress in *Letter to the World* belongs among these elemental figures also—as stern and puritanical guardian, nurse, death-crone. But whether menacing

Jane Dudley as the Ancestress in *Letter to the World* with Erick Hawkins and Martha Graham. Photograph by Barbara Morgan.

or benevolent, the cryptic females can be seen as other aspects of the passionate, "doom-eager" woman Graham herself played.

Graham also allowed the look of dances already made to alter as dancers came and went. For instance, May O'Donnell, on whom all but one* of the above-mentioned roles were created, was a tall, strong blond woman, but most of her parts were inherited in the 1950's and 1960's by Matt Turney, a black woman, who was also tall, but delicate, aloof. The sibyl gained power through mystery even as she lost it in terms of physical boldness and maternal warmth.

When Graham took men into her company, they at first appropriated the straightforward designs, the raw forcefulness that had been part of the style of her all-female company of the late 1920's and 1930's. Certainly some of her male characters were complex in their own right—the Revivalist in *Appalachian Spring,* for instance, who brings down hellfire and brimstone on all sinners, yet prances with his four followers and lets them rock him from one pretty lap to another—and many seemed to have her sympathy, her love. Yet often the Graham figure's perspective, particularly in the "Greek" dances, tended to reduce them to an essence of masculinity onstage. The movements that she devised for them—stiff-legged walks and jumps, bows bending like a V at the hips, hulking leaps, assertive gestures—imbued them with phallic significance. ("We're usually stiff foils, or something large and naked for women to climb up on" is how Paul Taylor put it.) During the 1940's and 1950's, the men she chose for her company were often tall—Taylor, Bertram Ross, Stuart Hodes, Robert Cohan—and they became imposing presences onstage. In those years, Graham's women dancers gradually became more willowy than those of the 1930's, and when they figured importantly in a drama, their movements reflected Graham's own, in terms of ambivalence and frustrated desire.

As early as 1948, though, with the rapturous *Diversion of Angels,* Graham had begun making dances that presented the men and women of her company as what she has many times called "celestial acrobats." In these dances, Graham did not appear. Although they might contain moments of indecision, patches of shadow, most celebrated the pleasures of love and the glories of human spirit and human prowess fused. The style was more open, less equivocal than that of her dark dramas. The Girl in Red (a role created on the gorgeous Pearl Lang) steps out onto one leg and—poised there strong

*Jane Dudley created the role of the Ancestress in *Letter to the World,* although O'Donnell took it over in the 1947 revival, since Dudley was no longer in Graham's company.

Diversion of Angels with (from left, faces visible) Miriam Cole, Linda Margolies (Hodes), Mary Hinkson, and Helen McGehee. Photograph by Martha Swope.

and confident, with her other leg stretching high to the side—contracts her body sharply; this contraction is no anguished gasp, but a profound shiver of delight. Graham dancers of the 1950's—Ross, Hodes, Cohan, Taylor, David Wood, Ethel Winter, Yuriko, Helen McGehee, Mary Hinkson, Matt Turney, Linda Hodes, and others—honed these images of powerful lyricism and estasy into indelibility.

During the late 1950's, when Graham was in her sixties and less able to generate new movement on her own body, the steps in her dances gradually became more codified, less particular to each dance. She drew ideas from the dancers she had trained, but only the most creative were able to give her back gestures that didn't come from the technique taught at her school. Beginning in the 1960's, an influx of lithe, ballet-trained male dancers reduced some of the distinctions between how women moved and how men moved in her works and created more yielding, voluptuous dancing. Instead of complex alternatives, the figure onstage seemed to be summoning up images of youth, beauty, physical prowess. Of necessity, she gradually reduced the scale of her own dancing, doing, with tremendous intensity, only those steps that still suited her, making astute use, say, of gliding little "Japanese lady" walks that accommodated her arthritic feet. By the end of

Martha Graham rehearsing Mary Hinkson and Bertram Ross in *Circe* (1963). Photograph by Arthur Todd.

her performing career (1973), the blaze of emotional energy issuing from her almost immobile figure seemed not just to summon up, but to animate and control her dancers, making them into the kind of appurtenances John Martin had long ago thought them, even though they were doing all the dancing.*

---

*Since retiring from performing, Graham has entrusted many of her roles to the women in her company. Whether they execute these roles with distinction—as many have—or not, the balance of the work is inevitably affected.

Graham's style influenced not only dancers and choreographers she had trained, but many who had only seen her work. Because of her longevity and artistic fecundity, most people cannot remember a time when she wasn't making dances. It is even difficult to remember a period when a supposedly narrative dance, with a high-keyed story to tell, wasn't free to travel rapidly back and forth through time, when three dancers did not usually portray separate aspects of a single character, or inhabit a landscape as violent, yet as distant as a dream. Before Graham, few defined the female dancer as passion-driven, yet intellectually complex; fated, yet capable of choice. In creating a theater of the mind where modern women and men jousted with their archetypes for the illumination of contemporary society, she construed herself as both celebrant and priestess, bringing Western theatrical dancing as close to ritual as it has ever come.

·6·

By Their Steps You Shall Know Them

Members of the New York City Ballet in George Balanchine's *Serenade.* Photograph by Martha Swope.

IN THE ROYAL IMAGE

In 1953 a new work for the New York City Ballet was being planned—a proposed collaboration between Igor Stravinsky and George Balanchine. The company's founder and director, Lincoln Kirstein, wrote to Stravinsky: "What he wants (as usual) is a ballet-ivanich." Stravinsky would have known exactly what Kirstein meant: Balanchine had in mind a suite of dances, or perhaps a ballet concerto, organized as formally as the poised and brilliant divertissements designed by Marius Petipa in St. Petersburg more than a half century earlier. In Balanchine's case, however, no scenario would be necessary to support the ballet, only what Balanchine called "musique dansante."

Kirstein's tone might be taken for exasperation. And perhaps a trace of that does figure. The predominant tone, however, is one of amused understanding: how ironic that it should be a Russian, with strong allegiances to ballet as it developed under the last tsars, who was formulating American classicism. Even though this was exactly what Kirstein had had in mind when he brought the young choreographer to America in 1933.

Many knowledgeable people back then thought he was crazy to do so, and John Martin, the powerful critic of *The New York Times* and champion of modern dance, saw nothing "American" about Balanchine's early work in the United States, although in 1938, when he saw Balanchine's choreography for *I Married an Angel,* he allowed as how Balanchine ". . . would seem to have found in the musical comedy field a more fitting outlet for his particular talents than those loftier fields with which he has been chiefly associated heretofore."

On the other hand, the general public in the United States, primed by the visits of Diaghilev's Ballets Russes and Anna Pavlova's tours, were accustomed to think of ballets as exotic or tragic or frolicsome tales and of ballet performers as frankly charming, soulful, sexy. Too Russian for Martin, an elegant, plotless work like Balanchine's *Serenade* might well have struck American ballet-lovers as not "Russian" enough.

Much the same thing happened in London in 1950, long after many Americans, Martin included, had come to understand and honor Balanchine, Londoners took great exception to the New York City Ballet. No plot, no acting, and no flirting by the dancers, occasional vague hints of drama—what was this? A review by P. W. Manchester suggested that British spectators' capacity for abstraction in dance was rather like their rationing-dulled stomachs' capacity for rich food: ". . . by the time we saw *Symphonie Concertante* we had all had about as much pure choreography as we could comfortably digest in a short period." Critic Richard Buckle, a convert, mocked his colleagues in a satirical playlet, "Critics' Sabbath," with a character designated as "Sheepface" pontificating, "They are mere callisthenics by automata . . . ," and the chorus howling, "We want kings with haughty glance/Wearing too much kit to dance. . . ."

The English public in the austere postwar years may have needed escapist tales. On the other hand, if, during the 1950's in America, Balanchine's ballets and his company's way of dancing became more widely admired, the young-adult generation's dissatisfaction with scenarios may have been partly responsible. As anyone who was an adolescent between the end of World War II and the Eisenhower years can remember, everything had a story line. A strong conservatism at home and a tough stance abroad were supposed to result in a peaceful world of happy democracies and a cowed Soviet Russia. On the home front, the discrepancies between reality and the well-worn plots—do this and you'll achieve this; do that and you'll end up in trouble—seemed more apparent than they once had. It's not surprising that the fifties saw the rise of Abstract Expressionism in painting and Balanchine's plotless ballets, that the most-touted new modern choreographer was Alwin Nikolais, and the most controversial one was Merce Cunningham—both of whom, in dissimilar ways, avoided "stories."

During those years, fans of the New York City Ballet would have known what Edwin Denby was talking about when he wrote of Balanchine in the *New York Herald Tribune,* "He has made our dancers look natural in classicism." Admiring spectators and hostile ones alike could perhaps have seen the truth behind Kirstein's flippant message about "ballet-ivaniches": of all the influences on George Balanchine, that of the nineteenth-century Russian classicist Marius Petipa seems to have cut deepest. And to look "natural in classicism," the dancers in the New York City Ballet had had to absorb not only the technique, but something of the decorum that was the rule in St. Petersburg when Balanchine was a child.

The old ballet master Petipa had been dead seven years when the little Balanchivadze* entered the Imperial Ballet School in St. Petersburg. But *Harlequinade, Don Quixote, The Nutcracker, Swan Lake, Raymonda,* and *The Sleeping Beauty*—all by Petipa and/or his assistant, Lev Ivanov—were in the repertory of what's now known as the Kirov Ballet when Balanchine was a student. He saw them, performed in them. His first stage appearance was as one of the children in the "Garland Waltz" in Act I of *The Sleeping Beauty,* and as a cupid in the last act. He must have cherished his memories of these ballets because he staged versions of them or excerpts from them for the New York City Ballet, "after" Petipa or Ivanov. An earlier Petipa subject, *A Midsummer Night's Dream,* became a Balanchine ballet too, and

Maria Tallchief and André Prokovsky in Balanchine's one-act version of *Swan Lake.* Photograph by Martha Swope.

---

*Balanchine was born Georgi Melitonovitch Balanchivadze. The frenchification of his name was urged on him by Serge Diaghilev when he went to work for the Ballets Russes in 1924.

it was said that he hoped one day to stage *The Sleeping Beauty* in all its grandeur. However, the Petipa that Balanchine most revered was *not* the court servant who devised five-act, five-hour dramatic spectacles with plots that inched forward, freighted with mime and divertissements. Balanchine thought ballet unsuited to complexities of plot, and he could be sardonic on the subject:

> A woman appears; presently another joins her, and the spectator, confused by his libretto, thinks it is her sister. But at second glance in the program notes, he discovers that it is her child by another husband.

Yet, as *The Prodigal Son, A Midsummer Night's Dream,* and a few other ballets show, he himself could tell a story well when so disposed, and he certainly knew the uses of spectacle. In *Union Jack,* he brought almost every dancer in the New York City Ballet onto the stage in a spine-tingling massing of kilted clansmen and clanswomen, marching in unison. Nor did he ever disdain the element of display intrinsic to the classical presentation of the dancer, although, unlike Petipa, he avoided calling attention to the difficulty of virtuosic steps. Both men made inspired use of the attributes and capabilities of various dancers that came their way, extended the vocabulary of ballet and—whether or not this was their intention—the expressive power of classical dancing.

Balanchine accepted on his own terms Petipa's vision of leading dancers as courteous and restrained in their dealings with each other. His ideal "princess" has never looked much like the charming, bejeweled ballerina of the Maryinsky Theater, with her hourglass figure, demurely rounded gestures, and sturdy legs, but she is as elegant. His "cavalier," even in a *pas de deux,* is more than a handsome ballerina-support with the concerns of a hero (as Petipa's princes often were); he stimulates his partner and complements her dancing with his. Nevertheless, he usually *is* a cavalier, whether he's wearing tights and a T-shirt (as in *Agon* and scores of other works), or is equipped with boots, tight pants, and a Stetson (*Western Symphony*).

What most inspired Balanchine in the ensemble dances, *pas de deux,* solos, and *pas d'action* of Petipa and Ivanov was the power of the dance phrase as a visual sign of the musical phrase, and the mysterious ways in which the steps and configurations of ballet can hint not only at human behavior, but at forces beyond human control. Some of Petipa's pure dance passages, like Balanchine's plotless ballets, are resonant with "unconscious images," as Edwin Denby put it, which are "suggested by devices of structure rather than by devices of gesture." Petipa's way of handling masses and soloists was a kind of dance symphonism embedded in the story ballets that were his prescribed format. In the brief period of wild artistic experimentation that followed the Revolution, long after Mikhail Fokine's reforms had supposedly proved the fustiness of Petipa, Petipa was invoked by one of the presumed radicals, Fyodor Lopukhov. In 1923 Balanchine performed in Lopukhov's controversial "dance symphony" *The Magnificence of the Universe*—a ballet set to Beethoven's Fourth Symphony that drew its substance almost entirely from Lopukhov's reading of the music. One of Balanchine's strongest memories is of Lopukhov's battle cry: "Forward to Petipa."

Marius Petipa arrived in St. Petersburg in 1847, his French training a gilt-edged security in a Russia bent on taking its cultural leads from Western Europe. A dancer of some accomplishment—he had once partnered Marie Taglioni—he worked under two French ballet masters in power in

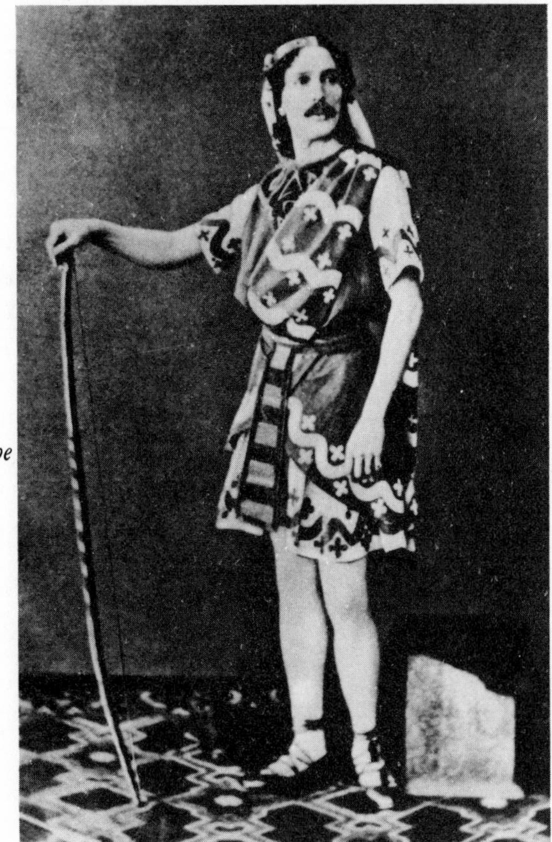

Marius Petipa in *Daughter of the Pharaoh* (1862)

St. Petersburg: Jules Perrot, an adept at romantic fantasy and naturalistic crowd scenes, and Arthur St.-Léon, a man who could deliver up spectacles quickly and smartly without worrying overmuch about dramatic coherence.

Petipa's first important ballets were produced during the period of restlessness and civil strife that followed the Crimean War. Alexander II's "great reforms"—the freeing of the serfs, the restructuring of local and town government, the legal system, and the military—led to an inevitable decline of the gentry and, along with economic growth, to the rise of a middle class. Yet the reforms ignited even more problems than they solved.

Ballet, however, as a virtual property of the tsar, didn't echo the populist sentiments of writers like Tolstoy and Dostoevsky or their later disillusionment and sadness. For the aristocratic and wealthy audience, ballet's spectacular fantasies and exotic dreamworlds and impossible visions were supposed to provide entertainment, beauty, and escape. Political awareness, if any, of balletmakers generally took the form of tactful tributes to enlightened power. In two of Petipa's most famous early ballets, *Daughter of the Pharaoh* (1862) and *La Bayadère* (1877), rulers were fair-minded and benevolent when faced with the truth—which was often revealed to them by heroines who carried on the tradition of righteous bravery established by French Romantic ballet. The scenario for Tchaikovsky's *Swan Lake,* which originally premiered in Moscow (reputedly with indifferent choreography by Julius Reisinger), offered the provocative dilemma of a future ruler who could not distinguish between good and evil, and suggested with the utmost poetic delicacy that without idealism, devotion to duty is nothing.*

During the reactionary period that followed the assassination of the reform-minded Alexander II in 1881 and the accession of his son, Alexander III, to the throne, the ballets demanded by Ivan Vsevolojsky, the director of the Imperial Theaters, emphasized pomp and diversion even more than had their predecessors—often at the expense of an even halfway interesting plot. Operas might be based on major works of Russian literature, as were Tchaikovsky's *The Queen of Spades* and *Eugen Onegin.* Not ballets. As

---

*In the story as it has come down to us, a royal maiden (Odette), along with her companions, is held in thrall by a wicked sorcerer. Allowed to appear in human form by night, the women spend their days as swans. Prince Siegfried falls in love with Odette; the formal plighting of their troth will break the spell. To the ball at which the Prince must choose a bride, the sorcerer, von Rothbart, brings his daughter, Odile, disguised as Odette. The Prince is smitten and vows to marry her. The real Odette is forever condemned to swanhood, and Siegfried, realizing his mistake too late, joins her in suicide. (Some productions now feature a happy ending, while others finish with a despondent Siegfried watching a crowned swan slowly glide past him.)

Russian cultural historian James H. Billington writes, *fin de siècle* Russian ballets provided "childlike interludes of graceful fancy for a harassed people."

Yet the swan was a traditional Russian symbol of purity and redemption. Given *Swan Lake*'s central image and its cautionary theme, it is an interesting coincidence that the rechoreographed ballet by Petipa and Ivanov was first performed in St. Petersburg early in 1895, around the time a new tsar took power. Vsevolojsky's choice of *The Sleeping Beauty* as a ballet scenario most certainly was based on the number of delightful divertissements it might occasion, but the story has a moral: a breach in royal courtesy, even to such nasty adversaries as wicked fairies, can allow chaos to upset the orderly flow of events. Did that really escape him, Petipa, and Tchaikovsky? Although such speculations may seem far-fetched, certainly Petipa formed a style not simply in spite of tsarist protocol, but in intuitive and creative response to it.

According to Russian ballet historian Anatole Chujoy, there were nineteen courts in St. Petersburg alone at the beginning of the twentieth century. The entourages of the tsar, the dowager tsarina, and the various grand dukes numbered in the thousands. Together with high-ranking military officers and foreign ambassadors, they could have filled all three of the Imperial Theaters in St. Petersburg and the three smaller theaters to which the general public was not admitted. Only at the Maryinsky (which seated 2,500) was ballet presented regularly—meaning about fifty performances a year, on Wednesdays and Sundays, only ten of them nonsubscription. Diaghilev's collaborator, Alexandre Benois, deploring the stodginess of the Imperial Ballet's presentations, said, with some truth, that they were created for "children, hussars, and ranking dignitaries."

When Marius Petipa worried about pleasing "the public," he was speaking of a power elite. Dancers on the stage of the Maryinsky could look out into an orderly assemblage, seated according to rank and prestige, with rich merchants, lesser officials, members of the theater or the press, students, and children filling the places not reserved for courtiers.

The spectators looked back at a stage world that flatteringly mirrored theirs in protocol, decorum, and elegance. The ballerina and premier danseur, like the tsar and tsarina, were framed by a select company of soloists (the grand dukes and duchesses) and demi-soloists (court officials, if you like) and by a further stratified corps de ballet. When Balanchine was staging the "Garland Waltz" from *The Sleeping Beauty* for the New York City Ballet's Tchaikovsky Festival in 1981, he joked with his dancers about the third echelon of the corps in prerevolutionary St. Petersburg. Barely able to stand on pointe, they were called "the fountain girls" because they were always placed at the back, near the apparently ubiquitous scenic water.

The parades, grand entrances, and large ensemble dances in the ballets affirmed the power of ceremony. The surviving works from this period—*The Nutcracker, Swan Lake, The Sleeping Beauty,* and *La Bayadère*—contain courts of their own. In last-act divertissements, the various dancers who entertained both the onstage and the offstage nobility are often presented before they dance their variations, and reenter together later to receive the royal thanks and applause. Soloists may walk ceremoniously to a position on the stage to begin dancing or be magically revealed by a shifting of the ensemble, as if living curtains were drawing aside.

It was natural that even the most frivolous of the spectacles would involve a manipulation of props and costumes as fastidious and elaborate as that of court ceremony and church ritual. Petipa's notes contain a plan for a dance during which horizontal lines of female dancers come forward alternately, displaying with each advance a different color of skirt. (While concealed by the line in front, the women in the next rank hastily tucked up the top layer to reveal the new shade.) His original plans for Act I of *Swan Lake* called for twenty-four couples and twenty-four red-and-green stools on which they could perch to create pleasing designs; the women were to carry flower baskets, the men batons that would suddenly sprout flowers. Fanciest of all: on the last beat of the "Grand Ballabile des Caryatides Animées" in *Daughter of the Pharaoh,* small children popped up out of baskets carried on the heads of dancers (who were presumably male and sturdy). Some divertissements we read accounts of may strike us as the nineteenth-century equivalent of half-time entertainment at football games; but in others, spectacle, symmetry, and repetition were used to profound effect. Of Petipa's mass dances, a contemporary, critic Akim Volynsky, wrote delightedly, "There are everywhere lines and figures which harmonize with one another and create the impression of one line and one figure." The dancing ensemble, which for Perrot and Bournonville had been a bouquet of individuals with common characteristics, became an impersonal force—garlands and frames and lakes of women, shifting the equilibrium of story and musical structure.

Gautier's scenario for the 1841 *Giselle* indicates that among the ghostly wilis could be distinguished girls of different nationalities, who performed wan echoes of their characteristic dances. After Petipa rechoreographed the ballet in 1850 and again in 1884, all that remained of that individuality were two exotic names, Moyna and Zulma—now owned by dance-alike

Members of the Kirov Ballet of Leningrad in Act II of *Swan Lake*

demi-soloists. Descriptions of the earlier *Giselle* suggest that the hapless gamekeeper, Hilarion, venturing into the woods at night, was waltzed around and around by one ghost-girl after another until, dizzied and exhausted, he was whirled into the lake by his last partner. "Ogresses of the waltz," Gautier called them. What has come down to us from Petipa is far more abstract: Hilarion is rapidly spun down an implacable diagonal line of wilis, each giving him a push, until he vanishes offstage; the image is of a glamorous and chillingly efficient machine for executing summary justice. And at the climax of Act II, the chains of wilis crisscrossing the stage with inexorable *arabesques voyagées** seem to cast a net of dancing to entrap the exhausted hero, Albrecht.

Petipa's memoirs give little clue to a brilliance that must have been largely intuitive. He appears to have concentrated on diagramming ever more

*Standing in *arabesque* on a bent leg, the dancers travel by means of tiny hops that scarcely leave the ground. Because they hold the position as steadily as possible, torso and extended leg almost parallel to the ground, they look almost as if they are being pulled across the stage by invisible cords.

Members of the corps de ballet of American Ballet Theatre in "The Kingdom of the Shades" act in *La Bayadère,* staged by Natalia Makarova in 1974, and incorporated into the full-length production in 1980

amazing spectacles, accommodating reigning ballerinas, attempting rather pathetically to curry imperial favor and win job security. Yet through his skill as a master mechanic, the most resonant of poetic images emerged.

The famous entrance of the dead temple dancers in "The Kingdom of the Shades" act in *La Bayadère* exerts its magic both as choreographic strategy and as metaphor. One by one, the white figures appear at the top of the ramp; each executes once the little phrase—step into *arabesque;* step back and arch deeply backward (*cambré*), arms wreathing overhead; straighten up and take two steps forward—before the next woman appears. Time seems suspended in a blanched eternity while, to the sweet, placid Minkus music, the women of the corps de ballet, one by one, advance to fill the stage with their snaking procession and, at last, assemble in four vertical lines to begin their smooth, slow balances. On a technical level, Petipa planned the dance as a long crescendo and an entrancing display of the unanimity of the corps de ballet. He had been struck, it's said, by Gustave Doré's illustrations for Dante's *Il Paradiso.* But the scene also reminds us that the entire act is an opium dream of the bereaved and anguished Solor, who, through opportunism and moral cowardice, has caused his beloved's death. The dance of the *bayadères* has a narcotized slowness and evenness; it suggests a blurring of the hero's vision: seeking one woman in the spirit world, he finds her endlessly multiplied.

Lev Ivanov allowed this kind of metaphor to shape the ensemble patterns and choice of steps in his choreography for Acts II and IV of *Swan Lake.* The white-garbed women don't imitate swans, they take on their habits—now gliding, as if over water, now flocking wildly, their wing-arms beating the air. In the "Snowflake Waltz" of his *The Nutcracker,* the leaping women swirl and drift past each other, giving the illusion of being caught up by imaginary currents of air. Such ensemble imagery had little place in the earlier French or Danish ballets, where dancing, for the most part, was "dancing," whether you were a sylph or a harem girl.

The late nineteenth-century dance ensembles, like the orchestra for a concerto or the court of a ruler, could echo the soloists' steps on a simplified level, or offer a step as a suggestion for the soloists to elaborate on. Sometimes these purely formal connections created an impression of human volition. Reviewing a revival of "The Kingdom of the Shades" act from *La Bayadère,* staged by Natalia Makarova for American Ballet Theatre in 1974, Arlene Croce wrote:

> My favorite moment comes in the final waltz, when the three principal Shades are doing relevé-passé, relevé-attitude cambré to a rocking rhythm, and the corps, seeing this, rush to join them on the repeat. They—the corps—remember those cambré positions from their big dance.

The solo dancers that Petipa and Ivanov had to work with were very different from the dancers of Marie Taglioni's generation. Years of interchange between ambitious dancers and ingenious shoemakers had gradually transformed the pointe shoe from a thin slipper reinforced by darning and buckram to something with a steel shank and a stiffened box of a toe. In Russia in 1858, in Perrot's *Eoline,* the Italian ballerina Amalia Ferraris managed jumps on pointe, a supported triple pirouette, and long balances, and was applauded by balletomanes for displaying only the tiniest of quivers. We can infer from a review of her performance that her shoes were fairly sturdy. Yet twenty-seven years later, when another Italian, the glamorous and dramatic Virginia Zucchi, made her Russian debut at the summer theater at Kin Grust, shoes and technique had developed to such a degree that she was able to perform an entire solo on pointe.

Most of the brilliant Italian ballerinas who came to Russia late in the century had been pupils of Carlo Blasis in Milan. Russian dancers who hadn't studied under Blasis in Moscow, where he worked between 1861 and 1864, picked up the dazzling new virtuosity on their own. Whether Petipa worked with the likes of the native Ekaterina Vazem (*La Bayadère*) or the foreign Pierina Legnani (notably in *Swan Lake*), he was assured of a ballerina whose "steel pointes" were a vital part of her equipment. The new virtuosity stimulated him. His many solos for women investigate all manner of illusions in the way that articulate toe can touch the floor—brushing, jabbing, piercing delicately, gliding, hopping, resting.

Solo *adagios* on the whole foot all but disappeared; the public wanted to see pointework. Since long, intricate balances on pointe necessitated a partner, a new duet form emerged—one that altered how men and women danced together. Instead of doing spirited steps side by side with an occasional lift or embrace, or taking turns—as did Bournonville's dancers—supporting and turning each other, the man stood behind his partner and, holding her waist or hand, revolved her smoothly on one toe through a sequence of constantly changing poses, as if to offer her to the audience in all her three-dimensional charm. Expanding the complexity of these earthbound patterns, lifts became more intricate too: the premier danseur might change the position of his partner in the air several times before returning her to the tip of one toe. Musical rhythms or an aura of narrative could render these *pas de deux* ardent, but they also occurred for sheer display, unoccasioned by the plot. In Petipa's remake of *Le Corsaire,* the heroine doesn't dance with the pirate-hero when they're safely aboard his ship; she dances *for* him with a handy and surprisingly amorous slave.

When Pierina Legnani made her debut in Russia, a critic complained that she "lacked lightness," but he was clearly a reactionary. Lightness was not the issue it once had been. Nearer to fingering contemporary taste was the critic who praised Legnani as Odette in *Swan Lake* for her "grace, art, precision, and confidence" and for "the extraordinary strength of her steel muscles and her beautifully shaped legs." Little bounding steps had lost precedence to large leaps, soaring to gliding. The sylph's emblematic steps had been airy jumps, quick little runs on tiptoe, and transitory balances; the swan moved by gliding and plunging. Her rippling chains of *bourrées* on pointe, her long, supported balances on one toe with the other leg probing floorward like a swan's neck or beating against the ankle like a quivering wing were all made possible by the changes in technique.

For the most part, photographs of the reigning ballerinas—Legnani, Mathilde Kshessinskaya, Olga Preobajenska, Carlotta Brianza, Varvara Nikitina, and others—show strong-legged, well-rounded little women with

Carlotta Brianza, the first Aurora in Petipa's *The Sleeping Beauty* (1890)

fashionably dressed hair and jewels (the gifts of admirers) prominently displayed. Leaning casually on a convenient pillar, or supported by a solicitous partner, the dancer establishes a plumb line from headdress to firmly planted pointe—and could stand there unflinching while the slow-shuttered cameras cranked out the seconds. For all her short skirts, she's a Russian princess and is meant to act like one. Virginia Zucchi was rebuked by Petipa for removing her headdress and disarranging her hair and clothes for the scene in *Daughter of the Pharaoh* in which Aspicia has been pursued by a lion. Lion or no lion, he told her, a princess ought *never* to appear without her crown.

Although cameras of the day were unable to capture motion, the photographs accurately reflect the fact that equilibrium, rather than ethereality and restless flight, had become the guiding image in the dance style, as well as in the overall stage picture. Mortal heroines, like Raymonda, show pas-

The "Rose Adagio" in a contemporary production of *The Sleeping Beauty* by American Ballet Theatre. Susan Jaffe is flanked by (from left) John Summers, Robert Hill, Ricardo Bustamente, and Clark Tippet. Photograph by Martha Swope.

sion through aplomb, their feet jabbing the floor in moody playfulness. Supernatural heroines may display their moral delicacy or their deadliness through how they assume a balance. When the spiritual Odette sits folded over and quivering on the ground, Siegfried leans over her from behind and gently pulls her up and onto one tentative pointe. When he is dancing with her double, Odile, the pawn of the evil Rothbart, she sits in something very like that position, then lifts her eyes, daring him; this time he faces her, and as he pulls her by one hand, she darts triumphantly into a *piqué arabesque* that's almost like the pecking of a fierce swan.*

A brilliant use of balance as combined feat and metaphor occurs in the "Rose Adagio" of *The Sleeping Beauty.* Sixteen-year-old Princess Aurora is supported by each of four suitors who have come to vie for her hand in marriage. The way in which she balances unsupported on one toe in an *attitude* while each prince comes forward in turn to offer her his arm, and

*It must be understood that *Swan Lake* as it is performed in Russia today, as well as those versions mounted in the West (many of them descended from the production that former Imperial Ballet artist Nicholas Sergeyev staged for the Vic-Wells Ballet in 1934), have been altered considerably. Nevertheless, many details of the choreography remain the same or similar in most productions, and we conveniently assume that in seeing these passages, we are seeing the Petipa-Ivanov choreography.

later to revolve her slowly in the same pose, showcased the virtuosic equilibrium of the original Aurora, Carlotta Brianza. Yet these balances can poignantly suggest the testing of a young princess's maturity and her ability to be calm, gracious, and balanced in her judgment under stress. A message any little grand duchess might take to heart.

In terms of excellence, women outnumbered men in Russian ballet—imported virtuosi like Enrico Cecchetti, for whom Petipa made the scissoring, flying solo of the Bluebird in *The Sleeping Beauty,** or eloquent Russian actor-dancers like Pavel Gerdt, who prolonged his career into late middle age by the device of dance stand-ins or assistants. (No one among Gerdt's fans would have thought it odd that during Siegfried's first impassioned duet with Odette, his friend Benno should be standing by to catch her as she swooned backward or to lift her into the air.) As in Western European ballet, the prominence given the ballerina sapped the strength of the male role and demoralized promising young student dancers. Training and morale had improved by the end of the century; no longer need Russian balletomanes suggest, as one had in 1879, that while men were clearly necessary to have around onstage, they ought not to be allowed to perform solos.

In the ritualized *pas de deux,* the male dancer was almost equal to his partner. He supported her in the *adagio,* danced a solo variation that showed off his ability to leap and spin, ceded the stage to the ballerina for her display of rapid pointework and secure balances, then spelled her and rejoined her in the fireworks of the coda. Yet his major function, as the sturdy physiques of the men affirmed, was as a *porteur,* and there were fewer opportunities for him to dance than for her. The character dances that Petipa also excelled at—a Spanish dance, a czardas, a mazurka—would bring on the men, but a male corps de ballet dancing without women was almost unheard of. The variation for four men in the last act of *Raymonda* (1898) attested to a renaissance of male dancing supposedly sparked by Cecchetti's performing and teaching.

The lives of the dancers who performed in the Imperial Theaters reflected some of the desired stability, decorum, and attention to protocol that characterized the roles they played in the ballets—roles that had to be affirmed in the increasingly precarious world of royal privilege that the majority of the audience hoped would never end. Children accepted into

*Cecchetti is often given credit for the choreography of his own solo. It is more likely that the two men collaborated—Cecchetti showing Petipa some of the fancy jumping he was capable of, Petipa weaving these into a pattern. In any case, it was certainly Petipa who conceived of the solo as aerial and an ideal vehicle for Cecchetti.

the imperial schools were wards of the court. Their clothes were provided for along with their schooling, food, and lodging. Uniforms distinguished the various levels, and boys and girls were kept separate, except in classes in social dancing and, later, in partnering. When a ballet required the participation of children, carriages with liveried coachmen and footmen carried them from the school to the theater. Teaching and playing character roles assured dancers of long careers. When too old or unsuited to do either, they were pensioned off. The jobbing around, the scrounging, the search for wealthy patrons was not a necessity for them as it was for dancers in most of Western Europe. If the women took archdukes and generals as lovers, it was because they liked the social cachet and the lavish support. Certainly those who attained stardom behaved like stars.

FORWARD TO PETIPA

"You discover," said George Balanchine, "that what stays with you are the essential things. You discover what you are doing is really Petipa." Petipa revitalized, expanded, stripped of all but the essentials, and delivered in the accents of modern America. The "ballet-ivanich" under consideration in 1953 turned out to be *Agon.*

When *Agon* was premiered in 1957, it was perceived as a dazzlingly contemporary work—dense and yet lean, like Stravinsky's score. No fancy costumes, only black-and-white practice clothes. No fat on the dancers, no fat in the steps. Rapid, propulsive, stinging. And hardly nineteenth-century Russian in tone. Edwin Denby wrote, "The basic gesture of *Agon* has a frank, fast thrust, like the action of Olympic athletes, and it also has a loosefingered goofy reach like the grace of our local teenagers."

Yet even as Stravinsky's music paid homage to the dance forms of Bach's day, Balanchine's steps acknowledged both these and the precepts of late nineteenth-century Russian ballet. The kind of "dissonance" that both men employed presupposes our ears' familiarity with consonance. "Dissonance makes us aware of consonance," Balanchine said. "We cannot have the cool shadow without the light." As in other Balanchine ballets to contemporary music—by Stravinsky or Hindemith or Ives or Webern or Gershwin—classical tradition may seem to be subverted, but a turned-in leg is understood in terms of the turnout that follows, a flexed foot in relation to a pointed one; a swing out of a centered posture imprints its in-balance counterpart on our brains.

The Stravinsky score opens with an explosive, discordant fanfare. To it, Balanchine sets a rapid sequence for four men that bursts from unison into canon and back to unison. Whirling, lunging, striking out from their separate spots in space, the men seem to be tugging the stage this way and that; yet the whole resolves as boundless symmetry. Later, in the second *pas de trois,* Stravinsky sets imitation as strictly as Bach might have, but although the precision is baroque, the dissonance and sharpness supply a pressure that's unmistakably contemporary. Balanchine's two men follow each other through a canon so compressed that one seems to be shadowing the other, daring him to widen the gap.

In *Agon*'s duet, a man bends his partner into positions far more extreme than any dreamed of by Petipa. Gently but firmly he lifts her leg up behind her and presses her head back, as if to see if they will touch. The rhythm of his manipulations suggests a pensive testing of his partner's range. Yet, although he holds her in intricately knotty ways and handles her with a matter-of-fact intimacy that would have seemed shocking at the end of the last century, his behavior refers to the supported *adagios* of that time. He promenades her in *arabesque* on pointe; only he is lying flat on his back,

Diana Adams and Arthur Mitchell in the *pas de deux* from *Agon.* Photographs by Martha Swope.

while she grasps his upraised hand, her leg sweeping above him like a compass needle. It is Petipa inverted, Petipa harried, Petipa atomized. Petipa with more than a nod to a vernacular of American cheerleaders, bred in studios where little girls studied ballettaptoeacrobatics. But Petipa nonetheless.

Of course, ballet scholars probing the origins of Balanchine's style can cite other influences. Mikhail Fokine's expansion of classical deportment and vocabulary surely had some impact on the young Balanchine. It is known that, as a budding choreographer in Russia, he was excited by the work of the classically trained Constructivist Kazian Goleizovsky, whose cool, erotic-gymnastic études and interest in American jazz stimulated the Russian art world of the twenties. Serge Diaghilev, whose last ballet master he was, exposed Balanchine to the Western European art scene, past and present. You can also postulate that his work for *Charles Cochran's Revue* in London and his stints on Hollywood movies and Broadway musicals during the thirties and forties had some effect when you note the jazzy hip thrusts and the one-knee-turned-in, pinup-girl stances that—exuberantly reconstituted—appear in certain of his ballets. Couldn't his occasional pretzely solo or duet (like that in *Agon*) be construed as witty and imaginative nods to ballet-acrobat Harriet Hoctor, for whom he choreographed a couple of dances in *The Ziegfeld Follies* of 1936?* You can remember that he once said that Fred Astaire was ". . . the most interesting, the most inventive, the most elegant dancer of our times."

Certainly Balanchine's take on Americans and America affected his style as profoundly as his background, as profoundly as the music that inspired his various ballets. Barely a year in the United States, he told Chicago critic Claudia Cassidy, "There is that love of bigness that is so important a part of the ballet. The skyscrapers, vast fields, gigantic machines, all make for thrilling spectacles." The scale may have reminded him of Russia, but he translated it into space less clogged and gestures larger in scope than anything on the Russian stages. He seems to have noticed the pace and complexity of our cities, and given them back to us as speed and density: into his choreography, he packed more steps per running foot than Petipa did at his most vivacious. Jean-Pierre Bonnefous, the French dancer who was a NYCB principal from 1970 into 1980, told writer Barbara Newman, "To dance his ballets, I think you have to learn to be part of New York."

Balanchine characterized the American spirit as "cold, luminous, hard as

---

*Some of Balanchine's very first ballets were created on Russian dancer Olga Mungalova, who was noted for her acrobatic abilities. It's conceivable that her image lingered with him.

light." Let us say that those were the qualities that attracted him. (Petipa too, it is said, liked a reserved performing manner.) He saw and liked something full-throttle and brisk and no-nonsense about the American character that he wanted to bring out. When he caught a dancer in his company mooning over the music and steps, he made dark mention of "Gisellititis." He wanted almost everything danced to the hilt in terms of energy and scale and precision. Tanaquil LeClercq, one of the first ballerinas to be trained from childhood in Balanchine's School of American Ballet, spoke this way about performing *The Four Temperaments* (1946):

> It should look maximum, 100 percent everything: move 100 percent, turn 100 percent, stop dead. Kick legs as much as you can, straight knee, pointed toe. Zip 'round. Fast. Nothing slow, no adaaahgio . . . Kick, wham, fast, hard, big. You have certain steps to do in a certain amount of time and the certain steps give it a certain flavor. But you can't interpret because you'll be late, you won't be with the music.

Not all Balanchine ballets have the fierce propulsiveness dictated by the score that Paul Hindemith wrote for *The Four Temperaments,* but the recommended lack of indulgence, the concentration on dancing and music are the basic equipment of Balanchine dancers, whether the ballet is *The Four Temperaments,* or an elegant essence-of-Petipa one like *Theme and Variations* (1947), or as poignantly romantic as the Brahms *Liebeslieder Walzer* (1960).

Petipa drew his images from plot or from pictures he collected or patterns he worked out with little figures on his dining table; Balanchine culled his almost wholly from music—working both from his profound understanding of musical structure and from his intuitive responses to musical atmosphere. Not that Petipa wasn't musically educated, but he planned his ballets before a score was composed to suit. Even the single great composer he worked with, Peter Ilyich Tchaikovsky, received instructions from him (these are for *The Sleeping Beauty*): "From 16–24 bars, which develops into another tempo. For Aurora's entrance—abruptly coquettish 3/4. Thirty-two bars. Finish with 16 bars, 6/8 forte."

Often Balanchine upheld the Petipa tradition of hierarchies, moderately ceremonious entrances, and patterns of ink-blot symmetry, but his rationale was different. He was not echoing the decorum of an elaborately stratified society, he was acknowledging the relations of parts within a musical composition. Quite often the members of the corps de ballet are indeed working behind the principal dancers, their dancing less complicated, their roles less individual, but this is simply because they are following a bass line or the chords that accompany a melody. And, even in such cases, their patterns tend to be more complex than those that Petipa made. As if Balanchine had taken to heart the equality implied in the American Constitution, he created the illusion of a company of equals, filling the available jobs allotted by the music.

Sometimes the ensemble assumes the prominent role. In the turbulent "Melancholic" section of *The Four Temperaments,* two demi-soloists propel the male soloist toward a far corner of the stage, and, as the string ensemble declaims the theme, four corps women advance toward the man on a diagonal; each swings one leg high before her, then jabs the toe down into the floor, swings the other leg up, jabs that toe down, thrusting her hips forward, as together the women bear down on the man. Responding to the powerful swing of Hindemith's music, Balanchine has given the women the force of avenging angels, women with thighs of molten iron. The theme invades the musical ensemble; the women take over the stage.

The little entourages of four who accompany each of the four soloists in the Stravinsky *Violin Concerto* onto the stage don't always maintain the respectful distance or the subsidiary role they might in a Petipa ballet. One woman (Kay Mazzo originally) prances with her four men, holding their hands, allowing them to hold her up and weave rings around her. They are

Kay Mazzo in *Violin Concerto* with (from left) Nolan Tsani, Tracy Bennett, David Richardson, and Michael Steele. Photograph by Martha Swope.

never anything but polite; their behavior suggests a witty hybridization of *The Sleeping Beauty*'s four cavaliers and a number in a twenties review. Four fraternity boys back up the prom queen while she does her song-and-dance.

In *Concerto Barocco* (1941), it is the job of the ensemble women to lay out the vocabulary before the two female soloists enter (with the violins of J. S. Bach's Double Violin Concerto) to begin their witty canonic bantering. It is the ensemble women who first fall from the stiletto line of fifth position on pointe into a sideways lunge, then swing back to fifth—the source of a myriad variations. It is they who affirm the rapid, even march of *piqué plié* that will turn out to be one of the ballet's most prominent motifs.

Balanchine's reading of musical structure, harmony, and dynamics causes odd and mysterious images to surface in his dances. For the first movement of *Concerto Barocco,* he embodied the two solo violins as two women, perhaps because the virtuosic speed and sharpness of the music made him see pointework, and the way Bach mingled and separated their lines made him see equals. In the tender and romantic *largo,* however, the two instruments often exchange roles in a subtle follow-the-leader; and in contrapuntal passages, the deeper-toned melody of the second violin flows under that of the first violin—as if supporting it. Therefore, at the beginning of this second movement, as the corps women matter-of-factly regroup, one of the female soloists leaves the stage so that a male dancer can take over her violin part. As the two violins play together, one may suspend a note as the other begins a melody, so that they seem to be twining unendingly around each other, never breaking free. In response to this, early in the *largo* movement, Balanchine choreographed a profoundly tender sequence in which the man and woman never let go of each other's right hands. As they slowly revolve and embrace, their linked arms become arch, barrier,

Tanaquil LeClercq, Nicholas Magallanes, and the ensemble in the second movement of *Concerto Barocco.* Photograph by Roger Wood, London.

support, love knot, and it is difficult to fathom what is a beginning for them and what an ending.

The music that began the *largo* repeats toward the end of it; and the woman from the first *vivace* movement enters and threads her way serenely through the group, simply in order to leave as she did before, and let the duet begin all over again. The device is sheer musical gamesmanship, but because of the gentle, dreamy passion of the *largo,* it creates an impression that the whole event is being rewound in memory, the way lovers recall their first meetings.

More often, it isn't gamesmanship but Balanchine's intuitive and subjective response to a musical fact that creates passages resonant with indefinable emotion. Charles Ives's *The Unanswered Question,* which Balanchine used for a section of his *Ivesiana* (1954), is a restless, yearning call of a melody that never resolves. Played first by a single horn, it seems suspended above a shimmering, featureless terrain. The choreographic response was a procession of four men who carried a woman aloft (originally the very young Allegra Kent, who appeared both innocent and mysteriously voluptuous). Standing at first, she lets herself fall backward into their arms. They walk on, bearing her as if she were an enigmatic harem beauty on a palanquin. But there is only air underneath her. They bend her gently, twist her into fantastic shapes, make her curve upward, or dip her down so that her unbound hair brushes against the solitary man who follows wherever she goes, oblivious to everything but this distant, carefully guarded beauty. Luminous as a beacon in darkness, like the melody, she never touches the ground.

In Petipa's day, style didn't wander far from what was taught in the classroom, although what was taught might reflect new discoveries. Even the various character dances employed a known lexicon of steps. Petipa spoke in one voice, Balanchine in many, partly because of the latter's interest in, and access to, a broad range of music. For Balanchine, musical style inevitably transformed the classical vocabulary in diverse ways; and some of the steps he created for the stage would be unlikely ever to appear in the classroom. Even in his most "abstract" ballets, his responsiveness to the music's texture engendered diverse images of people and society. (Beyond his initial delighted vision of American culture, any contemporaneity in his work seems to have filtered through his reaction to contemporaneity in music, rather than to have occurred in response to social or political events.) Stravinsky, he said, made ". . . time that shows how the small parts of our bodies are made." That response to the atomization in the music engendered the nervy pointillism of many of his ballets to Stravinsky's music.

The sparse, shrilling microtonal strings in Toshiro Mayuzumi's score for *Bugaku* drew from him a flow of serene yet contorted postures for the central couple that recall the elegant drawings of Japanese erotic art. *Who Cares* teases the purely classical steps out of equilibrium and syncopates them to evoke the carefree, impudent climate of orchestrated Gershwin tunes. In the first half of the no-longer-performed *Metastaseis and Pithoprakta,* great sloppy swarms and chains of people heaved about like a single convulsing organism, in keeping with the density and unpredictability of Iannis Xenakis's clouds and swirls of "aleatoric" music.

This last ballet would have fooled any contestant in a guess-the-choreographer contest. You would sooner have imagined it to be by Merce Cunningham than by Balanchine, and it certainly would have taken Petipa aback. Yet if a few of Balanchine's works seem unrelated to the tenets of Russian classicism, both in their designs and in some cases—such as *The Prodigal Son* and *Variations pour une Porte et un Soupir*—in their vicious, acrobatic heroines, dig down to bedrock in most Balanchine ballets, and you find Petipa. Chatting with Walter Terry in 1950, on the latter's interview series at the 92nd Street YMHA in New York, Balanchine remarked that the London critics had found the dancing in *Jones Beach* terribly acrobatic, but that everything they thought strange was derived from old Petipa steps.

The connections—and the differences—between the heroes and heroines of Balanchine and those of Petipa are nowhere more striking than in the Balanchine ballets that seem, wholly or in part, to be essays on Petipa. Arlene Croce has eloquently pointed out the kinship between *Theme and Variations* and *The Sleeping Beauty:*

> Evocations of the great Petipa role pass before us—pas de chat/pirouette, a rhythmic lunge striking tendue positions in fourth while the arms sweep through directional changes, and, in the finale, a pas couru ending in the dainty, girlish, tendue-front pose that Margot Fonteyn had her picture taken in a thousand times.

The ballerina of *Theme and Variations* is ". . . Aurora rewritten in lightning."

Abstracted, too, are the visions of womanly virtue and graciousness embodied by Petipa's good fairies. Balanchine shows us four demi-soloists, each flanked by two other women, who scroll around her with chains of *bourrées,* kneeling to support her *arabesques,* standing to build bowers over her head. Later the ballerina turns and dips and blossoms into new shapes at the center of a garland of eight women who support her. Petipa's choreographic rendering of the briar hedge around the slumbering Aurora was a vision of formal clusters and waves of attendant nymphs, framing and concealing a dream princess before the Prince's enraptured gaze. In Balanchine's ballet, the ballerina is as intimately connected to her attendants as a rambler rose to its vine.

At the beginning of the Stravinsky *Symphony in Three Movements,* women in white leotards arrayed along a diagonal whirl their arms, lunge, prance in place, paw the air, while a male soloist vaults into the air in front of them. As critic Nancy Goldner pointed out, they make you think of a Broadway chorus and a machine at the same time; it's the implacable destruction-machine in *Giselle* that they bring to mind most vibrantly, even though they're anything but wan and fated: "space-age Wilis," Goldner called them.

The "space-age wilis" of *Symphony in Three Movements.* Photograph by Martha Swope.

Gelsey Kirkland and Conrad Ludlow in the Second Movement of *Symphony in C*. Photograph by Martha Swope.

Images that recall *Swan Lake* surface in several ballets. In *Piano Concerto No. 2* (originally *Ballet Imperial*) and *Diamonds* (the last third of *Jewels*), they must have been summoned up for Balanchine by the Tchaikovsky music. In the case of the *adagio* in *Symphony in C,* he may have been struck by the resemblance between the haunting oboe melody that George Bizet made the heart of his second movement and the oboe melody that tells of the enchanted maiden Odette in *Swan Lake.*

In the *adagios* of all three ballets, the ballerina seems affected by something outside the circle of her partner's arms, something beyond the confines of the stage, whether it is a thing she appears to desire—as in *Diamonds*—or an inevitable force drawing her—as in *Symphony in C* and *Piano Concerto No. 2.*

The second movement of this last ballet might almost be a distillation and reassessment of one of the dilemmas of *Swan Lake*'s hero: to be world-weary and disinterested in the marriageable princesses presented for his approval and to desire an enigmatic and probably unattainable woman. But unlike Siegfried, this "prince" doesn't act polite boredom, restlessness, desperation. He lifts his arms, and small chains of women attach themselves to him on either side, as if summoned. When he presses his arms forward or back, the women *bourrée* in response. Because those farthest from him have farthest to go, he seems to be lashing ribbons of women about, or trying to use them as wings, scarcely aware of what he's doing. The woman we know to be his partner appears. After they dance a tender *pas de deux,* she *bourrées* backward away from him, down an avenue of women, and disappears. He lifts his arms, and the women re-form their chains. Although acting has played no part of this, you see loss, resignation, proffered consolation. As Balanchine once said of both music and his kind of dancing, "The dramatic elements are there, of course, but they are fused and transfigured."

And the dancers' "story" doesn't carry over into the first and third acts of the ballet, where they are every inch the happy rulers of this opulent kingdom of dancing.

When Balanchine presented his first work in America, for what was initially called The American Ballet,* he had to make do with the dancers he could get, choosing the best of those who were trying out his classes, recruiting from, say, Catherine Littlefield's new company in Philadelphia. The men had varied backgrounds: Lew Christensen came from vaudeville and Broadway; Charles Laskey had been in the Humphrey-Weidman company. There were strong and personable performers among these first "Balanchine dancers"—Leda Anchutina, Holly Howard, Gisella Caccialanza, Annabelle Lyon, William Dollar, Christensen, and others. But as for the corps de ballet, a snapshot of a spacing rehearsal for the first performance of *Serenade* in 1934 at Felix Warburg's estate in White Plains, New York, shows a group of predominantly chunky women in bathing suits and shorts,

*The original company founded by Kirstein and Balanchine made its official debut in 1935. From 1935 to 1938, it was the ballet company of the Metropolitan Opera. In 1941 it merged briefly with Ballet Caravan (1936–1939), a company founded by Kirstein to represent ballets on strictly American subjects, and, as American Ballet Caravan, toured South America. Between 1935 and 1946, Balanchine freelanced as a choreographer of opera, musical comedy, film, and ballet, was associated with the Ballet Russe de Monte Carlo from 1944 to 1946. In 1946 he and Kirstein came together again in Ballet Society, which, in 1948, became a constituent of the New York City Center, under the title of the New York City Ballet.

far removed in terms of training and self-image from today's lithe, long-muscled dancers. (The life was different too: today, NYCB dancers tour little and by jet; those earlier dancers had to drive to performances, maybe load and unload the bus, occasionally change the gelatins on the lights.)

During the forties and fifties, dancers who came to Balanchine from other companies began to adjust themselves to an image of what they sensed he needed. He wanted speed, flexibility, long lines, and no star manners. The ballerina he had most admired during his St. Petersburg student days had been the atypically slim and cool Elisaveta Gerdt. "You make yourself a Balanchine ballerina by dancing his ballets," said Melissa Hayden. "Your legs change, your body changes, you become a filly."

George Balanchine teaching a professional class at the School of American Ballet in the forties. Mary Ellen Moylan is first at the barre.

As school and company developed, it gradually became apparent that dancers could be chosen and prepared by the teachers to Balanchine's specifications—as if being selectively bred. People watching rehearsals are often surprised to find that the women are not *all* extremely tall and thin with long legs, a short torso, a long neck, and a small head ("pinheads," the critic R. P. Blackmur called them). Among the advanced students in the School of American Ballet, you can see already the widened rib cage, the high extensions, the slightly hollowed back (as if when the women lift their legs behind them in *arabesque,* one buttock will somehow fold up into that "pocket"). Their feet on pointe are strong but almost prehensile; Balanchine had said they should be like an elephant's trunk. He wanted "big girls with long legs. Not small girls with big heads." He got them. He wanted men with feet as quick, extensions as high, bodies as flexible as the women's, and as American society began to temper its disapproval of ballet as a career for men, he got boy students young enough to mold that way. As much as the principal male dancers vary, the men in the corps de ballet tend toward short, broad torsos and slim, articulate limbs—a look that's emphasized in Balanchine's "practice clothes" ballets by the black tights and white T-shirts that they wear.

When you watch the New York City Ballet dancing, the women's pointework doesn't come off as a stunt. It enables them to move more rapidly; it elongates their silhouette. Balanchine's appreciation of bigness, his desire to enlarge the scale of classical dancing involved other adjustments, some of which violate the strict academic principles that Petipa honored. Balanchine dancers are musically precise, precise in terms of energy, but not always precise in terms of "placement." They lift their hips in order to get their legs higher. They veer off the vertical plumb line. Their hands tend to flap, their chins to tilt. Their *port de bras* aren't always gracefully curved like those of the English dancers; Clive Barnes chided them for that once. But the strain of Russian classicism cultivated by such British choreographers as Balanchine's peer, Frederick Ashton, fits the English landscape and the English temperament; things are held slightly in check, moderately scaled. The British dancers seem to take up less space—to feel that they have a *right* to only so much space, however brilliantly they may fill it. As Balanchine has commented, English gardens and meadows are less expansive and more precisely laid out than waving Russian wheatfields. Perhaps the American sense of space reminded him of home.

It is no secret that George Balanchine adored women dancers. He told Lincoln Kirstein he'd come to America because it was the country that had produced Ginger Rogers. He married four of "his" dancers—Tamara Geva, Vera Zorina, Maria Tallchief, Tanaquil LeClercq—and was romantically attached to several others, including Alexandra Danilova. Muses were essential to him, but even dancers he didn't fall in love with, like Melissa Hayden, received roles tailored to their particular gifts or designed to extend them away from what was comfortable and toward what he saw as their essential nature.

Petipa, coping with royal favoritism, had to satisfy his leading dancers. Balanchine, answerable to no one on artistic matters, often did. "Which leg would you rather lift, dear?" "Which way do you want to turn?" Dancers new to roles often inspired him to alter choreography, or even reconsider issues of ballet technique. Suzanne Farrell thinks that an injury that made it temporarily impossible for her to work on half-toe, although she could manage full pointe, led him to expand his lexicon of pointework. Mary Ellen Moylan, who worked with Balanchine during the years he choreographed for the Ballet Russe de Monte Carlo in America, says that Marie-Jeanne's way of thrusting her hips forward, rather than simply arching her back above the waist influenced Balanchine dancers of the 1940's. Did Balanchine teach this to Marie-Jeanne or appropriate it from her?

Even though company style and choreography alter subtly over the years, Balanchine ballets still resonate with the image of the ballerinas who created the leading roles in them: the powerful and icily shimmering *Firebird* with Maria Tallchief, for example, or *La Valse* and its doomed, slightly decadent young girl with Tanaquil LeClercq, or the subtle coquetry of *Gounod Symphony* with Violette Verdy, or the elaborately erotic *Bugaku* with Allegra Kent (in LeClercq's affectionately irreverent view, "a rubber orchid"). The dazzling footwork of *Ballo della Regina* comes from Merrill Ashley and the springy dash of *Square Dance* from Patricia Wilde. The list could go on and on.

Balanchine's last major muse—the one for whom he created many glorious roles, the one whose influence on company style has been most obvious is Suzanne Farrell. During her first years with the company (1963–1968) and after her return from several seasons with Maurice Béjart's company in Brussels, Balanchine seemed to want to see what light her dancing would shed on almost every role he ever made. As Arlene Croce has pointed out, her success led first to eager, often clumsy imitations of her, and finally to a deeper influence. Farrell's musicality, her recklessness were inimitable, but the luxurious scale of her dancing, the alluring blend of speed and softness could be attempted.

Suzanne Farrell in *Don Quixote* (1965). Photograph by Martha Swope.

Farrell combines the insouciant, occasionally vulgar athleticism of a drum majorette with an elegance and purity that look native to her. She was up to any aspect of Balanchine's vision of woman: the spun-sugar ladies who dance the "Fairy Variations" in Petipa's *The Sleeping Beauty;* muse—angelically tender, but unattainable; enchantress; hoyden; precocious nymphet.

It is one of the curious and rather deplorable facts of the ballet world that performers are known as—and often call themselves—"boys" and "girls." In Petipa's day, dancers had long careers; today's accent on youth and ballet's athleticism usually force a dancer into retirement around the age of forty, so prolonged youth is cultivated. But Balanchine always adored very young girl dancers of brilliance—ones who have the appearance of beautiful women, but seem as yet touchingly unaware of their own power. (The revealing metaphor he fetched up in an interview with Simon Volkov was that of Parmesan cheese: he preferred it young and moist, not old and hard.) His appetite for dewy dancers was already strong in 1933 when he chose Irina Baronova, Tamara Toumanova, and Tatiana Riabouchinska—fourteen, fifteen, and sixteen respectively—for René Blum's Les Ballets Russes de Monte Carlo. But those three were baby divas, their worldly glamour combining piquantly with their presumed innocence. Tanaquil LeClercq, who made her debut in 1946 at seventeen, presented a different sort of girlish image: sometimes the impudent, crazy-legged tomboy came to the fore; sometimes she had the vulnerable elegance of a young girl at her first grown-up party.

It was perhaps LeClercq who made plain a kinship between Balanchine's image of woman and that of contemporary fashion designers. There were significant differences, of course: the designers were more interested in androgyny, but just as they thought that their garments hung most appetizingly on tall, boyishly thin bodies, so he thought that dancing draped on a lean frame became purer and more legible. (And many men in the audience found that look alluring, equating what may have been actual hunger on the part of the dancers with sexual appetite, or remembering the coarse old saw "The closer to the bone, the sweeter the meat.")

The men in the New York City Ballet are slim, but not unnaturally so. However, for the women, extreme thinness has become the norm, despite the relish Balanchine occasionally took in generously proportioned dancers like Gloria Govrin or Jillana. If anything, the sleek, pared-down look has intensified. As Violette Verdy says, "These days everybody's a greyhound." And when, according to a recent study conducted by Dr. L. M. Vincent, many ballet students' physical maturation is delayed because of a low percentage of body fat, it's possible for a five-foot six-inch dancer to weigh under one hundred pounds, and have no breasts or hips to speak of.

Yet, perhaps because dancers always look larger onstage than they actually are, the Balanchine woman in performances today rarely looks like a large, starving child. She tends to appear extremely slender, but also volup-

Balanchine rehearses Tanaquil LeClercq and Francisco Moncion in *Symphony in C*. Photograph by George Platt Lynes.

tuous. Her technical largesse imparts a kind of maturity to her stage persona. This bevy of contemporary nymphs is innocent but worldly, passionate but cool, soft but strong. Kent and Farrell were the prototypes. Their pale, sweet, blunt little faces and mousy hair could make their powerful sinuous limbs seem to be acting out passions beyond their comprehension. Yet they could move with great strength and speed and authority, their musicality imbuing them with a semblance of spiritual knowledge. They are, in a way, the fulfillment of an image of American dancers that Balanchine articulated in 1944 when he praised the way they could express "clean emotion angelically . . . a quality supposedly enjoyed by the angels, who, when they relate a tragic situation, do not themselves suffer."

* * *

Balanchine has been subject to some critical attacks from which Petipa was exempt. Feminists tend to froth at the mouth when discussing Balanchine's presentation of the male/female relationship—seeing in the expanded classical partnering only certain undeniable facts: women being manipulated, men in command. Remarks by Balanchine like "The ballet is . . . a woman, a garden of beautiful flowers, and man is the gardner," or (to his dancers) "You are ladies. Dainty. Beautiful," understandably do little to assuage them. Ann Daly could be said to express the feminist viewpoint in her analysis of the third theme in *The Four Temperaments*. Less angry than some, and acknowledging the prowess of Balanchine's female dancers, Daly notes that

> . . . the manipulated ballerina looks less like a dominant dynamo than a submissive instrument, both literally and figuratively. Her partner is always the one who leads, initiates, maps out the territory, subsumes her space into his, and handles her waist, armpits, and thighs. She never touches him in the same way: she does not initiate the moves. Metaphorically, she makes no movement of her own; her position is contingent upon the manipulations of her partner.

One frequently heard counter to such views is that the ballerina does not represent a real woman; she is an abstraction, a symbol of something. This only raises another question: since no one denies that her partner represents a real man, are we to believe that only real men get to manipulate symbols? The ambiguities in the ballerina/cavalier relationship and in the public's response to it are neatly expressed by this sentence of Claudia Roth Pierpont: "In his pattern of possession and release, her partner is a kind of falconer, taking his own freedom in her flight." Like the falcon, then, the female dancer is controlled by the male, in order that she may create an illusion of freedom *for him.*

Perhaps mannerliness and tenderness are the keys to the general lack of resistance to Balanchine's sexual politics. As an adaptation of the Petipa cavalier, the premier danseur in a Balanchine ballet doesn't grab his partner or maul her against her will. He offers a hand; she takes it. When this agreement has been reached, she may balance on one toe, or entrust her weight to him. The man displays, encourages, tends, adores his partner. Turning her into various angles to the audience, he's the jeweler rolling a

Kay Mazzo and Peter Martins in *Violin Concerto.* Photograph by Martha Swope.

diamond between his finger to show the facets. Yet often the woman leads her partner, challenges him, accepts his support even when she doesn't really need it. Quite a few duets, like that in the *Rubies* section of *Jewels,* suggest the bantering of equals. And it is certainly true that the mutual respect built into the Balanchine *pas de deux* makes his ballets seem less demeaning to women than some of the dramatic, erotic, muscular, lift-filled "modern" ballets. It is worth noting how sparingly Balanchine used lifts, compared to choreographers who present a woman draped over her partner's upraised arm like a platter of meat hefted by an enthusiastic waiter.

Balanchine's view of the male dancer as secondary strikes some people as discriminatory too. To be contented dancing in the New York City Ballet, a man must, I think, share Balanchine's view to some degree. Peter Martins, recalling how the boys he went to school with in Denmark hated partnering because they didn't like always being behind the girl, says, "I didn't think it was demeaning to stand behind. I never looked at myself as being *behind* the girl; I thought of myself as being placed accordingly and presenting her. What I did was very important to her . . . how I looked behind her and what I did to present her."

The leading male dancers of the New York City Ballet—many of them, interestingly, not products of the School of American Ballet—often seem in the ballets to represent Balanchine himself—tender, deferential, enraptured, but firm-handed. The gardener *vis-à-vis* the flowers. It should be remembered too that, although Balanchine made a number of well-publicized remarks about the preeminence of women in ballet, some of his greatest roles have been made for men: *Apollo,* for example, or *The Prodigal Son* (both built on Serge Lifar during Balanchine's Diaghilev years), *Orpheus* (made for Nicholas Magallanes), and the hero of *Baiser de la Feé* (in its last reincarnation, choreographed for Helgi Tomasson). Obviously interested by Bart Cook's unexpected ability to blend lyrical plastique with a forthright "American farm boy" persona, he inserted a new solo for Cook into a 1976 revival of the 1957 *Square Dance.*

In many of the ballets the autobiographical resonance is potent. In *Apollo* the young god woos Terpsichore, drops his head tiredly into the cupped hands of the three girl muses, then yokes them masterfully together. In the duet Balanchine made for Kay Mazzo and Peter Martins in *Violin Concerto,* the man helps his partner to strange, sinuous ways of walking; she might be a vine he's training. Nancy Goldner sensitively delineated the ballerina/choreographer image their dancing creates:

> At the end, he cradles her with one arm and slowly rocks her from side to side. His free arm reaches outward and draws in, in time to the rocking. It is an image of mutual love, until one perceives that she is totally passive. This love scene is, indeed, an image, and he is the image-maker.

Martins gently bends Mazzo's head backward, covering her eyes with his hand. She is both instrument and inspiration—inspiration because she is so superb an instrument. (These days, this ballet, made in 1973, can be seen as a mysterious presentiment of Martins's role as Balanchine's successor and ballet master of the company.)

Meditation (1963), a highly Romantic duet with Jacques d'Amboise as poet-dreamer and Balanchine surrogate, announced Suzanne Farrell's status as new muse—tender but elusive; in *Don Quixote,* Balanchine himself played the idealistic and impotent Don to Farrell's Dulcinea. *Davidsbündlertänze,* a dramatic ballet to the music of Robert Schumann, linked Balanchine with

Schumann and presaged the mental deterioration that was to be part of Balanchine's last illness, as it had been of Schumann's.

The notion of service in Balanchine's oeuvre is crucial and complicated. He saw his dancers as instruments, but he also saw himself as one—subservient to the music, and to the Russian Orthodox God that he devoutly believed in. He and Stravinsky both strove to undermine the nineteenth-century notion of the artist as "creator"; only God creates, they said, man discovers, assembles. The composer is a chef (Balanchine), a "pig snouting truffles" (Stravinsky). Lack of pretentiousness must be maintained. Balanchine coached Peter Martins for the mysterious and poignant finale of the *Violin Concerto* duet by telling him to reach out his hand "as if you're asking for money." To mention artists and muses in a rehearsal would have seemed to him fancy and absurd.

By study and intuition, the choreographer discovers forms and moods in a musical score. The dancer, in turn, conveys these without the interference of "interpretation," in its narrow sense, or "personality." This doesn't mean that the dancer is an automaton; he/she performs the steps fully and accurately, listens to the music deeply, and allows the moods that emanate from both to color the performing. To perform Balanchine's ballets demands not only obedience and craft, but the same, almost mystical faith in intuition and craft that the choreographer had. Writes former New York City Ballet dancer Toni Bentley, "The self is defined by the steps. The self is as large, as beautiful, as intricate, as curious, and as interesting as the steps." Some observers feel this; some do not. In 1958 R. P. Blackmur faulted Balanchine's female dancers for being "echoless technique," but to others, they display, as Clement Crisp said in praise of Merrill Ashley, a "selfless bravura"—not " . . . 'look at me dancing' . . . ," but " . . . 'look at this dance.' " Angelic messengers who reveal but do not comment.

Balanchine's often-quoted "Don't think, dear, just do" makes me wonder if he ever read Gordon Craig's essay "On the Actor and the Über-Marionette" or Heinrich von Kleist's "Puppet Theatre"—both of them obliquely admonishing actors for self-conscious and mannered performing. " . . . Grace has greater power and brilliance," said Kleist, "in proportion as the reasoning powers are dimmer and less active." It is just such thinking that the brilliant and rebellious Gelsey Kirkland attributed to Balanchine, and found demeaning. Yet Kleist's idea (and perhaps Balanchine's too) is not that simple. His mouthpiece in the essay, Herr C., goes on: ". . . When knowledge has . . . passed through infinity, grace will reappear. So that we shall find it at its purest in a body which is entirely devoid of consciousness *or which possesses it in an infinite degree;* that is, in the marionette or the god." Balanchine would not have said "god"; what he did say was "angel."

Of course, the humble choreographer is at the same time a dictator, a master puppeteer, the chief heavenly messenger; and many of the dancers, although trained to selflessness and docility, believe that they are special creatures, different from the rest of humanity. Lincoln Kirstein says, quite terrifyingly, that "ballerinas are kin to those mythic Amazons who sliced off a breast to shoot arrows the more efficiently."

As Toni Bentley made clear in her book *Winter Season,* few of her female colleagues think of themselves, the way Petipa's ballerinas did, as women who dance; they are dancers first, the better to create an image of idealized womanhood onstage.

Balanchine's vision of dancers has pervaded the American ballet scene and influenced many companies abroad. Now people in the United States proudly proclaim that the look of the New York City Ballet dancers is "American." We appropriate their lean, racy, long-legged look as an American ideal; we like to think of their boldness, their frankness, their speed, their cool absorption in music and dancing, their unselfconscious dignity and courtesy as attributes of American character at its best. Kirstein said once that classical ballet offered "a clear if complex blending of human anatomy, solid geometry and acrobatics, offered as a symbolic demonstration of manners." The residue of Petipa's decorum that clings to Balanchine's ballets tells us that we are all equal, and we are all royal.

·7·

Illusion of Choice— Acceptance of Chance

Meg Harper, Merce Cunningham, and Mel Wong in Cunningham's *Tread* (1970). Photograph by James Klosty.

Viola Farber and Carolyn Brown in *Summerspace*. Photograph by Richard Rutledge.

It's August 17, 1958—a hot Sunday afternoon in Palmer Auditorium at Connecticut College, where the last performance of the American Dance Festival is taking place. The audience is full of that summer's dance students, their teachers, dance buffs up from New York, and a few residents of New London. The week of performances has featured new and old works by Doris Humphrey, by her protégé José Limón, by Ruth Currier, Pauline Koner, Pearl Lang, the German-trained Inga Weiss, and Merce Cunningham. This afternoon, Cunningham and his dancers are presenting the second of two dances they've made this summer. Although he's been choreographing since 1942 and has headed his own company since 1953, he's still considered an outsider, and this is the first time he and his dancers have been invited to teach and perform in this bastion of modern dance. In his first new work of the summer, *Antic Meet,* presented three nights ago, Cunningham acknowledged his maverick status by including subtle and extremely witty parodies of the styles and concerns of his peers. Not a titter was heard by the performers.

Summerspace is second on the program, after Ruth Currier's lyrical *Quartet.* The seven dancers—Merce Cunningham, Carolyn Brown, Viola Farber, Cynthia Stone, Marilyn Wood, and Remy Charlip—wear rosy, spattered leotards and tights designed by the painter Robert Rauschenberg; when they stand close to his shimmering backcloth in the same pointillistic style and colors, they're almost camouflaged. The music, Morton Feldman's *Ixion,* sets up a glimmering and humming electronic climate within which the dancers maintain their own variegated rhythms. In their constant appearances and disappearances, the performers seem isolated from each other, bent on private errands. When they do come together, the encounters look almost accidental. Brown, leaping, with Cunningham running along in back of her, is repeatedly caught by him in midair, and eventually carried off-stage, but this is no climax of desire; it is a witty converging of two paths. As in all of Cunningham's works, his dancers play no roles, assume no emotions on demand, or pretend to a goal beyond the accomplishment of the dancing.

There is a sparseness, a delicacy about *Summerspace.* It's a veritable lexicon of turns and jumps. Charlip hops along a diagonal, one leg stretching behind him, arms out to the sides. His unchanging posture and calm, steady passage—hop, pause, hop, pause—give him the air of a hovering dragonfly. And although Cunningham is not telling stories, the darting, careering paths of the dancers, their energetic serenity, their long, erect spines and buzzing feet, their rapid flights and long pauses can remind one of the erratic activity in a field on a summer day.

It is doubtful that the Connecticut College audience sees any of this—although Merce has his admirers. ". . . being avant-garde is not always an open sesame to artistic achievement," snipes Louis Horst. And *Dance Magazine*'s tolerant Doris Hering hedges: ". . . a little sparse in movement invention—at first viewing."

The hit of the afternoon—and, indeed, of the whole festival week—is the dance that follows *Summerspace* on the program: Pearl Lang's *Falls the Shadow Between.* With her highly theatrical, emotionally taut recasting of the Persephone myth, Lang is thought to be making accomplished and imaginative use of her Graham heritage. This last concert of the summer concludes with the second performance of Limón's flawed but "sensuously lovely" *Serenata,* with dancers Betty Jones and Chester Wolenski singing Paul Bowles's setting of poetry by Federico García Lorca.

If that 1958 audience had been privy to the compositional methods that produced *Summerspace,* they might have been even more baffled. Students who had glimpsed Cunningham's rehearsals were shocked to observe him timing passages of dance with a stopwatch. Was this not heretical? It was bad enough not to set the movement *to* the music. But . . . a stopwatch; that seemed so calculating. Where did inspiration come in?

Whatever inspired Cunningham—and no one who has seen his dances can doubt that he *is* inspired—he has kept to himself. *Summerspace,* like all his works, acknowledges the status quo—the dancers he is working with, where they are working, and what physical or structural problem absorbs him at the moment. He had recently become interested in ways of turning, for instance, so turns were a big item. (Some of the dancers, he says, found this upsetting.) The dance was created in an immense empty parlor in one of the dorms—a beautiful, airy room, the dancers recall, with three sets of doors opening onto the hallway and French windows opposite them, looking out onto the summer lawn. Perhaps it was these many portals that inspired Cunningham to make entrances, exits, and the paths between them the structural principle of the dance. Adopting chance procedures similar to those that his close friend and colleague John Cage employed in composing music, Cunningham plotted twenty-one paths connecting the six possible entrance points on the stage and devised twenty-one phrases to travel along them. He also used chance methods (tossing coins onto charts is one possibility) to determine such elements as approximate tempo; overall time it took to accomplish each path; whether the movement happened in the air, along the floor, or on the floor; whether the paths were curved or straight or were bent into angles; how many dancers were to perform a particular action, and whether they were to do so alone or together. Years later, when *Summerspace* was being taught to the New York City Ballet, Cunningham mentioned—as he rarely does—the possible images that can emerge from the specific mingling of space and time in this dance: ". . . Like the passing of birds, hopping for moments on the ground and then going on, or automobiles more relentlessly throbbing along turnpikes and under and over cloverleaves." But such images are as likely to have been an unlooked-for result as the inspiration for the dance.

Now Cunningham is acknowledged as a master, the *éminence grise* of vanguard dance, yet his works still make some people uncomfortable. Although few today would say of his dancers what Hering wrote in 1958, "They are shapes in an image, not people in a world," they're often charged, as are George Balanchine's dancers, with not displaying much warmth or "personality" onstage. The world in which they move, however, unlike Balanchine's, lacks the harmony imposed by classical art. Lacks, too, its comforting hierarchies. With dancing, sound, and set functioning as separate, simultaneous experiences to be absorbed by the audience catch-as-catch-can, and with those elements in themselves embracing discontinuity and complexity, the spectator can feel himself in the presence of a chaos that he might rather not acknowledge in life—and certainly not in a theater. However, the many who follow Cunningham's work with excitement and great pleasure find the complexity heady and the lack of imposed scenarios or centers of attention liberating. He allows his audiences the responsibility, the privilege, of thinking for themselves.

To hear Cunningham tell it, he arrived at his distinctive movement style pragmatically. As a boy in Centralia, Washington, he'd had tap lessons with

Mrs. Maud Barrett; these were followed by a modest amount of training in modern dance at the Cornish School in Seattle—from Bonnie Bird who'd danced with Martha Graham—and two summer workshops at Mills College in Oakland, California. His second year at Mills, 1939, the whole Bennington Festival was transplanted there. In a home movie made that summer, you can glimpse him—tall, intense, and raw-looking—in the back of a Humphrey-Weidman class. Men were in demand, and raw or not, he was asked to join both the Humphrey-Weidman company and the Martha Graham company; it was Graham's he chose. While performing in her dances and taking classes with her during the thirties and early forties, he also—on her advice—studied at George Balanchine's School of American Ballet. It was perhaps while he was studying (and teaching modern dance) there that he observed that ballet dancers were lively with their feet, but fairly unmoving of torso. Modern dancers, on the other hand, used their torsos in vivid ways, but skimped on the feet. Why not combine the physical principles? Experimenting, he decided that the ballet dancer's modest *épaulement,* in which the hips remained locked and only the upper body moved, limited his capacity for speed; he reasoned that if twists and bends occurred from a deeper source, the hips would be freed, and the feet could move faster.

From the first, the dancing that he composed for himself went beyond any such rational notions. The critic Walter Terry said that ". . . five minutes of watching his dancing convinces one that 'to dance' is his equivalent of our 'to breathe.' " Although his performing inevitably presented a man of intelligence as well as a man passionate about dancing, the busy feet, buoyant jumps, level gaze, and long, erect, yet mobile spine were as much a result of his physique and predilections as of conscious thought. Edwin Denby captured the young Cunningham in his *Herald Tribune* review of the first joint recital that Cage and Cunningham gave, in April 1944—while the latter was still a member of Martha Graham's company:

> His build resembles that of the juvenile *saltimbanques* of the early Picasso canvases. As a dancer his instep and his knees are extraordinarily elastic and quick; his steps, runs, knee-bends and leaps are brilliant in lightness and speed. His torso can turn on its vertical axis with great sensitivity, his shoulders are held lightly free and his head poises intelligently. The arms are light and long, they float, but do not often have an active look.

Merce Cunningham, photographed in London in 1964 by Hans Wild

The picture Denby conjures up might almost be that of a great ballet dancer. Yet in Cunningham's case, the oddity of his coordinations, the unpredictability of his sequences, the awkwardness that he presents as another facet of grace belie any such notion, despite his use of turnout, pointed toes, and stretched knees. He has always differed from the modern dancers who were his teachers too, in that you never feel the presence of a force in the center of his body pulling the limbs awry, knocking him off-balance. The struggle with gravity, the debilitating whirlwinds—external or internal—that were so much a part of the modernist aesthetic did not apparently preoccupy him. A remark once made by Douglas Dunn, who danced with Cunningham's company for years, strikes me as crucial. "Merce," he said, "is dedicated to the image of a decision." Bold as his movements may be, he and the dancers who perform with him never throw themselves or any part of themselves so far off-balance that they cannot in an instant change their minds and go in the opposite direction.

From early in his career, Cunningham seems to have sought to understand the nature of his own body and at the same time to thwart its habitual patterns in order to put himself in touch with a larger "nature." He confesses to observing motion with a thieving eye: airplanes, birds, zoo animals, papers blowing in the wind, people in the street, his own dancers in repose—not to imitate these, but to discover unexpected rhythms, gestures.

"Nature." The word often crops up in writings about dance. Many innovative choreographers have believed it their mission to rectify the discrepancies that they notice between the artmaking of their day and nature as they define it. As they see fit to deal with it. Fairly early in his career, John Cage had been much taken with Ananda Coomaraswamy's statement that "the function of the artist is to imitate nature *in her manner of operation*." Cunningham, whose ideas have developed along similar lines almost from his first contact with Cage at the Cornish School, made a remark that might be considered a corollary: "I think of dance," he has said, "as a constant transformation of life itself." The subject, then, is life's processes, not tableaux and gestures from a particular life. And it is the ways in which he has attempted to bring his processes in line with his understanding of nature's "manner of operation" that make his dances and his dancers look the way they do.

Nature, as far as Cage and Cunningham are concerned, operates predictably only in a general sense. We might guess, for example, that in a given New England field, in a given summer, there will be black-eyed Susans, yarrow, Queen Anne's lace, that there may be less goldenrod because of last summer's early mowing. But we cannot divine where they will bloom; no formulas define the relations of wind, rain, sun, soil, and seeds. In our own lives, we think nothing of stirring sauce and glancing at a recipe while listening to a conversation. Today, human life is especially complicated and risky; we are hourly bombarded with more data than we can expect to absorb or make sense of.

Some artists have ignored life's hazardous or unruly aspects and emphasized an order that they sense or long for. It is Cunningham's belief that art which takes into account the unpredictability and cacophony of contemporary life may help both artist and audience to accept them with composure. After all, most of us accept both the randomness of meadows and the order of gardens. If a single sentence could be said to sum up his philosophy of dancemaking, it might be this, by Chuang Tzu: "Let everything be allowed to do what it naturally does, so that its nature will be satisfied." Let set, lighting, and costume design, music, and choreography be what they will be, as much as possible, and let any possible relationships generated between them by collision, intersection, or layering be what *they* will be. Let each movement or set of movements "find" its own time. Let each dancer be "himself" or "herself." If art is to operate like nature, it must decline further mimicry. Calvin Tomkins aptly summed up this viewpoint when he spoke for Marcel Duchamp, the expatriate Dadaist, an exemplar for many American artists: "Art . . . must be part of life—not an *interpretation* of life, not a description of life, not an attempt to comment or improve on life, but a piece of life itself."

Although in essence nature may not change, our understanding of it alters with new discoveries in science and new technological inventions. Isadora Duncan took account of the changes in thinking engendered by late nineteenth-century discoveries about electrical energy; the modern dancers of the thirties acknowledged the dynamism of machines. Cunningham's appetite for hazard, his eye for complexity and discontinuity may have been primed, in part, by developments in the field of high-energy particle physics. This isn't surprising. Cunningham, like Cage, has studied Zen Buddhism, read Taoist texts and other Eastern mystical writings which presaged the revelations of twentieth-century physics. If one can almost *see* Cunningham dance in this sentence by Radhakrishnan—"Buddha . . . reduces substances, souls, monads to forces, movements, sequences, and processes and adopts a dynamic concept of reality"—one can also see it in this statement by physicist Werner Heisenberg: "The world thus appears as a complicated tissue of events, in which connections of different kinds alternate or overlap or combine and thereby determine the texture of the whole." A former Cunningham dancer remembers that once when the company was performing at a college, a physics professor from India advised his students to see these people dance if they wanted to understand what he was trying to teach them.

Cunningham's work, then, acknowledges that nature does not operate in such logical and predictable ways as it once seemed to. Carolyn Brown pointed out in a landmark essay in *Ballet Review* that the universe which post–quantum-theory physics has revealed is one in which the rules of

Newtonian physics, the reassuring laws of motion and force, no longer seem to apply. In the particle-picture of nature, writes Sir James Jeans, the motion of particles ". . . can only be compared to the random jumps of kangaroos, with no causal laws controlling the jumps." Certainty has been replaced by statistical probability; the role of chance in universal processes has been acknowledged, along with the impossibility of verifying experientially exactly how the universe operates. A quantum, or subatomic particle, can better be described as a set of relationships than as a thing. According to the enthralling and unsettling Heisenberg uncertainty principle, we cannot know both the position *and* the momentum of a particle; we interfere with one when we investigate the other. And accepting discontinuity and hazard in a world we can't see somehow affects our interpretation of the world we *do* see.

On the other hand, one of the implications of Bell's Theorem on the behavior of photon pairs is that everything that happens—in the macrocosmic world as well as in the subatomic one—is dependent on something that is happening elsewhere, and that underlying everything is what physicist David Bohm calls an "unbroken wholeness"—a phrase that might have come from an ancient Buddhist text. Our choices may be the result of chance, but chance itself may be an illusion.

In 1950—during a period of extreme conservatism in America, when the Korean War and the House Un-American Activities Committee were attempting to impose the "American Way" at home and abroad—John Cage was looking for ways to impose as little as possible on musical elements. It was yet another instance of rebelling against prescribed scenarios, of refusing to accept that one thing had to lead to another. While composing the score for Cunningham's projected *Sixteen Dances for Soloist and Company of Three,* he got his first glimmering of how chance might be made to operate in the compositional process, thus demoting the composer from dictator to discoverer. Cunningham, in making the dance, followed suit—tossing coins onto charts to determine (at least, in part) the order of sections and the sequences, timings, and directions in space for the postludes that followed certain solos. It felt, he said, like "chaos has come again." If so, it was a chaos he determined to open himself to, and, like Cage, he began to utilize the hexagrams of the *I Ching* (just published in translation)—not, as they are intended, to obtain omens for future courses of action, but to ascribe to their patterns of solid lines (*yang*) and broken ones (*yin*) values that would yield dance action. For *Suite for Five in Space and Time* (1953–1959), he adapted another Cagean procedure to the composition of dances, marking the imperfections on a piece of paper and numbering them via coin-

John Cage and Merce Cunningham conferring during a stage rehearsal in 1977. Photograph by Lois Greenfield.

tossing to determine the path of a dancer through space, superimposing the pieces of paper to see where and if dancers came together. For *Untitled Solo,* with a piano score by Christian Wolff, he designed a range of movements for the various parts of the body, and tossed coins onto a stupefying number of charts to find the sequence, time lengths, and possible (or impossible) superimpositions. Working on it during the summer of 1953 at Black Mountain College in North Carolina, where he and his just-formed company were in residence, he sometimes doubted that he'd be able to perform it.

Charts for the 1953 *Suite by Chance,* which were published in his *Changes: Notes on Choreography,* give some idea of what he was up against. One of the many pages ruled into squares for the tossing of coins or the like lists fifteen possible leaps: leaps preceded by one step, two steps, three steps, or more than three steps; in each of these categories, leaps with both legs straight, both legs bent, or one leg bent and one straight. The last three squares are mercifully marked "free agent." There are similar charts for skips and jumps, for falls and rises, for seated positions and standing ones; for moves for back alone, for head, for arms. There is a chart of simple traveling steps ("neutral"), some couched in ballet terms for convenience: "pas de basque," "assemblé," "glissade," and "quick runs with stops" and "prancing in various manners." There must also have been a chart of more intricate phrases, because there is a chart to determine the speed at which these were to be performed. Charts for locations in space, for durations—there seem to have been about 155 in all, plus additional heads-or-tails instructions to find stillness-motion, onstage-offstage, and other things. Presumably chance procedures also determined the order of the charts to be used to create a given sequence for a particular dancer.

The Dadaists and Surrealists, too, employed chance procedures—dice-throwing, automatic writing—but for some of them this was a strategy against academies, for others a psychoanalytic ploy to get in touch with their unconscious urges. Cunningham simply makes dances that will be in step with the functioning of the world as he understands it, and chance is one way of making this visible. In his processes, however, the interplay of chance and choice acquires a mystical dimension: one consciously sets up situations which then open one to forces beyond one's conscious control. In 1955 he wrote:

> The feeling I have when I compose in this way is that I am in touch with a natural resource far greater than my own personal inventiveness could ever be, much more universally human than the particular habits of my own practice, and organically rising out of common pools of motor impulses.

Cunningham has continued to use chance procedures to choreograph, although rarely with the rigor of those early years. Having acquired the knack of alogicality, he no longer needs to create phrases step by step from charts. And once a dance is set, the choreography itself rarely varies, although for a few dances, like *Canfield* (1969), the order of sections or how many of them are to be performed may change from one performance to another. Seldom are the dancers given the opportunity to decide what steps to perform, even though during the 1960's, Cunningham—taking a cue from John Cage, Robert Rauschenberg, and the radical young choreographers of the day—did create several dances (notably *Story*) that incorporated a considerable amount of indeterminacy in performance. By all accounts, the ensuing near anarchy often made him uncomfortable.

* * *

In a review of 1951, critic Walter Terry wrote of Cunningham's dancers that ". . . their bursts of dynamic force appeared to come from nowhere and lead to nowhere." He meant the remark as a criticism; a Cunningham admirer might well have said it as a compliment. In fact, five years after Terry's little printed tantrum, a critic for the *St. Louis Globe-Democrat* was beguiled by the possibility that "a dancer may be standing still one moment, leaping or spinning the next." Time for the Cunningham dancer is not the time we associate with traditional theater and music. Events do not develop over its passage, lead inexorably toward climaxes and die away. The causality of Newtonian physics has been abandoned for a more discontinuous reality. Climaxes, which Cunningham has disapprovingly termed "privileged moments," seldom rise from the web of events. Relationships between dancers may occur fugitively, but, although they may seem tender of playful or contentious, they are not fraught with evolving drama.

Quite often, Cunningham wittily sabotages the audience's appetite for development and goal. At one point in *Walkaround Time* (1968), as it is preserved in a 1972 film, Douglas Dunn walks onto the stage and sinks into a deep lunge, facing the opposite side of the stage. Others enter, efficiently pick him up without altering his position, carry him across the stage, deposit him, still in his lunge, on a new site, and exit. After a moment's pause, Dunn slowly and intricately retraces their path and leaves the way he came. Many times, too, an action that suggests a follow-up will be repeated several times—as if a tape were being rewound and played again—until even the notion of follow-up is erased from the audience's slate of possibilities.

In one surprising sequence from the film of *Locale* (1979), Joseph Lennon plunges from the small stage in Cunningham's Westbeth studio and lands at the feet of Lise Friedman, who shoots straight up into Alan Good's arms. It makes you realize how seldom Cunningham's work displays such apparent causality, at the same time that you notice that the causality *is* only "apparent"; it is you who have imagined that of two events, consecutive in time and contiguous in space, one has caused the other.

As Cage has acknowledged silence as a necessary aspect of sound, so Cunningham has cultivated stillness as an aspect of motion. A dancer may walk onto the stage, begin calmly and pensively to stretch her arms and legs in the air around her, face in another direction to jump rapidly, stop with one leg off the ground and stand there for a long time. Another dancer, who has been leaping and spinning, runs up and touches her lightly on the shoulder; they burst into a dazzle of the rapid footwork that Cunningham is so interested in, then exit matter-of-factly together, while a clump of

Merce Cunningham and his company in an Event, 1972: (from left) Cunningham, Carolyn Brown, Susanna Hayman-Chaffey, Sandra Neels, Brynar Mehl, Barbara Lias, Meg Harper, Chris Komar, Ulysses Dove, and Valda Setterfield. Photograph by James Klosty.

three stride on in deep lunges, facing different ways, stepping at different times, yet sensing each other. This is no particular Cunningham dance, but it might well be. The extreme changes in speed occur without buildup; only a few steps can be seen as "preparations" for others. There is motion in the long stillness—the dancer seeming less to have stopped dancing than to have embarked on a different kind of motion—and there's an interior stillness in the most rapid passages of dancing. (The philosophers of the ancient East would have approved: "Only when there is stillness in movement can the spiritual rhythm appear which pervades heaven and earth.")

It may be, in part, the difficulty of keeping track of discontinuous sequences of movements and articulating their underlying connectedness that gives Cunningham's dancers their look of concentration. Although the choreography is set, the dancers often look as if each move were a product of individual will. The absence of a musical score that gives the beat affects this too. The dancers have to be intensely aware of one another. They seem to be listening to each others' rhythms, to know all that is happening in the space around them. They display the alertness of an animal in its native habitat, an animal's economy of motion, and its fullness in motion. Like animals, they do not pause to show worry over what is to come or what has been; they live in a perpetual present, accepting the world they have been given with intelligence, daring, and tranquillity of spirit. Man exercising his free will in a landscape beyond his control.

The critic Jill Johnston said years ago that one thing that she found novel and admirable about Cunningham and his dancers was their level gaze. Not that they didn't look down or up or out beyond the audience, but that "up" and "down" had been purged of the old Judaeo-Christian anxiety. She said, "He didn't have his head in the clouds and he wasn't hanging it between his legs either. I mean you didn't have to feel sorry for him on the one hand, or hope for his redemption from the powers above on the other."

Cunningham's chance-derived space patterns—complex and unpredictable—also contribute to the dancers' perhaps paradoxical air of making choices. Their paths, even though carefully plotted and rehearsed in the interests of safety, can be so eccentric as to seem spontaneously determined—unlike the neatly spaced-out, often symmetrical formations of much choreography which give the dancers the air of going obediently to preordained places. In traditional ballets, action in space creates hierarchies; there is never any doubt who the principal dancers are or where we are supposed to be looking. Classical theater theory ascribes values to various spots on a proscenium stage. Cunningham is dead against such obvious elitism. Unless he's staging an "Event"—an assemblage of sequences from the repertory—in a museum or a gymnasium, he works within the frame of the proscenium arch, yet he continually sabotages its traditions, treating it as an open field rather than as a hierarchichal arrangement of areas, setting action off to one side, or at the back, behind other action. Back in 1966, some members of the Cunningham company were dismayed to find that the New York City Ballet dancers, after performing *Summerspace* for a while, had instinctively altered the spacing, homing in on familiar spots like down left and center. One perhaps couldn't have expected the Zen Buddhist concept of a "multiplicity of centers" to have been picked up in ballet class.

A remark by Cunningham: "At this moment I read that sentence of Einstein: 'There are no fixed points in space.' I said to myself, if there are no fixed points, then every point is equally fluid and interesting." (It is, you see, no coincidence that readings in current scientific theory and Eastern philosophy keep throwing up sentences that remind you of Cunningham's dances.) And every point is interesting whether it is occupied or not: as stillness and motion are aspects of the same reality, so are form and emptiness, density and sparseness. "Being and fading of particles are merely forms of motion of the field." Like particles in a field, the moving dancers

Canfield with (from left with faces visible) Jeff Slayton, Sandra Neels, Meg Harper, Carolyn Brown, Mel Wong, and Chase Robinson. Photograph by James Klosty.

determine the structure of the space that surrounds them. Their frequent comings and goings, the lack of fanfare as to beginnings and endings can create the curious illusion that the action onstage is only part of the picture, the temporary framing of a larger dance that exceeds the space-time boundary of this performance.

The dancers and the audience also may have to contend with decor that invades the dancing space, decor that moves, decor that partially obstructs the spectators' view of the stage,* lighting that can glare in the dancers' eyes or leave them moving in near blackness. To his designers, too, Cunningham allows considerable autonomy, giving them the barest of clues to follow. It is on record that he told Robert Rauschenberg, who was design-

---

*Certainly true when Robert Rauschenberg was the company's resident designer (1954–1964); and when Jasper Johns, as artistic adviser (1965–1973), commissioned designs from various major artists, it became apparent that such decor was being phased out by 1977, during Mark Lancaster's tenure as principal designer.

ing the lighting for the 1965 *Winterbranch,* that he envisioned the dance happening in "night light"—not the romantic night light of moon on water; something more like headlights striking a wet highway. Rauschenberg's design involved a large, brilliant lamp that, when on, shone directly in the audience's eyes and a chance-determined shifting between patches of darkness and light that in no way depended on where the dancers happened to be at the time.

The glinting, gently bobbling stacks of helium-filled pillows that Andy Warhol designed for *Rainforest* (1967); the row of tall standing electric fans (one or two of them whirring) that Bruce Naumann arranged along the front of the stage for *Tread* (1970); Robert Morris's vertical, light-filled pillar that during *Canfield* moved back and forth across the proscenium arch, scanning the dancing with a white column of light—such designs inevitably affect the way we perceive the dancers.

Cunningham was not the only artist of his day to treat space as an open field. The painters were there before him, and some of them shared other of his concerns. He had known the work of Mark Tobey and Morris Graves in Seattle; Josef Albers was at Black Mountain College when he and Cage did their various teaching stints there; in New York he met Willem de Kooning and Marcel Duchamp, as well as younger painters like Robert Rauschenberg and Jasper Johns. He seems to have found talking with dancers less interesting.

The years just after Cunningham left the Graham company to focus on his own choreography were important years in the history of American painting. One by one, painters of the New York School were abandoning figurative art: Clyfford Still, Barnett Newman, Franz Kline, Mark Rothko, Jackson Pollock. Different as were their concerns and styles, they began to cover their canvases with paint, to avoid hierarchical structures that would guide the eye to a single point.

Although Pollock's overt preoccupation with his own unconscious distinguishes him from Cunningham, their work has something in common, and Cunningham has spoken admiringly of Pollock's efficiency in defeating his sense of central focus by laying his canvases on the floor and walking around them, skeining paint on them from above. The way you look at a painting of Pollock's is similar to the way you watch a Cunningham dance. They're both roomy: the eye wanders at will, caught now by this, now by that. Because of the complexity, you see the work differently each time. Their processes, too, have some kinship. Pollock walked around his paintings; Cunningham makes his dances as if they were to be viewed from all sides.

That Pollock never actually touched the canvas with his brush slightly distances—undermines—the almighty power of the artist's hand, as chance mildly subverts Cunningham's choreographic taste. Pollock downplayed the factor of "accident" in his work (but it's there); Cunningham tends to downplay control (but it's there).

A handwritten note found in Pollock's files after his death says:

> Experience of our age in terms of
> painting—not an illustration of—
> (but the equivalent)
> Concentrated
> fluid.

John Cage, emulating nature, has attempted to evade all notions of "purpose" in the composing of music, beyond the fact of doing it at all. To this end, he has refused to discriminate among sounds on the grounds of tonal beauty. Any sound (noise) can figure in a musical composition. Further, sounds ought not to be subject to undue manipulation, but be allowed to reveal their own natures.

Cunningham has never gone as deeply into nonintention. His compositional strategies may undermine his authority as a "creator" and give his works a semblance of self-determining mechanisms, yet there is a fundamental difference between his aesthetic and Cage's. The equivalent of "noise"—which might be everyday, unregulated human activity—appears only rarely in a Cunningham dance. In the course of the 1965 work *Variations V,* for instance, Cunningham did rollbacks on a miked mattress, pulled the leaves off a plastic plant, and rode a bicycle; in *Objects,* three people mimed playing jacks; in *Signals* (1970), the dancers jockeyed for position in a line-up and held up fingers, as if to determine who did what next. Similar events come to mind. But for the most part, Cunningham's "naturalness" derives from the unpretentious demeanor of the dancers; everyday movement is mostly a matter of walking and running. The dancing—far from being ordinary—is extraordinarily beautiful and requires a high degree of training. He has made no bones of the delight that the act of dancing gives him or of his pleasure in unabashed performing as a part of life. He has remembered his astounded and delighted first glimpse of his first dancing teacher, Mrs. Maud Barrett, a member of the Centralia, Washington, community and of the same Catholic church where he served as an altar boy. The middle-aged Mrs. Barrett welcomed the audience for her

Sandra Neels, Chris Komar, Nanette Hassall, and Merce Cunningham in *Borst Park* (1972). Photograph by James Klosty.

pupils' recital while juggling Indian clubs, then, fastening a band around her legs to hold her skirts modestly in place, turned several cartwheels. Cunningham doesn't object to dancing to the full extent of one's powers and entertaining an audience by so doing; it is forcing dancing to be about some subject beyond itself—like Clytemnestra's revenge—that doesn't suit him. Although he'd be likely to say that if you, the spectator, chose to *see* it as a Clytemnestra's revenge, that was your lookout.

It's possible to relate the notion of everything being allowed to do what it "naturally does, so that its nature may be satisfied" to Cunningham's elegant and virtuosic dancers only if you imagine that he defines the "nature" of his dancers *as dancers,* rather than simply as people (and never as "characters"). Their individual humanity is revealed through how they perform the steps he has chosen to make for them or that chance has dealt them. ("Enlightenment," the saying goes, "consists merely in becoming what we already are from the beginning.") As meditation frees the mind, the discipline of daily class, ideally, drains the body of mannerism and everyday tension, yet doesn't neutralize it. Cunningham is not interested in the uniformity that characterizes the corps de ballet, in which all the dancers are, as he puts it, "an image of the same thing." In being themselves-dancing, however, his dancers are expected to strive and at the same time not be consumed by the striving. One of Cunningham's most enigmatic statements—maddening, no doubt, to all but the most courageous dancers—goes like this: "Perfection is something you should aim for and dismiss at the same time. It robs you of a certain spirit."

To him, as to George Balanchine, it is the simple act of concentrating on the movement itself, rather than on any scenario lurking within it, that frees dancers to be themselves. No Cunningham dancer onstage can be identified as "just recovering from a love affair." One can imagine Cunningham, like Balanchine, reading and approving of the essay on the marionette written by Heinrich von Kleist.* In an essay of his own, written in 1955, he likens the dancer performing at his fullest and most transparent to a "nature-puppet . . . mother-nature and father-spirit moving his limbs, without thought." So while he does not, in his oblique terms, "cultivate nonexpressiveness," neither does he direct his dancers how to perform the movements.

In the essay quoted above, he eloquently states his principles:

> So if you really dance—your body, that is, and not your mind's enforcement—the manifestations of the spirit through your torso and your limbs will inevitably take on the shape of life. We give ourselves away at every moment. We do not, therefore, have to *try* to do it . . . In one of my most recent solo works, called *Untitled Solo,* I choreographed the piece with the use of "chance" methods. However, the dance as performed seems to have an unmistakable dramatic intensity in its bones, so to speak. It seems to me that it was simply a question of "allowing" this quality to happen rather than of "forcing" it.

If you watch the film of *Walkaround Time,* your eye is riveted by a solo that Carolyn Brown performs, once in brief, once in expanded form. It is

---

*It is interesting to note that Nancy Wilson Ross pointed up the kinship between Kleist's ideas and Buddhist ones by including his essay in her anthology, *The World of Zen Buddhism,* published by Random House in 1960.

Carolyn Brown shown in *Walkaround Time* with Jasper Johns's set, conceived in homage to Marcel Duchamp's *The Large Glass*. Photograph by James Klosty.

very serene—the eye of a storm. She balances for a very long time on tiptoe, feet drawn together, then lifts one leg slightly in front of her, now straight, now unfolding, as if she were testing the air. She travels across the stage with soft, jumped turns, one leg raised behind her, then strides forward and, balancing on one toe, swings the other leg smoothly around. The quite classical motions have the force of some unknowable, yet comprehensible statement—private and universal at the same time. It thrills you very directly the way you can be thrilled by light hitting a tree, without your having to know its motive or ascribe to it symbolic value.

Not all of Cunningham's dancers, however, especially in recent years, have been/are able to "allow" the power of the instant to speak through them, and may instead express their nervousness, their preoccupation with being correct, or even their desire to remain neutral. On the other hand, how tricky it must be for even the most daring of them to remain flexible and not attempt to decide on the "meaning" of a gesture or how it is best played, and so thwart the idea of dancing as responsive to the instant. Cunningham himself takes more liberties with the steps than any of the dancers, and powerful images of feeling emerge through his dancing.

* * *

Many people have commented that despite the absence of narrative, many of Cunningham's dances conjure up vivid, often very dramatic images of places and people. Part of this derives from the accidental intersections of decor with lights, sound, and dancing. I remember thinking, when I first saw *Scramble* in 1967, that Toshi Ichiyanagi's abrasive, rumbling score combined with Frank Stella's bright-colored, movable horizontal strips of canvas made the dancers sometimes appear threatened, like innocent picnickers on the edge of an active volcano. The soft, quizzical voice of John Cage and/or David Vaughan reading Cage's extremely droll stories while the dancers performed *How to Pass, Kick, Fall, and Run* made many audiences perceive a gaiety in the sportive steps that was not necessarily intrinsic to all of them. The many falls of *Winterbranch,* taken with La Monte Young's piercing score (two different constant sounds coming at high volume from two speakers), the dancers' dark, drab clothing, and Robert Rauschenberg's indeterminate lighting with its extremes of glare and darkness certainly produced strong images of disaster in the minds of audiences.

At the party after the first performance of *Roadrunners* in 1979 at the American Dance Festival at Duke University, the dancers expressed surprise at the loud laughter that the performance had evoked and asked to know what was so funny. Besides the wit of some of their friezelike poses, or the spectacle of Cunningham trying to put on a pair of pants, shoes, and socks while a determined nymph danced all over his territory, what had especially amused the audience was the way composer Yasunao Tone's heavily accented readings from ancient Chinese geography—not heard by the dancers prior to the performance—colored the events onstage.

Structure and movement, however, are the most powerful determinants of a dance's climate. Cunningham has said that in choreographing he is always asking himself a question or two, investigating some principle that is new for him (perhaps about space or time or movement, perhaps about landscape, perhaps, I suspect, about human behavior). The kind of question determines quite a lot about the effect of the finished work, as, for example, the way the many individual entrances and exits along chance-plotted paths in *Summerspace* created a sense of beings who were largely independent of each other. If in *Inlets* one often gets a perhaps fanciful impression of dancers temporarily marooned when a tide of movement ebbs, it may be because he established four points on the stage where any of the six dancers might be required to stop and move slowly in place. If *Winterbranch* strikes people as drastic, it is certainly in part because falling was his principal subject, and because—wishing to get the fallen dancers off the stage without having them get up and run—the dancers fell onto pieces of cloth that others had rushed in with, and were dragged away. The 1975 *Torse,* on the other hand, is a fast, bright, rather dry virtuosic piece, and no wonder: stressing five basic positions of the torso—straight, twisted, tilted, arched, and bent—Cunningham made sixty-four sequences of dancing, containing one to sixty-four units. However, the dancers don't proceed metrically from number to number; no matter how much bending and gesturing is going on, unit twenty-one doesn't become twenty-two until the dancer's center of gravity is raised, lowered, or shifted to another foot. Relate the sixty-four phrases to sixty-four subdivisions of space (in accord with the sixty-four hexagrams of the *I Ching),* and you get a dance that is primarily about rigor and complexity and flexibility at top speed.

Some of Cunningham's game plans may themselves arise out of feelings that he is not about to impose on his audiences. Sentences dropped here and there provide intriguing clues. *Rainforest* (1968) he has characterized as "a little community," and although the ambiguity he delights in is there, so are startling images of territoriality and mating. Of *Rebus,* he has remarked casually, "a dramatic dance . . . with me as protagonist"—which might also be said of the comical *Gallopade* and the somber and mysteriously poignant *Quartet.*

The dancers working with him at any given time, the day-to-day life of the company are inevitably acknowledged in the dances. When Cunningham made *Rainforest,* the company *was* still a little community, nine strong (only six appeared in the dance), still enjoying picnics together on the road. After Carolyn Brown left the company in 1972, he made no more extended duets for himself and a woman, although he later "partnered" others—Karole Armitage in *Squaregame* (1976), Catherine Kerr in *Duets* (1982), for instance. (Brown's solo in *Scramble* must have been intimately bound to Cunningham's perception of her as a dancer, because until recently the solo was omitted from revivals of the piece.) By the time Cunningham made *Rebus,* he had become interested in working with a larger ensemble of dancers—dancers whose strength, speed, elevation, and high-swinging legs obviously fascinated him. Inevitably the widening age gap between him and the other dancers affected his relationship to them. I think that he has never really perceived them as the "beautiful machines" one rather rueful recent company member feared, but somehow either it was harder for them to reveal their individuality through the steps he gave them, or harder for them to reveal to him the individuality that would enlighten him when he was making up the steps.

In *Rebus,* Cunningham bustled among the vibrant young performers, helping here, holding up there, surveying the scene from the back, well

A stage rehearsal of *Travelogue* (1977) with Merce Cunningham and (rear) Karole Armitage, Chris Komar, Julie Roess-Smith, Meg Harper, and (front) Ellen Cornfield and Robert Kovich. Photograph by Lois Greenfield.

out of the way of the powerful legs. Sometimes, especially in his solos, he looked almost aghast at what he had set in motion. He struck me as a kind of Catcher in the Rye, watching to see that this flock of beautiful creatures, lost in their dancing, didn't fall over the edge of the world. And in *Quartet,* he—a fifth dancer not even acknowledged in the title, increased age and old ground-in injuries inevitably limiting his movements—stays withdrawn from the activity, occasionally snared by his own creation.

Seeing the act of choreographing as an expedition into uncharted territory, Cunningham lays his plans, keeps his eyes open, and, with luck, discovers something unforeseen. (The work he has done, beginning around 1974, with film and video* represents, in part, an attempt to teach himself new ways of looking.) Delighting in ambiguity, wanting always to be surprised, he tries to cultivate flexibility in himself and his dancers and his audiences. Here's one rehearsal episode: out of some unconscious instinct, or following through on the movement flow he has established, he asks a

*Cunningham's initial projects with video and movie cameras were done in collaboration with Charles Atlas. Elliot Caplan is now the company's resident videographer.

female dancer to lean over a male one; in doing so, she turns her face to one side and lays her cheek against the man's chest; suddenly, it looks like "an intimate connection" (in Cunningham's carefully reticent assessment), and he is pleased as Punch—far more so than had he tried to design a tender moment.

In a fairly long sequence in *498 Third Avenue,* a documentary film about the Cunningham company made in 1967, Cunningham works on a lift for Sandra Neels and Gus Solomons, Jr. It's a scorching summer day in the big, shabby studio. Staying close to the two dancers, supporting, prodding gently, talking, demonstrating, he gets Sandra facing Gus with her arms around his neck (no one uses the word "embrace"). He asks her to lift her feet off the ground and bend her legs behind her. She's supported by her own arms and Gus's left hand, locked under her right knee. Now Merce wants Gus to bend slowly backward from the knees, sinking until he catches his weight (and Sandra's) on his right hand. It's difficult and scary: Gus plots the whole thing out; he and Merce both reassure Sandra. They do it, talk some more, do it again; it works. And suddenly in the sweaty studio, this extraordinarily intimate and drastic event has been created: a woman entrusts herself to her partner and, with no show of alarm, holds calmly onto him while he falls backward, making himself into a table on which she rests momentarily. Later, in the dressing room, Neels jokes with an interviewer about Cunningham's inscrutability: "What's this exotic fall supposed to mean?" She'd never dare ask him, she says. But when you see her perform it, you know that, in her bones, she knows all that it can mean.

As the episode described above shows, there is nothing radical about Cunningham's view of the relationship between men and women. One suspects he sees female dancers with a courtly and tender eye. They are often supported, lifted, carried by men. Sometimes, the moves have an air of camaraderie—boisterous or solicitous. At other times, you may see a woman arched backward, draped over her partner, leaning the weight of her body against him. It is in passages for a man and woman that Cunningham's steps are most resonant with mystery and poignancy. In the 1987 *Shards,* Victoria Finlayson arches her body and jumps backward, straddling Alan Good's side, catching her feet with her hands; Good gently puts an arm around her and rocks slowly from side to side, touching his cheek to the top of her head. There is something both intimate and alienated about the lift that makes it strangely moving. In *Place* (1966), a dance that most concede has a decidedly ominous atmosphere, there is an ineffable little duet for Cunningham and Carolyn Brown. It begins quite wildly: he spins her, then sidles along while she, holding onto him, hops in *arabesque.* Then, very slowly,

Douglas Dunn and Susanna Hayman-Chaffey in *Landrover* (1972) with Louise Burns and Sandra Neels in the background. Photograph by James Klosty.

standing close together, the two of them look up, down; again, up, down, and briefly back to their left. With their arms around each other, they begin to walk slowly toward the right side of the stage, both of them bending forward and lifting one leg behind them with every step. Several times, they stop and lift their heads again; several times she looks back to the left and, leaning against him, lifts her left leg like a pointer. Then while he holds her, she slowly wrenches her leg across her body, places her foot back on the path, and together they resume their walking. It takes them a long time to vanish from the stage. Had a choreographer deliberately chosen the Expulsion from Eden as his theme, he could have made nothing more eloquent than this.

Watch Cunningham at work—on film, tape, or in the flesh—and you're struck by his calmness and cheerfulness. Demonstrating or sketching out steps, molding a dancer into the desired position, giving corrections, he works with the matter-of-fact, untemperamental efficiency of a master carpenter. Fixing a tricky spot or two in a passage for Rob Remley, Meg Eginton, and Lise Friedman (in 1979): "You have to . . . she takes your hand; that is, you've turned around and Meg takes your hand. Just do that, Meg; now you're there. Now *you* must start to come up at that point, Lise.

Otherwise, you see, you can get hit . . . Now there's one other place where you . . . could I just see that a moment? There's something wrong with the position. You're bending over, but you must curve your back. I don't know which foot is front. No, it should be more like this, you see—rather than this way, it should be more like this. You may have to be further away from them to pull them out. That's right, *that's* right. Now do the next thing. Up. That's right. Now this is straight: the knees are bent, but the back is straight, right? Now continue. Up, curve again, *really* curve your back."

Merce Cunningham and Carolyn Brown in *Place.* Photograph by James Klosty.

Concentrating on, laboring at dancing is for him and, he hopes, for his dancers the only possible task. Asked about seeking for the Buddha nature, Zen master Po-Chang replied, "It's much like riding an ox in search of the ox." And Cunningham, astride his chosen mount, says:

> Dancing is of divine origin, and to try to express that divinity is like pinning jelly to the wall. It only escapes you. It expresses itself if one gives one's life to dancing out of love for it, out of reverence for the nature of its action and the discipline necessary to allow for that action.

The dancer is no more, and no less, than "he who dances," "she who dances"; and what they do is the dance.

Merce Cunningham.
Photograph by James Klosty.

·8·

Everyday Bodies

Trisha Brown in her *Primary Accumulation* (1972). Photograph by Babette Mangolte.

Steve Paxton in *Flat*. Photograph by Peter Moore.

Three Dances of the Sixties

Flat (1964), by Steve Paxton. A man wearing a shirt, a suit, shoes, and socks walks impassively in a circle. Once, twice. He crouches, one hand lifted higher than the other; he might be wielding a racket, an oar, a spear. He holds the pose. Rises and circles again. Almost sits on a plain wooden chair; pauses; does sit. Walks another circle. Sits on the chair. Removes one shoe and one sock, freezing momentarily in mid–sock-removal.

The action of the piece consists of the circling walks, interrupted by sits and by repetitions of five simple, attentive poses, at least four of which appear to be derived from sports (an up-at-bat stance is the most obvious), and all of which Paxton performs with a minimal yet efficient use of energy and a noncommittal facial expression. He may hold a pose for as long as fifteen seconds. During the course of *Flat,* Paxton removes most of his clothing, one piece at a time, freezing now and then in the act. By the time he has stripped to his underwear, he's got the jacket hanging from a hook taped to his chest, the shirt hanging from a hook behind his left shoulder, the pants from one on his back. The walking and the posing continue with these encumbrances: clothing as totally nonfunctional adornment. Bit by bit, he re-dresses.

At least, this is what Paxton does in a revival of *Flat* that was videotaped in 1980. The dance is, in part, indeterminate, the order of the "photographic score-catalogue" of poses not set in advance.

Rulegame 5 (1964), by Trisha Brown. Seven parallel rows of masking tape are laid out on the floor to define a space twenty-one by twenty-one feet. Five performers wearing any old clothes (at the 1965 performance at the

Trisha Brown's *Rulegame 5* in a 1966 performance by Walter DeMaria, Red Grooms, Tony Holder, Simone Forti, and Olga Klüver. Photograph by Peter Moore.

Judson Memorial Church in New York: Trisha Brown, Walter de Maria, Alex Hay, Steve Paxton, Robert Rauschenberg, Joseph Schlichter) follow each other along a track that runs up one line and down the next, and so on. Each row is associated with a prescribed level of activity—ranging from erect to prone. A performer working along one row may pass partner(s) on neighboring rows only after all parties have ascertained whether their comparative levels are correct in relation to each other. Is the person crawling along row 6 lower than the one crouch-waddling along row 5? Okay, proceed. If not in the proper relationship, the performers tell each other what's wrong and make the necessary adjustments. The manner of achieving the correct level is up to them. As the "players" progress through the pattern, the audience, understanding the game plan, can predict or silently referee any hitches. Or simply concentrate on the slanted design created by the performers' graded paths and the changes in speed and density that the rules of the game have generated.

Trio A (1966), by Yvonne Rainer. At its first performance (when it is billed as *The Mind Is a Muscle, Part 1*), three performers (Rainer, David Gordon, Steve Paxton) execute a single long phrase of movements with almost no repetitions. Some of the moves are mundane, others are unusual—odd in shape or coordination; the dancers lope sideways, head rolling slightly; they step onto one foot, the other stuck out behind them, and, gazing down, crook one hand to their foreheads, not unlike a person indicating a unicorn's horn. The phrase contains turning steps, somersaults, balances on one leg, lunges, squats, moves for arms alone, steps in place, and steps that cover ground; yet the inherent contrasts are offset by the performers' deliberately neutral rendition. Using the minimum amount of energy required to execute the steps, they stitch together the modules that make up the sequence without the dancerly phrasing that would shape the material into peaks and transitions—substituting an uninsistent delivery that gives every moment equal importance and irons out even the dynamic inflections that inform everyday movement. The performers are rather like dancers "marking" or people speaking in a clear monotone.

Yvonne Rainer's *Trio A* with David Gordon, Rainer, and Steve Paxton. Photograph by Peter Moore.

The three are not in exact unison; over the course of the dance (approximately four and a half minutes long), their individual senses of time pull them into what might be termed an informal canon. Too, Paxton's well-oiled muscles and composure; Gordon's soft, shambling delivery; and the bluntness that always characterizes Rainer's performing make subtle variation one of the essential features of the work.

At that first performance, in New York's Judson Church, Alex Hay, stationed in the balcony, pushes strips of lath, one by one, over the edge and lets them fall. By the end of the dance, a large, untidy pile has accumulated to one side of the dancers. The irregular bangs made by the laths as they hit the floor function, in a sense, as "accompaniment" to the dance, yet the pieces of wood are accorded a status equal to that of the dancers, and their dropping is a simultaneous performance—a performance analogous to the dance in workmanlike quality, but a contrast to it in terms of rhythm and riskiness.

It's clear by now, I hope, that some radical notions about who was a dancer and what constituted a dance emerged during the 1960's in America.

As these works show, "dancing" might consist of postures and activities drawn from sports or everyday life, isolated from their usual context, but re-created as accurately as possible (*Flat*); ordinary occurrences, like dressing and undressing, given an unusual twist but *performed* as if nothing extraordinary were happening (*Flat*); highly imaginative "games" played with attentiveness and all the skill that can be mustered (*Rulegame 5*); complicated and demanding sequences of movement matter-of-factly executed (*Trio A*). The performers might appear as neutral as their props and costumes (*Flat, Trio A*), or they might display all the natural impulsiveness traditionally pruned out of dance performance (*Rulegame 5*). A "dancer" might have had years of studio training, as Paxton, Brown, and Rainer had, or, like some of the performers in *Rulegame 5,* be more experienced with, say, paintbrush or chisel than with *pliés.* Whether trained or untrained, however, the bodies tended to look like those you'd meet on the street, rather than those of slim, lithe, well-muscled dancers trained to affirm their specialness with every look and gesture.

A Decade of Revolt and Redefinitions

There are other dances I could as easily have chosen to be paradigms of sixties vanguard dance as the three described above: Lucinda Child's *Geranium* or David Gordon's *Mannequin Dance;* Deborah Hay's *Victory 14,* Robert Morris's *Waterman Switch,* Meredith Monk's *16 Millimeter Earrings;*

maybe Simone Forti's *Huddle* or Ann (later Anna) Halprin's *Parades and Changes.* Or any number of others. Many, like *Flat, Trio A,* and *Rulegame 5,* emerged from what came to be known as Judson Dance Theater—named for the church in New York's Greenwich Village where, on July 6, 1962, the people who had been taking a stimulating and unorthodox composition class with musician Robert Dunn gave the first of many performances. From that historic three-and-a-half-hour concert, which featured twenty-one works by fourteen choreographers (including a film by choreographer-filmmaker Elaine Summers), Judson excited some spectators—especially visual artists and musicians who saw their own concerns echoed—and horrified others—especially those members of the dance establishment who mistook the belligerently alternative approach for a state of siege.*

It's not hard to understand why they felt baffled and threatened: the value of the skills they had arduously acquired was being questioned. Virtuosity—and with it almost anything that looked like "dance technique"—was out. So were role-playing and acting and glamour. So, at first anyway, was creating illusions with lighting, costumes, or props. So were climaxes induced by phrasing or determined by an accompanying piece of music. The ultimate put-down "it's not art" was heard as heavy punctuation to the applause of aficionados and the qualified approval of discerning critics.

Any attempts to analyze the period are fraught with peril. Although most of the radical choreographers of the sixties were a part of Judson, a few—notably Ann Halprin, Simone Forti, and Twyla Tharp—were not; Meredith Monk and Kenneth King appeared on the scene after the first wave of experimentation. The 1962 Judson concert was not the beginning of the new consciousness, nor were the Nine Evenings of Theatre and Engineering (1966)—a last collaborative bash in which many Judsonites and one-time Judsonites participated—the end of iconoclasm. The sixties† was a period of almost continuous experimentation, of anarchy channeled into ebullient creative statements. An immense number of dances were made and performed; many, having made their point, were then discarded. To these choreographers, maintaining repertory was not of interest. As Twyla Tharp put it in 1971, ". . . it's like bubble gum . . . You can keep on

---

*It should be noted that not all of the choreographers whose work was presented under the auspices of Judson Dance Theater were radical. Those who were not only radical but extraordinarily gifted came to dominate the public's imagination and to characterize Judson for historians of the future.

†Cultural historian Richard Kostelanetz thinks that the sixties began in 1958, and the ferment in the art world offers some justification. Although the most radically experimental and collaborative period of Judson was over by 1964, the search for new definitions of dance and dancer that characterized the sixties lasted at least ten more years.

chewing for ten hours but after about a minute and a half you've got all the good out of it."

Change was a constant. Yvonne Rainer's *Three Seascapes* (1962), with its ironic use of the slow movement of Rachmaninov's Second Piano Concerto and its "screaming fit downstage right in a pile of white gauze and black overcoat," was much rawer and more impulsive than her pristine, minimalist *Trio A* of 1966. And by 1968, *Trio A* had become just one component of *The Mind Is a Muscle,* an evening-long work for proscenium stage that juxtaposed film to live performance, people to objects, words to motion, unpredictable or commonplace behavior to structural finesse.

Rainer also taught *Trio A* to students and anyone else who asked to learn it, giving them the right to pass it on: "I envisioned myself as a postmodern dance evangelist bringing movement to the masses, watching with Will Rogers-like benignity the slow, inevitable evisceration of my elitist creation. Well, I finally met a *Trio A* I didn't like. It was 5th generation, and I couldn't believe my eyes." The quote, from Rainer's *Work 1961–73,* nicely points up a minor dilemma of the sixties: the tussle between pride of creation and noble intentions about relieving the creator of godlike control over the finished work.

As diverse as the new dances were, however, as individual their creators and as subject to change, one of the achievements of the period as a whole was a redefinition of the dancer as "doer" and the dance as whatever was done—whether that meant performing an impressively choreographed piece of offbeat dancing, sitting still, climbing a wall, reading a text, munching a sandwich, or not showing up to perform.

The controversial new view of the dancer stemmed in large part from aversion to the status quo, to the New York studio dancers whom artist-choreographer Carolee Schneemann pityingly described as follows:

> Their eyes reached into space without touching it, They were alone. The distance between "art" and "life." Space was anchored in their bodies, space was where they felt their spines. They didn't realize that a radiator behind them equaled their mass, asserted verticals against their legs.

Schneemann was deprecating what many in the dance world praised: that by the fifties, modern dance had built conventions as elaborate as those of ballet. The most popular and most copied styles tended to present the dancer as tragic hero, suffering victim, pawn of passion, celestial acrobat. In 1958 the two reigning giants of modern dance created works that enshrined them in roles they had been polishing for years. *Clytemnestra,* arguably the last significant work that Martha Graham produced in her "Greek" vein, offered Graham as the tormented heroine and her dancers as the passion-driven characters who peopled the Trojan War of her memory. In *Missa Brevis,* José Limón again cast himself as a man torn between opposing forces—in this case, the character's lonely, privileged agnosticism and the rapt faith of people in a war-ravaged community (Steve Paxton was one). Anna Sokolow, a powerful voice for 1950's angst and kin to the Abstract Expressionists in her avoidance of narrative and intensification of emotional gestures, presented dancers scrabbling on the brink of disaster, menaced by huge forces beyond their control.

Such roles, the Judsonites felt, had had their day. There were, of course, alternatives. Alwin Nikolais, in a landmark concert of 1953, had offered dancers as components of a landscape molded equally by their fluid motion, by light, and by color. Erick Hawkins, expunging his history as Graham's "dark lover," performed serene, soft-muscled incantations of poeticized nature. James Waring's dancers could be mysterious or glamorous, but they were likely to seem eccentric as well. Merce Cunningham's dancers played no roles, displayed no high emotions, appeared as alert and unselfconscious as animals.

Yet although some members of the upcoming generation of rebels found things to admire about these New York choreographers and other mavericks of the fifties, they could not wholeheartedly accept any of them as models. There was a high degree of artifice, of contrivance in the work of Nikolais and Hawkins. Waring, always sympathetic to the Judson dancers, a sort of floating component of Judson himself, was drawn to illusion, pathos, camp, and other elements the new choreographers wished to avoid.

Cunningham, as the most radical figure on the scene, was the most admired. His ideas were discussed; it was to his classes that insurgents most often went for their daily workout; Judsonites Steve Paxton and Judith Dunn danced in his company—Paxton through 1964. Yet there was one tradition that Cunningham had never cared to break with: his dancers were experts and looked it. Technical marvels doing beautiful and difficult things. They might have the air of decision-making individuals rather than obedient executants, but they cultivated the same ideals of flexibility, strength, and control that ballet dancers did. The new generation noted the discrepancy between Cagean theory and Cunningham practice: if any noise could figure in a musical composition, why wasn't any human movement a fit ingredient for dance?

A single, extremely atypical concert given by the young Paul Taylor on

October 20, 1957 (while he was a member of Martha Graham's company) prefaced the radical presentation of the performer that was to become common during the sixties. The movement for some of Taylor's seven dances consisted mainly of walking, running, simple changes of position. One, *Duet,* might be considered an analogue to John Cage's famous musical composition *4′33″,* in which pianist David Tudor sat before the keyboard without playing a single note. Taylor—who had been with Cunningham at Black Mountain College in the summer of 1953 and whose concert featured some of Cage's music—offered stillness as the equivalent of silence: he stood, and Toby Glanternik sat for the entire duration of the dance. The query of *Dance Magazine* critic Doris Hering, "Does nothing lead to something?" was not at the time taken as prophetic.

The Antielitist Ideal

Demythologizing the dancer was more than a revolutionary tactic within the dance field. It was a considered stance—idealistic even—one of many strategies devised to bring dance in line with aspects of contemporary life.

Liberation emerged as the most potent theme of the sixties, Americans having begun to notice that reality and morality were out of sync, especially within those elite groups that decree codes of behavior or determine political or economic policy. The Vietnam War made visible to many the breach between high-minded rhetoric and self-serving action. Volunteers in the Peace Corps, established by John F. Kennedy in 1961, found that competitiveness was not a universal value, that technology, efficiency, and happiness did not always go hand in hand. One important modernist dream, the gleaming, sterile, formidably equipped hospital, was discovered to have flaws: it was not invariably conducive to healing. By the late sixties, a revival of interest in home births and homeopathic medicines attested to public disenchantment. The flower children who began to blossom on street corners and set down roots in communes refused to accept their parents' definition of a successful life, and their increasingly fanciful clothing made bright-colored mockery of the idea that there was a "right" way to dress. Radically minded artists were among the first to become aware that our institutions had generated problems that they were unable to solve. Historian Sally Banes, in titling her definitive book on Judson Dance Theater *Democracy's Body,* aptly identified the revised image of the dancer as an embodiment of the revolutionary spirit of the day, a sort of relief map of cultural unrest.

In the various social and political movements of the sixties and early seventies, communality was another motif—a counterpoint to liberation.

The demonstrations against American involvement in Vietnam, the prison riots, the marches and rallies to protest discrimination against blacks, phenomena like the Woodstock Festival of 1969 created new if temporary communities.

Wrote Yvonne Rainer in 1981 of her early work and that of her colleagues, "If it was an explicit assault on then-current artmaking, it was also a response to the same economic/cultural pressures encountered in other areas of our competitive, atomized society, that turn one victim against another, although both may have the same interests at stake."

In Judson workshops and performances, a collaborative and consensual spirit—strained though it may have been at times—went hand in hand with the iconoclasm. Unlike the pioneers of modern dance, these choreographers at first neither established companies around themselves, nor founded training techniques. They performed in each other's works. The rejection of elitism and hierarchies and the attempted democratization of dance's processes, ingredients, and nature became focal points—with one goal being to free the dancer from the tyranny of rules, ideals, and "technique," as it had come to be taught.

Art-World Allies

In pursuing the goal of liberation, it was inevitable that the dancers would seek alliances and exemplars in other disciplines. A network not only of shared ideas, but of relationships, both personal and professional, linked this adventurous corner of the dance world with equally radical musicians and artists. There were obvious parallels between the choreographers' disinclination to model themselves on Graham and Limón and the painters' repudiation of the dominant Abstract Expressionism. The latter move was already well under way by 1958, the year of *Clytemnestra* and *Missa Brevis.* Early that year Jasper Johns unveiled his flag and target paintings in his first major one-man show at the Leo Castelli Gallery. A few months later, Castelli introduced gallerygoers to the shock of such Robert Rauschenberg "combines" as *Bed*—a pillow mounted on the wall above a narrow rectangle of patchwork quilt, both of them streaked with paint. (The image of reality dislocated has obvious parallels with Paxton's later clotheshooks-on-the-body in *Flat*). It was in 1958 that Allen Kaprow showed his first "Environments" at the Hansa Gallery—a concept that eroded the barrier between observer and observed and sent viewers threading their way through the work, inevitably altering it by their presence.

It was also in 1958 that New Yorkers with adventurous musical tastes could enjoy a major retrospective of the compositions of John Cage. Like it or not (and he occasionally didn't), Cage, by implication, stood godfather to many works produced during the sixties by composers, choreographers, artists, dancers, playwrights, and directors. His book, *Silence,* came out in 1961, disseminating a wealth of unsettling ideas. To the composition class that he taught at the New School between 1956 and 1960 came not only composers like Toshi Ichiyanagi, but writers and painters, among them Jackson MacLow, Dick Higgins, Al Hansen, George Brecht—one of the founders of the Dada-ish Fluxus group—and Kaprow, creator in 1959 of *18 Happenings in 6 Parts* (from which a raucous genre of performance art took its name). It's small wonder that La Monte Young said in the course of a lecture he gave at Ann Halprin's Dancers' Workshop in San Francisco in 1960, "It is often necessary that one be able to ask, 'Who is John Cage?' "

Two of the key issues for Cage and those who gathered around him had to do with dismantling or ignoring distinctions between the various art forms, and with allying the materials and processes of art with those of daily life (an idea the dancers were to grab and run with). Performance was seen as a way of making art immediate, verifiable, and "real." The Beat poets had already made the public reading of their poems in coffee houses as important as the writing and publishing of them; in 1952, at Black Mountain College, Cage himself had staged what is generally regarded as a proto-"Happening." By 1960 Fluxus artists were presenting what they called "Events," and Red Grooms, Claes Oldenburg, Jim Dine, Robert Whitman, and others were orchestrating "Happenings"—going beyond assemblage to incorporate human performers, sound, light, and a time element into their work.

Since dancers were part of this larger scene—performing in Happenings, Events, and music concerts, attending others, debating the heady and uproariously unconventional ideas—it was perhaps natural that the styles and techniques that dominated the art world of the sixties—junk-art assemblages, minimalism, Pop Art—would be literally embodied in the "new dance," and that the artists and composers would take a serious view of the dancers' work. Some made dances themselves. Counted among the choreographers and performers of Judson Dance Theater were musicians Philip Corner and John Herbert McDowell, and artists Robert Morris, Alex Hay, Robert Rauschenberg, and Walter de Maria (the last as performer only). The dominant image in Morris's small, witty sculpture *I-Box* (1962) was a nude photograph of himself. Factually offering that body in motion at a dance performance was a logical extension of presenting it as an art object

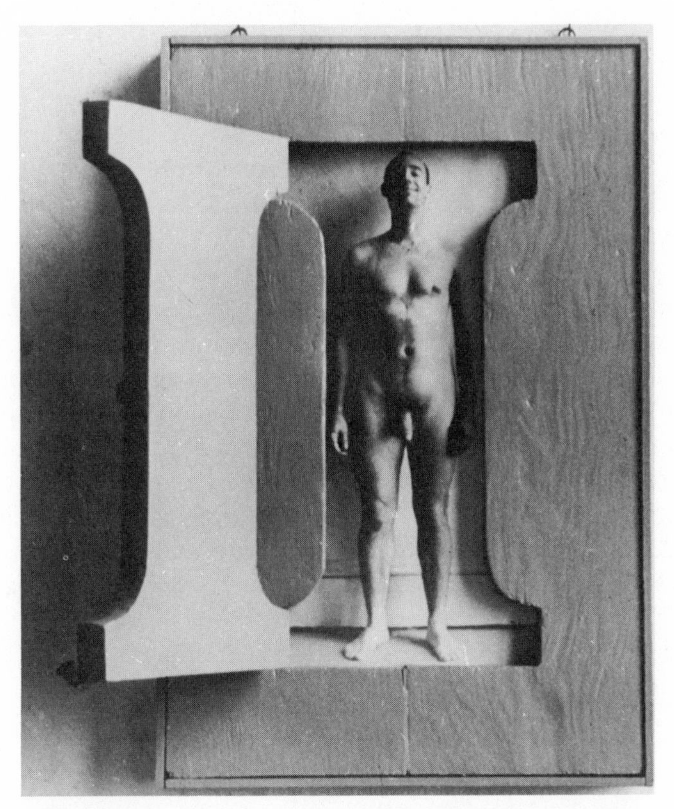

Robert Morris's *I-Box,* 1962 (open). Photograph by Bruce Jones.

in a gallery. Ironically, the choreographers, trained for years in how to *mold* the body, their prime material, had to discard a lot of ideas to arrive at that point.

The mingling of artists, writers, musicians, and dancers recalled the heyday of Futurism, Surrealism, and Dada in Europe—evenings in the Zurich of 1916 when dancers such as Sophie Täuber contributed to riotous performances at the Cabaret Voltaire and its successor Galerie Dada. The iconoclasm, the emphasis on performance, the serious playfulness, the use of chance and indeterminacy to thwart the ego of the creator also linked the two eras. Marcel Duchamp's "readymades" predated the incorporation of found objects into collages; and Kurt Schwitters, who in the 1920's avowed that any sound could be put into music—"violin, drum, trombone, sewing machine, grandfather clock, stream of water"—might have considered as a logical extension of his work La Monte Young's *Piano Piece for David Tudor*

#1, the score for which begins, "Bring a bale of hay and a bucket of water on stage for the piano to eat and drink."

Yet in the sixties, the cross-fertilization between the arts was more intense than it had been during the first quarter of the nineteenth century, and it yielded some startling hybrids, especially at Judson. Amid the fracas of Dada and Futurist performances, the image of the professional dancer had remained polished: bizarreness, rather than naturalness, replaced prettiness. Given that in dance, beauty—balletic or Duncanesque—was a norm, it is to be expected that the first and most obvious form of counterstatement would involve distortion. Too, while painters and poets could engage in outlandish acts in a theater, and musicians bang on peculiar instruments, a dancer who walked onstage and performed an everyday act in an everyday manner might not even have been recognized as a dancer, or have been dismissed as incompetent rather than argued over as radical. Prior to the sixties, it's doubtful that any self-respecting dancer would have described himself as David Gordon has done: "When I began performing in other people's works, I had a fairly good-looking body (though not a flexible one), good height and good balance, and no training . . . My leg never went very high and turning still makes me vomit." And then there is Simone Forti, who decamped from the Graham School because ". . . I *would not* hold my stomach in."

The Body

Although the presence of so many brilliant nondancers in dance performances was one catalyst for the change of body image, well before the sixties certain dancer-choreographers had already begun to be dissatisfied with the status quo and to think differently about the body and how to train it.

Here's an exercise that Simone Forti remembers doing during her first class at Ann (Anna) Halprin's San Francisco Dancers' Workshop:

> We must have been a group of about fifteen people. Ann had us walk together in a circle, not single-file but all moving in the same direction around a general circular path. She asked us to follow the speed of the flow around the circle, and not to initiate any changes. We started out very slow, and over a period of an hour gradually picked up more and more speed until we were running. We ran for some time and then started to slow down. The slowing-down process was much faster

> than the speeding-up process. Within this general speeding up, running and slowing there were several minor speed-ups and slow-downs. We finally came to a stop and collapsed on the ground.

Forti studied with Halprin between 1956 and 1960; so, on occasion, did her then husband, Robert Morris. Future Judson choreographers Trisha Brown, Yvonne Rainer, and Ruth Emerson arrived for the 1960 summer session. Halprin was developing dance training that, rather than seeking to mold The Body, considered each individual body and its potential. In her classes, as in her choreography, she utilized improvisation to get past dance clichés to more basic human responses. Such improvisations—"tasks"—whether complex or as simple as the one described above, had a forthrightness that helped set the tone of the sixties; by casting the dancer as a decision-maker, intent on solving a particular problem, they inevitably presented him/her as a wily, alert individual, perhaps athletic, perhaps not. A dancerly body and mind-set were not only *not* prerequisites; they might actually get in the way.

The seminal dance-composition class that Robert Dunn taught with his wife, Judith Dunn (1960–1961, 1961–1962, and 1964–1965), further honed this image of the dancer. Dunn, who had participated in John Cage's New School classes, advocated the use of scores or game plans—his own, Cage's, ones the dancers devised for themselves—to generate performance. This certainly had a practical side: those with no access to studio space could bring in a dance in the form of instructions to be interpreted on the spot. But, more important, scores could undermine habit, artifice, premeditation, and present both choreographers and performers in the role of problem-solvers. A score could push art-by-inspiration out of the picture and still foster an individual approach.

For example, using the number structure that Dunn extracted from Erik Satie's *Trois Gymnopédies,* Rainer designed a complex of intersecting graphs to ally composed actions with musical counts. (The result, *Three Satie Spoons,* was one of her finest early works.) On the other hand, Simone Forti, as "classmate" Remy Charlip recalled in an interview with Sally Banes, approached the same number structure in a more fundamental way: she touched a predetermined part of her body to the floor, depending on whether the count was a five, a four, or a three.

Like Cage, Dunn had studied Zen Buddhism: the nature of the person, material, situation must be allowed to be what it is. He attempted to maintain Cagean notions of nonevaluation during discussions of work in his own

Yvonne Rainer's *We Shall Run,* performed at the Wadsworth Atheneum in Hartford by (from left) Rainer, Deborah Hay, Robert Rauschenberg, Robert Morris, Sally Gross, Joseph Schlichter, Tony Holder, and Alex Hay. Photograph by Peter Moore.

classes, to emphasize that there was no "best" way to fulfill an assignment. If Steve Paxton wanted to offer as a dance the act of moving furniture out of the studio office piece by piece, fine. As Paxton himself was given to remarking, why not? The others would respectfully question and analyze.

Many dance fans and dancers, looking at a picture like one taken of Yvonne Rainer's *We Shall Run* in 1963, might be baffled. This was a dance? (Why not?) Little pattern or rhythmic unity is discernible. Some of those in the herd look mildly athletic, some do not. Alex Hay is wearing a suit and tie; Rainer and Deborah Hay are dressed for action in loose pants and T-shirts. No one is presenting an unusually beautiful or powerful physique. Were it not for the bare feet, you might almost mistake this for a street scene or a curious adult party game.

The point is that to this generation of dancers, the body in all its states was acceptable. Clumsiness could figure in dance as well as adroitness, plumpness as well as trimness. Even weakness could play a part. (In 1967 Rainer performed *Trio A* while in the shaky state that followed a serious illness and called it *Convalescent Dance.*)

The kind of clothing worn in performance made an aesthetic statement too (although some critics found the clothing itself distinctly unaesthetic): anything that glamorized the body was to be avoided. During the first part of the sixties, many of the "dancers" who were dancers continued to wear the standard classroom attire of leotards and tights. But at the same time—influenced perhaps by the nondancers among them, perhaps by their own changing ideas—they began to perform in everyday clothes: jeans, T-shirts, dresses. The athlete's baggy sweatsuit, the workman's coverall became popular: these garments allowed for mobility, didn't enhance the body, and buoyed up the image of the dancer as efficient and workmanlike. Given the fact that the lofts these people rehearsed in were often underheated in winter, sweatpants, sweaters, and thermal underwear constituted their usual attire. What could be more honest than wearing your workclothes to perform in?

If the dancers occasionally appeared naked, or nearly so, the purpose was not to titillate the audience, but to reveal the human body in its most unembellished and vulnerable state. Too, clothes often covertly exaggerate sexual features. Nudity was a way of defusing gender differences by calmly focusing attention on these—as well as being a commentary on the guilt-ridden eroticism of much then current dancing. The new dance wished to point out that male-female relationships were not invariably marked by sexual tension, that there were situations in everyday life in which gender was not an issue. *Word Words,* a one-shot, clear-the-air piece made and performed by Paxton and Rainer in 1963, gave viewers the opportunity to take in a ten-minute movement sequence executed by a male performer wearing only a posing strap and then compare this with its rendition by a female performer similarly clad. A third run-through by both dancers could further stimulate thoughts about the effect of gender on dance style and the effect of nudity on an audience. Did Rainer and Paxton perform the dance very differently? If so, was this due to differences in physical structure or differences in temperament? Did gender have a decisive effect on either or both of these factors? And, given our cultural conditioning, did the nudity prevent the audience from focusing on the dancing?

The 1960 Supreme Court decision against censorship fueled such assaults against restrictive prudery in many areas of American culture, from movies to swimming parties. Over the decade, the naked human body (highly visible in now legalized pornography) became less of a taboo in daily life, particularly among the younger generation. It is ironic, though, that while Rudi Gernreich's topless bathing suit was a high-fashion gimmick of 1964, three years later in New York Anna Halprin was threatened with arrest because in parts of her *Parades and Changes* (composed in 1965), the company danced completely naked.

It seems to me that dancers working in different decades are often trying to make their audiences focus on similar things about the body. However, because they *are* working in different decades, their methods and viewpoints are not at all alike. Isadora Duncan, associating breath with a kind of electrical force animating all nature, built rise and fall, ebb and flow into her movements and patterns. Doris Humphrey and Martha Graham made breathing the basis of physical systems that could express the drama of human existence. Yvonne Rainer, wishing to make a more factual acknowledgment of inhaling and exhaling as a function of the body, taped to her throat a contact mike that amplified the sound of her breathing (*At My Body's House,* 1964).

During the sixties, objects or media were often used as a way of extending the body through space or heightening the audience's perception of its functions. The nine dancers in Judith Dunn's *Last Point* (1963) appeared at Judson Church both live and on films which showed them in other sites—via three projectors and seven screens. In that same year, Paxton, who had become intrigued by clear plastic inflatables, inflated one in which he stood, then deflated it by means of a vacuum cleaner; what started out as a room shrank into a suit. At one point in *16 Millimeter Earrings* (1966), Meredith Monk put a paper Japanese lantern over her head, onto which was projected a movie of her face, eerily distancing her from her image. Alex Hay, with *Leadville* (1965), created a wry and bizarre melding of the human and the technological: wearing coveralls spray-painted silver and in a silver wig, he progressed slowly forward with a working tape recorder strapped to his back (the sounds included crickets chirping and his own voice). There was no take-up reel, and the plastic entrails spilled onto the floor. When the tape ran out, Hay fell.

A paradox? Such uses of media involve illusion—an aspect of theater that most dancers of the new generation were leery of. However, the various apparatuses were used in ways that were consistent with the rawness and plainness with which bodies were presented. They weren't concealed; there was an almost ingenuous directness about how they were utilized, albeit immense sophistication too. (During *Fantastic Gardens,* Elaine Summers's beautiful full-evening intermedia piece of 1964, spectators were given hand mirrors in which they might capture fragments of the films being projected on walls, ceiling, floor, dancers, and themselves—playing at being artists or imaginative children in the garden of technology.)

Meredith Monk in *16 Millimeter Earrings.* Photograph by Charlotte Victoria.

Alex Hay in *Leadville.* Photograph by Peter Moore.

To these choreographers, the status of media devices as contemporary artifacts was a given. The space exploration necessitated complex life-support systems. Trailing wires, amplifiers sprouting knobs and dials were the indispensable paraphernalia of rock concerts. The long-distance images of television and the myriad pieces of equipment required to transmit them were likewise extending the human body in ways both beguiling and dangerous. There was some question whether scientists could control the flow of data they were generating. New and recent technology seemed as apt for artistic commentary as, early in the century, had been the machines that ironically enlarged human capabilities while atrophying certain human skills.

In focusing on the unadorned moving body as an art object—something to be considered in terms of its volume, its mass, its nature, its spatial relation to other object-bodies—the dancers juggled another apparent contradiction. Their respect for their material allied them with the Bauhaus artists of the twenties and thirties, but they shunned the Bauhaus notion that the artist had to mold the material in order to bring out its essence. Determining the "essence" of something involved meddling with its essential nature, with its naturalness. La Monte Young cautioned composers against "enslaving" sounds; the dancers were wary of anything that might interfere with what Rainer called "unenhanced physicality." A different brand of idealism: in theory anyway, Utopian democracy in art. Wrote Robert Morris of conventional dance techniques: they ". . . give the body definition and role as a dancer, but qualify and delimit the movement available to it." Or, as Trisha Brown put it years later, "There's a kinaesthetic reality that's only encumbered by skeletal artifice"—referring to the stretched legs and lifted chest that constitute the Western dancerly stance. If illusion and transformation were to be shunned, the thinking went, then anything that promoted the image of the dancer as superperson or gravity-defying whiz was out. In this crowd, a trained dancer had to keep a certain amount of physical versatility under wraps to avoid the onus of showing off, and the amateur performer was valued for rough-and-readiness, for lack of preconceived muscle-set.

This isn't to say that some polish wasn't inevitable, and some expertise not desirable. Morris and Alex Hay were naturally capable movers. Rauschenberg, who later denied having any physical gifts, affirmed that the choreographers ". . . wouldn't just stand around and let you screw up their work." The dancers continued to take technique classes. During the first part of the sixties, at least, many of them could be found at the Cunningham studio; Judith Dunn, and sometimes Yvonne Rainer, appeared in morning technique classes at Robert Joffrey's American Ballet Center.

Trisha Brown and Steve Paxton in Brown's *Lightfall* (1963). Photograph by Peter Moore.

However, they viewed class as a way of staying in shape, not as a system for designing bodies or a factor in the finished work. "Technique" to them meant the skill to do what you wanted to do as simply and convincingly as possible—whether that was standing on one leg or howling.

Where the movement approached discernible technical difficulty, a blunt, antibravura execution (what Trisha Brown called "not buttering it up") cut it down to size. And Rainer's criterion for rejecting certain moves, such as a *grand jeté,* was that you couldn't "just do" them; without a heroic lift-off and a stretched leg, without a push toward climax, you couldn't do them at all.

The Dancing

As with any choreographers who disassociate themselves from traditional movement styles, one issue for the Judson dancers and other radicals contemporary with them was, what *do* you do? How do these bodies "dance"? Visual artists with similar antielitist views had already begun to use found material in their collages, assemblages, and Events—often material too banal or dilapidated to win any aesthetic seal of approval. It was one way of blurring the boundaries between art and nonart, of querying the hegemony of the skilled hand. In George Brecht's *Three Chair Event* of 1960, for example, one ordinary chair, a white one, was displayed, suitably lit, in the Martha Jackson Gallery; a black one stood in the gallery's bathroom, a yellow one on the sidewalk outside. Only one component of the artwork was certain to be regarded (and questioned) as "art"; the others might be sat on or overlooked altogether.

For the dancers, the most obvious forms of found material were pedestrian movement and everyday gestures. Actions like running, walking, crawling, sitting, lying down affirm their connections with day-to-day living. Rainer's *We Shall Run* (1963) was just what its title implied: twelve people jogging around in patterns for seven minutes to the (ironically) heroic "Tuba Miram" from Berlioz's *Requiem.*

In *Silence,* Cage averred that the sight of a horn player shaking spit out of his instrument was as interesting a part of the performance as the music being played, and why not make art that took that fact into account? In line with this, the dancers often did not, in performance, suppress any nose-wipes or hitchings-up of pants. And, given the reluctance to limit dance to certain forms of physical activity, it was also reasonable for a dancer to talk, paint, grunt, scream, sing. Rainer wrote in her diary in 1962 that her sources could be

> . . . as diverse as the mannerisms of a friend, the facial expression of a woman hallucinating on a subway, the pleasure of an aging ballerina as she demonstrates a classical movement, a pose from an Egyptian mural, a hunchbacked man with cancer, images suggested by fairy tales, children's play, and of course my own body impulses generated in different situations—a dance classroom, my own studio, being drunk at a party.

All such acts, of course, were to be regenerated as authentically as possible, although a new context and altered timing—an unusual amount of repetition, slower than normal performance—might alter how they were perceived.

The creative act as a matter of ordering and framing reality has received no more ardent tribute than the review that Jill Johnston of *The Village Voice,* exegete of the "new dance," wrote of Steve Paxton's 1967 *Satisfyin' Lover:*

> . . . thirty two any old wonderful people . . . walking one after the other across the gymnasium [at St. Peter's Church on West Twentieth Street] in their any old clothes. The fat, the skinny, the medium, the slouched and slumped, the straight and tall, the bowlegged and knockkneed, the awkward, the elegant, the coarse, the delicate, the pregnant, the virginal, the you name it, by implication every postural possibility in the postural spectrum, that's you and me in all our ordinary everyday who cares postural splendor . . . there is a way of looking at things which renders them performance.

Paradox: making art more like life doesn't necessarily make it more accessible or more popular. In fact, the reverse is likely to be true. To Johnston's thrilled response might be countered that of the viewer who believes he/she's being cheated unless the dancers are doing something certifiably difficult: "Call that dance?"

Everyday actions, however, were not enough to fuel a career for dancers with a lust for movement. Many of the choreographers found sports to be an appealing source. Unlike their revolutionary counterparts of the thirties, they weren't attracted to an ideal of prowess or the beauty of a streamlined body; it was the intent-on-the-goal efficiency of ballplayers, runners, boxers, pole-vaulters that they focused on. It was how to get there first, get

Yvonne Rainer performing "Corridor Solo," with Deborah Hay, Steve Paxton, Tony Holder, Judith Dunn (between mattresses) and Robert Rauschenberg performing "Crawling Through" from Rainer's *Parts of Some Sextets.* Photograph by Peter Moore.

the ball, and get away with it. Such athletes, observes Trisha Brown, weren't "on tiptoe with their shoulders up." The dancers seemed particularly drawn to the unguarded body—the catcher hurled off-balance, the hitter splatting into third base—although even a poised and shapely movement like a dive was preferable to an *arabesque.* (A good part of Twyla Tharp's brief and meticulous 1965 *Tank Dive* was devoted to a slow preparation for a dive.) In addition, there were the satisfyingly raw windups and follow-throughs to regenerate: the boxer's jigging feet, the lope-to-place, the on-your-mark crouch, the dust-yourself-off-and-walk-away. Such moves might be transformed by context. Steve Paxton used sports photos to generate movement for *Flat,* and he employed them again in making *Jag ville görna telefonera,* a 1964 collaborative duet with Rauschenberg; the pictures, arranged in a "score" and subjected to chance ordering, produced dancing with a continuity unlike that of any known sport and unlike any hitherto-dreamed-up dance.

During the latter half of the sixties, when American interest in Eastern art, philosophy, and culture was escalating along with the war in Vietnam, many dancers were drawn to Eastern martial-arts forms—the Chinese Tai Chi Ch'uan for Rainer, Forti, Brown, Deborah Hay, and others; the Japanese Aikido for Paxton—both as a source of movement ideas and a technique for working the body. These people were not, of course, interested in the possible goal of success in deadly combat, but in the forms, and the principles behind them.

At the heart of Tai Chi lies the Taoist notion of *wu-wei.* As defined by Chinese-American dancer and Tai Chi expert, Al Huang, the best translation of this term is "to act without forcing—to move in accordance with the flow of nature's course . . ." If there was one sense in which the sixties dancers were idealistic about the body, it was that of being in tune with natural processes, of being able to feel alive in passivity as well as in action. In Tai Chi the performer refrains from sharp angles and sudden force, pursues paths as subtly modulating as those of water. The weight is rarely locked on both feet, but is in the process of transference: the dancer is grounded, but not rooted; alert, but not tense. Soft in the joints. One part of the body may lead another part into motion or ripple into peaceable opposition. The Tai Chi forms, like "Needle at Sea Bottom" or "Step Back and Repulse Monkey," create a complex three-dimensional interplay of the parts of the body, and in this interplay, obliqueness and directness, pressure and yielding can exist almost simultaneously. The view of the performer is very different from that projected through almost all Western dance styles—less single-minded, less aggressive. If you leaf through photos of sixties dance, you get a general impression of people who care more about rolling with the punch and deflecting the blow than about shooting from the hip. Blunt as it is, Rainer's *Trio A* has more in common with Tai Chi than it does with American Modern Dance.

Equally liberating as a way of thinking about the body were the rock-and-roll dances of the day. Prior to the sixties, black vernacular dance styles had entered the overall social-dance scene primarily as passing fads. In the sixties, both because of the allure of the music and the rise of a militant black consciousness, the images they generated were more pervasive. Nineteen sixty-four, the year of the tragic Sunday school bombings in Birmingham, Alabama, was also the year the Beatles and their danceable tunes hit the big time. Like the martial-arts and tap—another black-based form which enjoyed a renaissance in the late sixties—rock-and-roll encouraged ways of showing the body that differed from accepted Western norms of dancing. Instead of moving the torso in one piece, dancers in discotheques imitated the mobile hip action of Afro-American styles. The professional dancers took note of the individuality and liberated physicality that rock-and-roll promoted. As with Tai Chi, the finished pose—that image of expertise—was subordinated to shifts in impetus; follow-throughs eroded into new windups.

Persona and Performance

Trisha Brown recalls that *Rulegame 5* was performed in two different ways. It all depended on who the participants were. One cast followed the rules as efficiently as possible, while another more ebullient set of performers took pleasure in faking each other out and prolonging the adjustments. Many of her colleagues considered anything but the "pure" version a betrayal of minimalist ideals; others (Brown herself was one) relished the individuality that was a feature of the rowdy version.

As I look back on the sixties, it strikes me that Brown's story epitomized two notions about performing that coexisted during the sixties. On the one hand, it was felt that the performer should go about his/her "task" of performing in the most neutral, efficient way, interposing no comment between concept and art object. On the other hand, wasn't it a suppression of nature to conceal feelings of glee or discomfort or puzzlement? That such apparently contradictory aesthetics could feed each other was perhaps possible only because the Judson choreographers so assiduously attempted to quell rivalry among themselves.

Carolee Schneemann's *Lateral Splay* (1963), which had the performers tearing across the space, bouncing off the walls and various objects, colliding with each other, collapsing in heaps, was one of a body of dances that celebrated raw physicality and playfulness. Describing a performance of *Lateral Splay,* Jill Johnston said that Alex Hay ". . . pitched forward with such speed—his body laterally off center . . . that he looked like a bat out of hell with one wing scorched and the other left behind." Suppose, for example, that you had been one of the two performers in Simone Forti's 1961 *From Instructions:* you'd have been told to lie on the floor for the duration of the piece; your partner would have been instructed to tie you to the wall. From before the Cuban missile crisis through the Vietnam War, in what seemed an increasingly bellicose America, these dancers deliberately projected the image of peaceful and cooperative people. But it should be noted that this did not preclude wit and rambunctiousness.

As the choreographers gradually modified their severity toward anything like showing off, such fearless and exuberant use of the body engendered a new, unpolished sort of virtuosity. In 1968 two women—Trisha Brown and Barbara Dilley (then known as Barbara Lloyd)—stood on a smallish mat and took turns falling. This was Brown's *Falling Duet.* The faller might daringly reverse directions in mid-lunge, causing the catcher to race around, dive to the floor, and, at the last minute, cushion the impact. The audience—gasping, laughing, crying out—experienced a kinesthetic excitement different from, but no less potent than that roused by a male ballet dancer doing his Sunday-best leap.

To those who approved of reckless physicality, one value of using everyday movements in a dance was that their simplicity and concreteness rendered them transparent; spectators could not only take in, say, the act of running, but could discern individual human ways of doing it. The runner pretended no emotions, assumed no disguises, but neither did he or she attempt to conceal any resultant fatigue or glee.

Those who favored a detached performing style viewed quotidian actions in a different light. Theorist of the avant-garde Michael Kirby remarked in his *The Art of Time* that the use of everyday movement and the elimination of dance technique in themselves objectified the art; with no "exciting" movement to provoke a kinesthetic response in the spectators, they would focus on what was being done more than on who was doing it. This is why Noël Carroll called Rainer's *Trio A* "expressive in the broadest sense . . . it calls attention to hitherto unexplored, even suppressed, movement possibilities of the dance medium." It is this objective "expression" that Rainer, and Paxton in *Flat,* thought could be enhanced not only by a nonheroic performing style, but by deliberate impassivity.

Deborah Hay's *Group I,* which Kirby made the subject of a chapter titled "Objective Dance," illustrates his points admirably. Hay deliberately created a structure that would inhibit spontaneous behavior, and which did not entail a loss of control on the part of the performers. In one section of the work, a group of people wearing suits or dresses stood on a platform wielding long wooden poles, which they raised, lowered, clattered together in response to signals from a "conductor." Five others, also dressed in neat street clothes entered, exited, formed a line, and executed a number of rudimentary patterned actions—for example, putting their arms around each other's waists and swaying slowly from side to side, kneeling, stepping, stamping a foot. Throughout the sequence, they maintained a relaxed posture and an expressionless face. When *Dance Magazine* critic Doris Hering wrote of the performers in a later Hay piece that they ". . . did not have much physical magnetism," she didn't realize that her negative criticism accurately reflected Hay's goal.

Deborah Hay's *Group I* (1968). Photograph by Peter Moore.

A noncommital manner was yet another antielitist strategy. It appealed to those who didn't wish to manipulate the audience or seduce them with the power of the dancer's personality. That's why Rainer averted her gaze in *Trio A,* why Judith Dunn and some others adopted a motionless face—"glazed," as Paxton put it. They wished to elicit little more empathy than the ropes, clothing, boards that shared the stage with them. Neutral delivery commented on the overemotive dances of the preceding generations, just as the simple forms and flat surfaces preferred by many sixties artists can be seen as manifestos against the thick paint buildups of the Abstract Expressionists—which came to be construed not as a particular use of the material, but as emotional outbursts.

Too, objective performing could comment ironically upon the loaded political, social, or artistic issues that interested some choreographers. As Marshall McLuhan's influential books *Understanding Media* (1964) and *The Medium Is the Message* (1967) pointed out, hot messages can be cooled down by the way in which they are presented or by the medium in which they're couched. Lucinda Childs seemed to be addressing this issue in her 1965 *Geranium.* While a tape played the noises of a crowd at a football game, Childs, wearing dark glasses and a fur coat, touched points on a white pole to indicate rises and ebbs in the crowd's excitement. Later, she padlocked herself to a hammock attached to the wall in such a way that she could carefully assume and hold the pose of a football player upended in the melee, without any rush of impetus or expenditure of energy. The cool commentator, the nonparticipant charting emotional and physical highs through visual signals.

An objective performing style could produce what Barbara Haskell, in referring to Pop Art, called ". . . a new kind of subjectivity, one that did not rest on the artist's expressive gesture." For those choreographers who were (or became) interested in emotion, persona, or narrative, but were disenchanted with the strong acting component of the modern dance they'd grown up with, detachment was a useful ploy. The performers could appear not to own the emotions being dealt with, or to function as what Rainer in the seventies termed "stand-ins for personae." When Meredith Monk and Kenneth King performed Monk's *Duet with Cat's Scream and Locomotive* at Judson Church in 1966, their actions remained calm and deliberate, while a tape loop of a rushing train with horn blowing and the screech of a cat created an aural picture of motion and emotion. Various-sized crescents painted with smiling mouths were rocked on the floor by means of invisible strings. To "talk," Monk held up a mouth. The neutral actions and cool style set up a tension with the loaded or "hot" sounds.

More paradoxes: studied inexpressiveness, carried to extremes, could mask humanity as thoroughly as conventional dance technique was thought to do. Too, complete blankness in performance, instead of taking attention away from the performer, often had the reverse effect—making audiences wonder just what sort of person would want to look so unresponsive. Of course, the human body is rarely completely inexpressive. At the very least, in some of the more severe sixties dances, it expressed the fervent wish for neutrality or detachment. It is also worth noting that, as important as it was for Rainer to eradicate her considerable charisma as a performer, she was never entirely able to do so.

The Object

Over the sixties, objects proliferated in dances, as they had in the junk-art assemblages of visual artists. Handling objects was something you could do instead of "dancing," which might, until the late sixties anyway, have branded you as a reactionary show-off. The tons of plastic, wood, mattresses, fabric, wires, water, food may not have been intended as an ironic comment on the affluence of Lyndon Johnson's "Great Society" in which, in 1966, the average American spent forty-eight cents out of every dollar on luxuries, but they can be read that way. Carolee Schneemann's 1964 *Meat Joy,* in which a glut of sausages and chicken carcasses was dumped onto the performers, scandalized some spectators by its gleefully messy sensuality—not, I think, by its "wastefulness."

Objects, like the strips of lath in *Trio A* or the garments in *Flat,* by juxtaposing their factual, ready-made natures to that of the created dance, further reinforced art's connection with life. They could also affect the quality of movement by impeding or altering the body's priorities—sometimes simply calling the spectators' attention to interesting physical facts, sometimes suggesting that possessions can control their owners, or that an acquisitive and

Carolee Schneemann's *Water Light/Water Needle* (1966) with the floor of Judson Church covered with newspapers for the audience to nest in. Photograph by Peter Moore.

ambitious society sets up obstacles for itself. The audience at Judson was moved to hilarity by the spectacle of Alex Hay (in his *Prairie*) attempting, with the aid of three pillows, to rest in some remarkable positions on top of a bar supported by slim metal poles, while gravely answering questions posed by his taped voice: "Are you comfortable?" "Are you comfortable *now*?"

To emphasize the objectlike aspect of the dancer, performers could literally function as objects—be lugged from place to place, set up on display, contrasted to other objects. The title of a dance that Kenneth King choreographed in 1964, *cup/saucer/two dancers/radio,* made the status of performers and objects quite clear. Indeed, as King and Phoebe Neville, staring expressionlessly straight ahead, raised the cups to their mouths with both cups and hands shaking furiously, it was hard to tell whether the humans or the objects were initiating the action. The issue of responsibility was still at stake when the cups slowly tilted and spilled colored liquid down the dancers' bodies.

The manipulation of objects was just one of several devices that could provide the performer with a desirable new role—that of "worker." (If for some of the dancers, this was a way of avoiding dancing, for the artist-choreographers it was a natural extension of their usual work, now transformed into an act of theater.) The worker, like the gamesplayer of Brown's *Rulegame 5,* subverted theatrical time by working in real time, continuing a task for a predetermined period or until some goal had been achieved. Any climaxes had the virtue of being unplanned.

Presenting the dance to be made as the task to be accomplished was perhaps the ultimate act of honesty. Referring via taped words to what might be included in this dance under other circumstances or in future performances, describing what had been left out, outlining elaborate dance scenarios while shuffling around—these were devices utilized by a variety of choreographers. When Twyla Tharp was asked to reconstruct her *Excess, Idle, Surplus* for avant-garde dance's first Broadway season in 1969, she and her company performed to a rehearsal tape of their muttering voices trying to remember what came next. Counting aloud or clapping, marked passages, and going blank became features of the new dance.

Incorporating process into the finished work—mixing the raw with the polished, the improvised with the rehearsed—not only monkeyed with audience expectations about what was art and what was not, it undermined the image of both dancer and choreographer as aloof in their certainty and accorded them the status of explorers.

The Space

The sixties dancers were abetted by economics and other circumstances that landed them in galleries, lofts, and churches more often than in theaters. The traditional picture-box stage, a seventeenth-century invention, was designed to promote illusion, to frame fictional images of time and space, and, as in academic painting, foreground and background are not just locations in space, but hierarchical distinctions. These vanguard choreographers wished to democratize space, and, unlike their radical counterparts of the thirties, they were more interested in inhabiting it than in dominating it.

The Judson Church gym made up in adaptability for what it lacked in equipment. Chairs could be made to face in any direction. When performances took place in the church sanctuary itself, walls, pillars, the choir loft, and the recess beneath it were often made use of. And, like the Happenings artists, the dancers designed works for existing sites—from woods to public buildings. The audience for these might be solicited in the usual manner, but it might simply consist of passersby. There is something irresistible about coming upon performance unexpectedly, as Paxton discovered in 1963 when his *Afternoon* was done in a forest in New Jersey and he got his first fan letter.

Works performed in notable sites not usually associated with dancing fiddled with the borderline between art and life in quite beguiling ways, providing bright illustrations of the Heideggerian view of art as a bridge to life—something that facilitates the understanding of nature without violating it. In environmental pieces, the figure can merge with the ground or call attention to it, losing some self-importance along the way. Lucinda Childs's *Street Dance* (made in 1965 for one of Robert Dunn's composition classes) neatly formulated the issue. Members of the class, advised by Childs's taped voice, went to a window and looked out. There was Childs down in the street, anonymous (except to them) among cars and pedestrians, pointing out permanent aspects of the scene, like architectural features, and temporary ones, like the weather.

Childs's piece was minimal in movement, but environmental works or ones that used unconventional indoor spaces could also enlarge movement possibilities and play with audience perception. In *Planes,* the first of a series of "equipment pieces" that Trisha Brown made beginning in 1968, two dancers crawled smoothly over a white "wall," putting their hands and feet into barely discernible holes. Even without the film that sometimes

Trisha Brown's *Planes* in a 1975 revival. Photograph by Lois Greenfield.

played over the action, the piece could befuddle one's vision. The climbers appeared to be in free fall; we were looking at them from above? From below? Illusion, a taboo in the early sixties, returned as an inquiry into itself. A year later Meredith Monk made *Juice,* the first of several spectacular theater pieces that raised questions about the relationship of scale and distance to our view of the performer—who might be now a distant figure in a vast pageant (1st installment, in the Guggenheim Museum), now an "actor" in a wacky "play" (2nd installment, Minor Latham Playhouse), now a disembodied head on a television screen (3rd installment, Monk's loft).

Monumental works that, like *Juice,* appeared in the late sixties and early seventies might involve many performers functioning in peaceful, mutually supportive ways. They can be considered not only in relation to large-scale pieces made by visual artists, but to the unplanned mob scenes of the times: friendly ones like the Woodstock Festival of 1969, unfriendly ones like the 1968 Democratic Convention in Chicago. So primed were the police for public demonstrations that during a performance of Twyla Tharp's 1969 *Medley* in Central Park, when (as planned) some fifty young dancers jumped up out of the audience and raced to join others dancing over the Great Lawn, a mounted policeman galloped into the center of the dance, fearing a riot.

It seems to me that the adventurous approach to performance spaces was a crucial factor not only in how the work was perceived, but in how careers were managed and audiences won. During the 1950's, young modern-dance choreographers (and many not so young ones) usually maintained themselves by teaching, saving up money for perhaps a single annual concert at a rented hall. Working under those conditions, it was difficult for any artist to maintain creative momentum and build an audience.

Because the nontheater spaces favored by the choreographers of the 1960's often came cheap, more performances were possible. Audiences excited by the pace of experimentation didn't have to wait a year to see what Yvonne Rainer would come up with next.

In the early sixties, the audience for "new dance," was, somewhat ironically, an elite one: artists, playwrights, poets, dancers with similar concerns seemed to form the bulk of the audience. (Even so, people regularly booed, walked out, made rude comments.) However, as choreographers became increasingly daring in their choice of space, they drew spectators who were intrigued by the idea of seeing a dance happen on a rooftop, in a parking lot, on park benches. Going to a contemporary dance performance hadn't been so controversial since Merce Cunningham started working with chance in the early fifties, and it hadn't been so much fun in who knew how many years.

The Aftermath

Even before 1970, the communal spirit that had produced the Judson Dance Theater was waning. After 1966 the painters made no more dances. Once again, choreographers formed small companies and began to refine and build upon their discoveries. The distinction between dancer-choreographers and those who were primarily performers had eroded during the Judson days; it resurfaced. During the seventies, Yvonne Rainer gradually abandoned choreography for filmmaking, although David Gordon, who had stopped choreographing for several years, returned to the scene with greatly developed skills.

But many of the values and roles developed during the early sixties proved durable: no grand manners, no pretense, no showing off, no body set that announces, "I am a Dancer." In the various approaches to movement and form that developed in the movement-and-form-oriented seventies, the notion of the dancer as a mundane and matter-of-fact "worker" persisted. The thrifty, no-wasted-effort approach was well suited to a decade marked by the energy crisis, by our first serious inkling that we were drastically depleting our natural resources.

Impersonation remained a rarity. Even when the "task" was performing fluid, complicated dancing like that which Trisha Brown devised, and keeping track of witty structural permutations, the performers' relaxed demeanor anchored art in the real world.

In a sense, the styles of the seventies were constructed on the image of the sixties. Contact Improvisation, the "art sport" duet form that Steve Paxton fathered, appeared as a formalization of the sixties values of peacefulness, cooperation, and daring: a person might be catapulted into the air, not through any manipulative laying-on of hands by his/her partner, but as the inevitable result of leaning, sliding, twisting to follow the current of movement and play with the tug of gravity. The gamesplayer emancipated by a new and highly flexible set of rules. Twyla Tharp, never one to eschew full-out dancing and always the head of a company, began to meld black dance moves with ballet, athletics, and who-knew-what to produce a distinctive style that refined antielitism into a quizzical principle. Her dancers, obviously pros, executed demanding and rigorously structured choreography while retaining the spontaneous edge, the casualness, the occasional inelegance of people dancing for pleasure. In David Gordon's dances, process and polish were linked in witty paradoxes.

The stringent rhythmic and geometric patterns of simple, minimal movement that Lucinda Childs began to make in 1969, and newcomer Laura Dean in 1970, were clearly related to minimalist styles in music and the visual arts. But they, too, were related to earlier sixties convictions about the body, even though the look of the pieces was unashamedly dancier. (Dean's works, especially, were not unlike Western European folk dances in their emphasis on small, lively steps.) Through their complicated yet clear rhythms and formations, they crystallized the image of the dancer as a member of a peaceful, vigorous community—not part of an elite species, but, as Tharp put it in 1969, "a grown-up person with a skill." Deborah Hay took the idea in a different direction—abandoning professionalism per se for a while, and creating simple *Circle Dances,* meant to be performed by all present, not watched.

Simone Forti's study of animals—the primal motions of crawling, rolling, swaying, the impulses of polar bear and heron—fed into the ongoing search for nontechnical movement and basic sources. And the obstreperous improvisational company The Grand Union (at its height: Brown, Gordon,

Paxton, Barbara Dilley, Nancy Lewis, and Douglas Dunn) trekked into the seventies bearing ideas of cooperation, in-performance daring, spontaneity, wit, and anarchy hatched ten years earlier.

Even as choreographers in the vanguard began again to make polish, virtuosity, and spectacle components of their work, they still wanted to affirm that the act of dancing was simply something that some people did with their lives, and tried to do well. They might not be making dances that, like *Flat,* featured the putting on and taking off of clothes; they might not be as concerned with looking nondancerly, as Rainer was in *Trio A;* their games might be more complex and less apparent than *Rulegame 5,* but as *dancers,* they wanted to retain vestiges of that rough-around-the-edges look, that lack of a high-powered manner, that affirmation of human vulnerability once considered so shocking.

Judy Padow, Susan Brody, and David Woodberry in Lucinda Childs's *Reclining Rondo* (1975). Photograph by Lois Greenfield.

· 9 ·

The Allure of Metamorphosis

Wigman dancers in "Schatten" from *Der Weg.* Photograph by Herbert Vohlwahsen.

"I think I'll be a ghost," says the Hallowe'en child, cutting eyeholes in an old sheet. Then: "Do you think anyone will guess it's me?"

On a social level, the masking of face and body has the charm of sanctioning behavior that's "not-me," although the idea of surrendering identity, of perhaps being totally transformed is a scary one. In African rituals or Asian dance traditions rooted in religion, possession or transformation—often an integral part of masked dances—are known to be dangerous. The Balinese *Topeng* dancer prays before assuming his mask, in order to put himself in tune with its powers; performing those *Topeng* solos that are sacred, he becomes a conduit between humanity and the gods. In African rites, a performer wearing a particularly potent mask often covers his entire body to shield himself from the forces that his performance attracts and seeks to control.

Few instances of ritual disguising remain in the traditions of Christianized Europe and America. Certain American Indian tribes still honor it in their dance-rites. The Wild Men of Oberdorfer, decked in fir branches, perform every seven years. The demonic *Perchten* show their horrid faces at house windows in Austrian villages at Epiphany and must be bribed away. But complete concealment of head and body are uncommon, and the extraordinary art, complex symbolism, and exacting rituals connected with African masking have no parallel in Christian tradition.

For the Western stage dancer, too, some element of transformation often figures in a performance: the mask, whether it belongs to one of the disputatious "diplomats" in Kurt Jooss's *The Green Table* or to the demonic creature in Mary Wigman's *Witch Dance,* leeches the personal and familiar from a performer, making him or her either more recognizable as a type or more potent as a symbol. Artists who use disguises to alter or hide the dancer completely are less common, but they are almost always hugely popular with the audiences of their day. Working in different decades of this century, Loïe Fuller, Oskar Schlemmer, Alwin Nikolais, Mummenschanz, and Pilobolus have all made transformation an important part of their art; and even though their aesthetics, as might be expected, differ considerably, they have generated quite similar responses in their audiences: wonder and delight, an undefinable sense of mystery, even terror.

". . . Human beings will always love bright games, disguises, masquerades, dissimulation, artificiality . . . ," wrote Schlemmer in a diary entry of 1926. But the notion of *total* disguise seems to offer a more particular kind of excitement, that of vicariously escaping something that is with us always: our own shape. Whether the transformation turns a human performer into a shimmering blossom, or a creature with feet where its head ought to be, or one component of a many-limbed structure, it meddles with the accepted reality of the human body in relation to the world and offers something more slippery, more magical.

Too, although technology is often an important part of the illusions, transformations that occur on stage, live, have a power that's very different from that of the special effects of film and video, and more like that of ritual or Hallowe'en games. The child at the costume party enjoys the tension between the disguise and the reality beneath it. The *komo* dancer of Ghana—a mound of swaying grasses topped by a huge fantastic face—may symbolize the indwelling spirits of natural things, but it is crucial that a member of the community be able to do this. Stage dancers who traffic in similar illusions understand that the spectators' enjoyment depends in good part on the knowledge that the phantom visions are being engineered on the spot by living human beings. In *Brainwaving,* by the currently popular Pilobolus spin-off group Momix, there isn't a dancer in sight. Standing loops travel along a white rope stretched across the stage; they appear to follow each other, to give chase, to turn tail and run, to get pooped. Yet you know that only the nuances of timing and effort of two invisible dancers could bring such an effect to life. More unequivocally even than puppeteers, they translate their muscular action into another form of life.

Schlemmer presented the dancer as a fantastic automaton, yet he wrote, "I never created a 'mechanical ballet,' intriguing though it might be to construct figurines and sets directed by an automatic mechanism . . . the fluid human element always forms part of the game."

Loïe Fuller in her *Danse Blanche,* a precursor of *Lily.* Photograph by Taber.

Silken Lily, Mechanical Man

The "game" of transformation is one that, like the dancerly roles already discussed, alters with the viewpoint of the creator, the state of stage technology, and the taste of the times. Loïe Fuller was a contemporary of Isadora Duncan, Ruth St. Denis, and Maud Allan (although Fuller was born more than a decade earlier). Like them, she reinforced, expanded, and challenged prevailing images of women.

Writing of Fuller's solo dance *Lily* in 1905, Roger Marx took full literary advantage of the delirium with which Parisians responded to the American performer's metamorphoses through lights and fabric and hidden wands. She didn't simply assume the outlines of a plant the moment the curtain rose, he said. Standing on a transparent pedestal that made her seem immensely tall, illumined from below, she seemed to him more like an archangel spreading great wings. Then as she began to turn, her long, long silk robe spiraled up around her up, swelling and compressing into the shape of a gigantic lily bud, with her body dimly visible as its pistil. More spinning, and the lily gradually opened at the top to display a glowing, quivering calyx. The spectator's eye was ravished by this "fantastic and passional vegetation, which sets the human being and nature side by side in a single symbol, and which illumines with a woman's smile the flower's fragile flesh."

Marx was one of many writers and artists who raved about Fuller, dedicated works to her, captured in bronze or marble or paint the swirl of her hundreds of yards of silk. The Goncourt brothers, Georges Rodenbach, J. K. Huysmans, Henri Toulouse-Lautrec, Jean Lorrain, James McNeil Whistler, and Stéphane Mallarmé numbered among her admirers. The curvilinear lines and wave patterns she created epitomized the exotic yet organic shapes of Art Nouveau. The artists animated metal, paper, and stone with subtle curves that created a semblance of vital growth; Fuller's shapes fulfilled that vision—actually swelling, subsiding, spiraling through space. Louis Tiffany's workers blew opalescent glass vases in the shape of budding lilies; she *became* molten glass. To the Symbolist poets and painters, she was superior to the mere dancer doing steps in a scenic world: she had fused with that world, was that world—the single glowing object on a black stage, in Mallarmé's words, "the visual embodiment of the idea." Her visions of clouds, flowers, flames epitomized the imaginative transformation—"deformation"—of nature that the Symbolists espoused. Albert Samain's sonnet "Dilection" might have been written in her honor:

J'adore les indécis, les sons, les
couleurs frêles,
Tout ce qui tremble, ondule, et
frissonne, et chatoie,
Les cheveux et les yeux, les
feuilles, la soie,
Et la spiritualité des formes
grêles . . . *

*I adore things indistinct, frail sounds and colors,/ All that trembles, undulates, and quivers, and shimmers,/ Hair and eyes, leaves, silk,/ And the spirituality of fragile forms.

Fuller took Paris by storm with her debut at the Folies-Bergères in 1892. In 1900, at the apogee of Art Nouveau and her status as its emblem, she had her own Art Nouveau bijou of a theater, designed for her by architect Henri Sauvage, on the fairgrounds of the International Exposition. Her "Danses Lumineuses" complemented both the changing colored lights playing nightly over the fountains in front of the Palace of Electricity, and the swirling lines of the furniture and decorative objects displayed in S. Bing's gallery, La Maison de l'Art Nouveau. Visitors to the American pavilion could ponder Charles Courtney Curran's painting *The Spirit of Roses*—women in diaphanous gowns frolicking among piles of outsized blossoms—and then go to Loïe's theater and watch a woman turn into a flower.

Fuller's thoroughly modern technology was also in tune with the electrical and mechanical marvels of the exposition. Although Mallarmé acknowledged the pairing of ". . . an artistic intoxication and an industrial achievement," most of her poetic admirers didn't dwell on the fact that it took a small army of sweating electricians to engineer her magic—that because each lamp required its own handler, fourteen men were necessary to create the eerie interplay of red and gold for her *Fire Dance.* A few of her fans, like Pierre and Marie Curie and astronomer Camille Flammarion, must have been impressed by her experiments with electricity, her interest in the phosphorescent properties of radium, or by the devices she came up with, such as the transparent mirror-lined platform and sheets of heavy glass embedded in the stage floor which made it possible for her to be lit from below, or the color wheels and slides that dappled the light, or the cunningly devised mirror room that multipled and scattered the dancing image. Those in love with her illusions didn't need to know that in order to protect herself from imitators (there were many), she prudently patented her discoveries. One patent application, filed in 1893, shows the stitched-together triangles of fabric and the long bamboo or aluminum wands (preferably curved at one end like a shepherd's crook) with which she whipped herself into the airy, luminous semblance of natural phenomena.

Her story of her evolution from small-part actress to dance soloist is this: she was flitting about the stage, in a play called *Quack, M.D.,* in response to the commands of a hypnotist. Because her white silk skirt was too long, she had to hold it up to avoid falling. She must have made a rather good thing of it because, said Loïe, a thrilled audience began to cry, "It's a butterfly! . . . It's an orchid!" This was in 1891, and, according to her, lights went on in her brain, and the idea for her "serpentine dance" was born. Fuller's autobiography, *Fifteen Years of a Dancer's Life,* is kin to the later writings of Isadora Duncan and Ruth St. Denis. She wasn't quite as lofty in her sentiments as those two were, but she was as anxious as they that sudden inspiration and total originality be understood as the base and substance of her work.

Original she undoubtedly was, but she was also shrewd enough to blend ideas she learned in the popular theater with artistic visions of the day. Born in 1862 near Chicago, she got her start in plays and musicals. Her well-rounded little body and pert, dimpled face fit her for soubrette roles (although she also played young boys). Historian Sally Sommer has suggested that Fuller picked up some ideas of what could be achieved with stage machinery when she performed in two different American productions of the Orientalist fantasy *Aladdin* (in 1877 and 1887), which teemed with effects produced by veils and magic lanterns and carbon arc lights.

The popular skirt dances of vaudeville may have given her an idea or two also. (In the tongue-in-cheek *Carmen Up-to-Data,* at the Gaiety Theater in London in 1890, she had replaced Letty Lind, who was known for her skirt dancing.) They certainly gave audiences a context in which to place her. "It is not a skirt dance, although she dances and waves a skirt," wrote a critic, reviewing *Uncle Celestin* (1892), a play in which she performed her "serpentine dance" as an entr'acte. By making the skirt out of lightweight silk and increasing the amount of fabric, Loïe gave birth to an image infinitely more mobile and mysterious than that of a pretty woman swishing her dress fetchingly and showing her legs to an audience.

Some descriptions of Fuller's solo dancing give an impression of purity and impersonality. She could ". . . evoke the dream of pure form," and transcend her own womanly presence. On the other hand, that presence was a vital part of her art, and not just as a motor to churn up the silk. Many writers allude, as Marx did, to the vision of her body beneath the fabric. Mallarmé wondered whether she was not ". . . half element, half human, mingling—as they are apt to do—in the floating world of reverie." ". . . Distinctly visible as though she were in her bath," wrote a British critic, less poetic than his French counterparts.

In abstract solo dances as in her *Salomé,* Fuller capitalized on *fin de siècle* male fantasies of women. In *Butterfly* (1892), her head topped by an antennaed headdress, she fluttered about the stage, waving huge wings created by wands and silk, arching back to open them, bending forward as she folded them in—the picture of mindless, weightless beauty. In *Lily* and *Violet* she could become fused in the spectators' minds with the deadly flower maidens of Wagner's *Parsifal:* glowing alone in darkness, the huge flower buds, swells, unfurls. No contemporary paintings of women half-buried in flowers, affirming their alluring yet mind-sapping fertility, spelled

Watercolor of Loïe Fuller—one of a group painted by Beaux Arts students and presented to her to commemorate her 550th performance in Paris on March 24, 1895

out the message of erotic danger as vividly as Loïe's gleaming patterns of tumescence and detumescence.

Some of her dances seem to have had the cruel beauty so admired during the Decadence, like her famous *Fire Dance,* performed to Wagner's "The

Ride of the Valkyries." In an 1896 critic's scenario, flames lick at her hem, rise, set her on fire; she beats at them with a scarf, which also catches fire; she crumples down and vanishes into blackness. Yet, as the flames begin to rise, she crosses her wands, opens a slit in the flaming veils and shows her eyes to the audience. Some say she smiled. "She stood in embers and did not burn," marveled Jean Lorrain. But whether she was on fire herself or luring others to the flames, the metaphors of ecstatic martyrdom or consuming passion would have been equally fascinating to her audiences.

Her ideas, extended to a group, also emphasized the playful, the macabre, or ebbing and flowing visions of nature. Light and fabric could be chased and gamboled with by her all-female company of "natural" dancers. In the 1922 *Ombres Gigantesques,* an immense shadow hand, then a foot menaced the dancers; three performers became a six-armed witch. The most abstract of her group creations kept the dancers invisible: in *La Mer*—set to Debussy's music of that name and conceived for the 1925 Paris Exposition des Arts Decoratifs et Industriels Modernes—they animated a shimmering ocean of blue silk.

Fuller not only fed the troubled late nineteenth-century view of women, her career epitomized its ironies. She was a brilliant theatrical inventor and innovator, an astute and nervy businesswoman-impresario. The carefully plotted waves and spirals and blossoms were brought to life by the considerable muscular strength of her arms and back. Yet onstage she offered, in Martin Battersby's words, ". . . exotic visions of Woman as a sensuous but intangible dream metamorphosing into a flame, a cloud, a moth, or a flower." And one who could be menaced, drowned, extinguished by her own creations.

No such romantic view of stage illusion or double-edged fantasies of woman figured in the choreographic works of the German visual artist Oskar Schlemmer. Although to him the history of the theater was ". . . the history of the transfiguration of the human form," he wasn't interested in transforming human beings into clouds or flowers. In September of 1922, he wrote in his diary: "Life has become so mechanized, thanks to machines and a technology which our senses cannot possibly ignore, that we are intensely aware of man as a machine and the body as a mechanism." The costumes that were the motivating force of his dances might exaggerate the body's spatial designs ("the egg shape of the head, the vase shape of the torso, the club shape of the arms and legs, the ball shape of the joints"). They might emphasize basic laws of motion in space (one costume, emblematic of spinning, presented the performer as a top), or reconstruct the body in cubes in order to relate the performer to the architecture of the surrounding area, or emphasize such symbolic forms as "the star shape of the spread hand . . . the cross shape of the backbone and shoulders." In Schlemmer's creations, man became literally "the Measure of All Things," a fantastic geometer, diagramming inner and outer space with his gestures, a moving embodiment of the design principle that had shaped his costume.

The disguises that turned Schlemmer's performers into "art figures" didn't have a life of their own, an afterflow, as Fuller's fabrics did; they were static shapes—wired, padded, molded—cages that confined the dancers at the same time that they redefined their outlines. Sexuality or glamour scarcely figured, and gender was often obliterated. Even the props that might extend the body in space tended to be made of stiff materials—wood or metal or papier-mâché.

Just before the premiere of Schlemmer's *The Triadic Ballet* in Stuttgart in September 1922, he contributed an article to the *Stuttgarter Neues Tagblatt,* announcing that "in the past we have neglected Form and Law in favor of the cult of mood and individualism. Today they are emerging again . . ." It is probable that to many in his audience he seemed to be redressing the balance with a vengeance.

Imagine "The Black Series"—the last of *The Triadic Ballet*'s three sections. The stage is entirely hung with black curtains, and the three performers are clothed in black; against this darkness, their metallic masks and costume pieces seem to float in space. First onstage is The Abstract (Schlemmer himself, under the stage name of Walter Schoppe). His right leg is white—padded into a shape like a turkey drumstick—while the left, in black tights, is almost invisible. His right arm ends in a bell within which is a purple ball with a silver point; his left hand is a metallic club. A round breastplate covers his chest, and his pear-shaped head is painted half pink, half red. To the adagio from Haydn's Piano Sonata, Opus 50, he faces the audience and begins to move. The bell rotates, the point within it stabs, the club swings, the inflated right leg leads off in wide steps and jumps.

In the last dance, to the passacaglia from Handel's Seventh Suite, in G minor, a "ballerina" performs, flanked by two cavaliers. But in black leotard and tights, she is seen mainly as a gold headdress and a glittering gold tutu—a coil of wire loops that roll from the bottom of her torso to her armpits. When she whirls, the loops fuse in a single gleaming mass. The men, wearing stylized gold face masks, appear to have no arms, and only a gold ball for an upper torso and a gold disk for a lower one. Graduated

Oskar Schlemmer: Study for *Triadic Ballet* (ca. 1921–1923)

black balls cover their legs, and white strings stretched from the disk to their ankles cage these preposterous limbs. What can they do? Walk and lean solemnly.*

The audience at the first performance seems to have been divided in its opinion: some people pleased by the audacity, others deploring the lack of interesting steps, most marveling over the artistry of the designs. You can imagine them asking afterward, as Arthur Michel did when he saw some other Schlemmer works in 1929, "What is this? Mechanistic cabaret, metaphysical eccentricity, spiritual tightrope walking, ironic *verité*? Is it perhaps all of these together . . . ?" That it was definitely "modern" would have been too obvious to mention.

Loïe Fuller was still producing performances (although not, at sixty, still dancing) when Schlemmer first mounted *The Triadic Ballet,* yet a vastly different vision of art separated his generation from hers, just as in America, it eventually separated Martha Graham and Doris Humphrey from Ruth St. Denis. Those born around 1889 (the year of Schlemmer's birth) were too young to catch the *fin de siècle* malaise. They came of age in a new century, and that knowledge fired them up. Schlemmer was only one of many artists who wished to define contemporaneity through "abstraction, mechanization, and the new potentials of technology." When in 1912, as a young art student, he first began to conceive of *The Triadic Ballet,* the Cubist painters had already begun breaking down natural forms into semi-abstract shapes and reorganizing them in space to make several facets of an experience simultaneously visible.

And, in the exalted and irreverent ferment of Futurism and Dada, painters had already begun to work in the theater and to design costumes as art objects. In 1912 Wassily Kandinsky published in *Die Blaue Reiter* his scenario for a production, *The Yellow Sound,* to be presided over by fantastic and monumental giants. Kasimir Malevitch dressed performers in the Russian Futurist opera *Victory Over the Sun* (1913) in cardboard costumes, painted in a Cubist style, with oversized papier-mâché masks. In 1917 Picasso designed the towering costumes for the Managers in the Ballets Russes production of *Parade.*† By the time *The Triadic Ballet* was finally produced, Oskar Schlemmer was on the faculty of the Bauhaus School in Weimar.

---

*In 1977 Gerhard Bohner produced a new version of *The Triadic Ballet* at the Berlin Akademie der Kunst. While Bohner did not attempt to duplicate Schlemmer's choreography, his production utilized the original costumes and copies of these, and gave a good idea of the movement capabilities of the figures.

†Léonide Massine restaged *Parade* for the Joffrey Ballet in 1974, and it has been in the company's repertory ever since.

Gifted as a painter, sculptor, and designer (he had already created costumes for two one-act operas by Hindemith), he was first put in charge of the mural workshop, then of the stone and woodworking workshops, and finally of the Bauhaus Theater. In an atmosphere that stressed both idealistic schemes and hand labor, his talents bloomed. The director of the Bauhaus, architect Walter Gropius, understandably approved of Schlemmer's ". . . magic of transforming dancers and actors into moving architecture," and Schlemmer noted shrewdly in a diary entry on June 14, 1921, that since the postwar economic crisis made building difficult, theater provided an outlet for the Utopian fantasies of the Bauhaus architects: "We must be content with surrogates, create out of wood and cardboard what we cannot build in stone." At the Bauhaus, even the parties had to be masterpieces of design.

Judging from recent reconstructions of some of the short dances that Schlemmer made at the Bauhaus, the tension between the human performer and the disguise that he or she wears is as striking as it was in Fuller's work, although very different in nature. In *Stick Dance* (1927), the performer wears dark leotards, to which are attached white wooden slats at knees, elbows, shoulders, ankles. He wields a stick in either hand. As he lunges, steps, stretches his arms, you see a dance of sticks, it is true, but the sticks refer to the joints of a human body—extending their inner functioning into visible space. In *Space Dance* (1926), three masked and padded figures of indeterminate gender travel along the geometrical paths taped to the floor: Blue ponderously stalks on half notes; Red advances by quarter notes; Yellow scuttles along in triplets (the relationships between color and form and sound were a subject of hot discussion at the Bauhaus). As the dancers progress along their diagonal and right-angled paths, they come to resemble imperturbable commuters neatly avoiding collisions. One sympathetic critic pointed out that ballet, "with its static stereotyped forms and formulas," often made dancers look mechanical, while Schlemmer's wit humanized his art figures and made them surprisingly lively. Fuller's wizardry lay in transcending human shape and behavior, Schlemmer's in affirming them in abstract form.

Kaleidoscope, Mask, and Mythical Beast

To Oskar Schlemmer, the development of the sober, stark *Ausdrucktanz* and its concomitant "cult of strength and beauty" justified presenting the antithesis—the "bright games, disguises, masquerades . . ." he thought eternally appealing. He died in 1943, having seen his artwork and theater pieces excoriated as "decadent" by the Nazis, but in America, something over five years later, Alwin Nikolais was beginning to think about brightly colored masquerades—this time as an antidote to what he saw as the overliterary state of current modern dance and the self-important, virtuosic princes and princesses of ballet. And also, perhaps, to a certain grayness endemic to the Eisenhower years. In 1956 *The New York Times*'s John Martin, with what he called "fairly bold prophetic enthusiasm," praised a concert at the Henry Street Playhouse,* where Nikolais and his dancers taught and rehearsed:

> In these days, when there has been so much cause to lament the sterility of the modern dance, it comes as a balm to the spirit to find a spot where it is lively, in full health and aware of its true substance.

The substance he was referring to was motion—motion that didn't proceed from a codified vocabulary—whether personal like Graham's or academic like ballet—motion that spoke for itself without dramatization. A man who took the future of dance seriously, Martin didn't lavish many words on the illusions that Nikolais created with lights, props, and costumes, but it was these that particularly delighted audiences and drew them down to the little theater on New York's Lower East Side.

One of the works that Martin saw was *Masks, Props, and Mobiles* (1953), Nikolais's first exploration in the vein he's still working, as of 1987. When the lights go on for "Noumenon"—a section of that early work that's rarely been out of the company's repertory—the audience sees what might be three large abstract statues, blobs of red sealing wax, each on its pedestal. They begin to move, to tip and rock stiffly from side to side in time to the electronic score composed by Nikolais, to change shape, as if motion were rendering them softer and more malleable. Perhaps because of their rhythmic coordination with each other and the taped sounds, the dance strikes people as funny as well as bizarre. At the end, the blobs elongate and re-form, so that the heads and arms of the dancers inside are identifiable.

Nikolais says that he never saw Schlemmer's pieces; even so, there are some striking similarities, say, between Nikolais's *Structures* and Schlemmer's *Flat Dance*, in which colored screens sidle about by invisible human

---

*Nikolais's twenty-year association with the Henry Street Settlement gave him an opportunity (a unique one in modern dance) to work out his ideas on a stage. The playhouse had been the original Neighborhood Playhouse.

agency, and the limbs and head of an apparently gigantic figure appear at the edges (several dancers do the trick). Schlemmer's widow thought enough of Nikolais to offer him her husband's costumes. Unlike Schlemmer, however, Nikolais uses a complex vocabulary of movement. He studied dance extensively—first in Hartford, Connecticut, with Truda Kaschmann, who had danced in Germany with Rudolf von Laban, then at Bennington with the major American modern dancers, and most particularly with Hanya Holm. From Kaschmann and Holm, he inherited the German modernists' tradition of space awareness and an analytical approach to the body that facilitated his artistic fantasies.

The idealistic and prescriptive moral tone of early modernism was in short supply during the fifties; the Cold War and the escalation of atomic weaponry provided fodder for a pessimistic appraisal of human endeavor. Man the master had become as vulnerable to his own devices as the rest of creation. Whether Nikolais articulated such an idea to himself or not, it did not please him to emphasize mankind as the measure of the universe, any more than as a mute tragic actor, holding the center of the stage with his inner agonies. In Nikolais's words:

> I began to establish my philosophy of man being a fellow traveller with the total universal mechanism rather than the god from which all things flowed. The idea was both humiliating and grandizing. He lost his domination but instead became kinsman to the universe.

The dancers he trained became but one aspect of vivid worlds of color, light, shape, and sound—with all the elements composed and designed by Nikolais. At their most visible, perhaps wearing only leotards and tights, the dancers function as explorers in strange worlds, or primal celebrants taking on the semblance of rocks and beads of water. Atavism and science fiction often mate in the works. "Shaman dances for our time," a critic called them. The dancers may wear costumes that alter their shapes, carry huge props that conceal them, or become moving surfaces to catch the patterns of light. In some of the dances, particularly the evening-long suites that Nikolais created during his twenty-year residency at Henry Street, the dancers might play all these roles—now displaying their range as dancers, now scuttling about in the dark, turning flashlights on and off at carefully timed intervals, now making a piece of fabric shudder into ghostly life.

To the community on the Lower East Side whose children took the classes

Members of Alwin Nikolais's company in "Mantis" from *Imago* (1963), Murray Louis in foreground. Photograph by Robert Sosenko.

that Nikolais, and later his dancers, gave at Henry Street, these first dances must have seemed a logical extension of the homemade magic that was an ingredient in his early dance-plays for children. In the late sixties, with the development of Nikolais's innovative work with large hand-painted slides, the effects became more spectacular. This, coinciding with the National

Endowment for the Arts's first support for dance touring, brought him a second wave of fame. Work that had been seen in the fifties as a bright antidote to "modern dance"—entertaining enough to be regularly featured on *The Steve Allen Show* in 1959—now, like Op Art, stoked a swelling interest in beauty that needed no explanation, in sensory delights, psychedelic effects, and drugs that heightened perception. The opulent patterns and dream-country electronic sounds, the metamorphosed bodies in Nikolais's dances provided the public with a theater of hallucination that was almost as good as an acid trip, and built audiences around the world.

Watching a Nikolais dance, you can imagine that you're peering through a microscope into a world of tiny organisms, gaily anthropomorphized; or perhaps it's a kaleidoscope you've glued your eye to; at other times the images might be inside a dreaming mind. Nikolais's creations can have the lollipop brightness of a popular-science demonstration, like *Prism* (1957), in which the dancers, costumed in the colors of the spectrum, shone flashlights on their legs—blacking themselves out and popping into view in another spot—and, at the end, raced up the back wall of the theater and bounced off in an orgy of refraction. In the recent *Video Game,* to one of Nikolais's quirky electronic scores, the dancers, in black light, with luminescent space-invader patterns on the front of their black leotards and nothing on the back, create a witty semblance of sidling across the space of a computerized screen and disintegrating on impact.

Luminous, bewitching, antic as Nikolais's work can be, it is also frequently pessimistic. There's something macabre about the triple totem figures in *Temple* with great black holes where their insides ought to be; the pretty, light-mottled dancing scrolls in *Count Down* have a habit of swallowing dancers, or turning one dancer into another. Quite a few of his dances end in apocalypse, with the environment obliterating the humans. In the finale of the macabre *Gallery,* the invisible dancers, holding phosphorescent, luminous moon faces above a partition, glide along like automata in a shooting gallery; in time to electronic explosions in the score, holes and gaps begin to appear in their faces, until there are almost no targets left to carry on the game.

Fuller's body within her silk creations roused her viewers to consider metamorphosis. Was this woman gradually becoming a flower? Disguise as a fact, rather than metamorphosis as a process, is usually the focus of Nikolais's works. (It was of Schlemmer's too.) He creates an image, manipulates it through ingenious variations, and then abandons it for another.

While Fuller's mechanics were not supposed to be seen, we are often aware both of Nikolais's illusion *and* of how it works. Each performer in the group dance from *Sanctum* is encased in a white loop of stretch fabric. We can see that it is the way people press their hands against the fabric and bend forward that makes a tall oval turn into an irregular trapezoid; we know how a human is making the white shape tip and rock; yet we can also *not* watch the mechanic-dancers and see only the strange runic dance.

In his famous essay on the marionette, Heinrich von Kleist spoke of the puppet's "symmetry, flexibility, lightness." Nikolais's designs often stress these qualities, and the hallmark of the Nikolais dancer is an ability to isolate any part of the body, as if an elaborate electrical circuitry were monitoring the flow of energy, causing small twitches, shudders, ripples to pop out now in a leg, now in a fingerip, now in surprising combinations. A leg can appear as boneless as an arm. And often, the dancers float and skim through their worlds, never articulating a struggle with gravity like the archetypal modern dancer. Their encounters with each other lack emotional weather; they merge and separate as effortlessly as drops of pond water. Even when Nikolais dancers are at their most unadorned and sensual, mating is apt to be a matter of harmoniously blending designs. The costumes rarely emphasize gender differences. Nikolais has been known to remark during lecture-demonstrations that he figures the locations of crucial bumps on dancers are a dead giveaway, and that any spectator with decent eyesight can tell which dancers are men and which women.

Nikolais's style was built to a large degree on Murray Louis, his leading

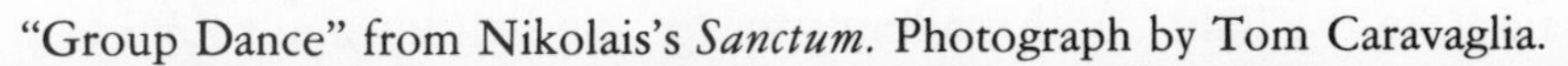
"Group Dance" from Nikolais's *Sanctum*. Photograph by Tom Caravaglia.

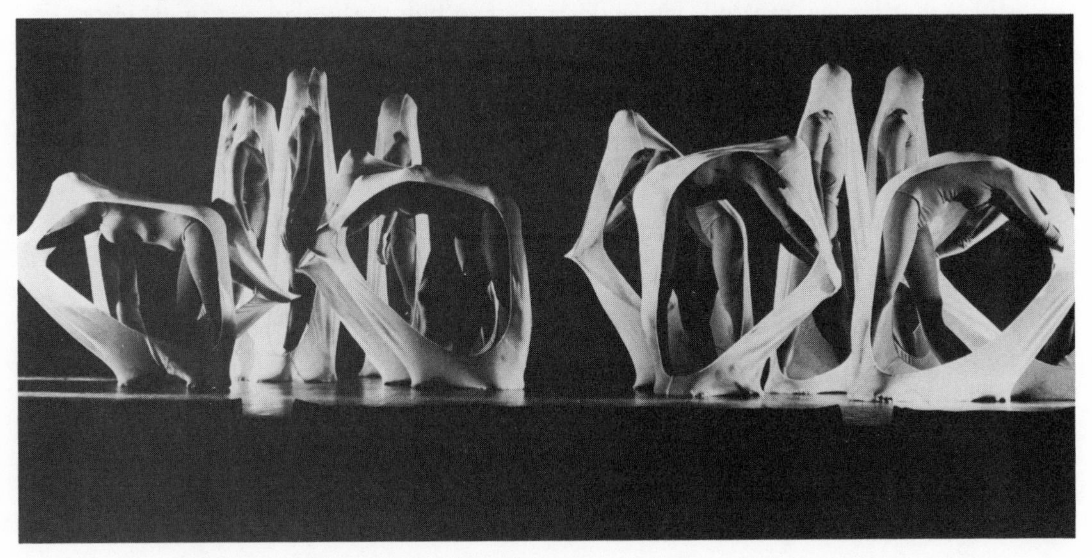

dancer for many years. Louis's uncanny deftness of coordination, his control of dynamics created an image of carefree wit—a rather androgynous latter-day imp delighting in the curious interplay of his own joints and muscles. In solos in which he was fully visible, his motion often resembled that of a puppet: elbow and knee might behave as if an elastic string connected them, one arm might fly up and surprise him. Over the years, the best of Nikolais's dancers have had something of this flexibility—from Phyllis Lamhut and Gladys Bailin of the Henry Street days through the lithe, crazy-limbed Carolyn Carlson to Gerald Otte, Suzanne McDermaid, Lynn Levine, and others of more recent times.

There is quite a lot of actual puppet imagery in Nikolais's work. It has a practical source: Nikolais was a puppeteer, entertaining children in Hartford, Connecticut, parks before he was a choreographer. Sometimes you see the dancers as hand puppets, molded from within into strange shapes; at other times, they are less complex stick puppets, skimming through space with invisible gliding steps that make them look as if they're on wheels, or being wafted along. They can project an image of marionettes, either as a by-product of the style, as described above, or through disguise. In *Guignol,* two dancers manipulate a third by means of strings attached to him, cheerfully cruel in their demands. The puppet image pokes sardonic fun at the notion of humans as masters of their destiny and of the natural world.

Nikolais has never danced in his own works, and his presence in the theater reinforces his persona as a puppetmaker or magician—sitting, as likely as not, in the light booth, calling cues for the lighting and sound and slides that he has created, while on the stage below the dancers swim into existence, merge with each other or the backdrop, now sentient creatures, now quivering shapes. Ironically, these dancers, highly trained, often contributors to the choreography, by the very nature of the work often appear will-less onstage—performing their fluid actions as if they were unconsciously, or instinctively, fulfilling their part of a large pattern, whether it be carnival show or mechanical organ or the life cycle of an amoeba.

It is not surprising that a group formulating its aesthetic in Paris during the late sixties and early seventies should make social criticism a potent ingredient in its disguises; on the other hand, two of the three creator-performers of Mummenschanz, Andrès Bossard and Bernie Schurch, were trained in mime (by Jacques Lecoq in Paris), and mimes have a long history of mocking human foibles and the social status quo.

Although Floriana Frassetto, who joined Bossard and Schurch in creating

Mummenschanz in 1972, was trained in dancing and acrobatics as well as in acting and pantomime, the choric dance element we see in Nikolais's compositions—a sea of disembodied legs waving up from the floor, for example—plays no part in the group's work.

The Mummenschanz aesthetic is centered around the mask. You don't see the faces of the performers until the curtain call. The mask is not a caricature of a specific character; it's a bizarre, often highly abstract creation. In the show that Mummenschanz brought to New York in 1973 and again in 1977 (for a three-year run at Broadway's Bijou Theater), the performers often *were* their heads. That is, you hardly noticed their bodies,

Mummenschanz: The performer is the mask, the mask the performer. Photograph by Christian Al Torfer.

while their masks—highly visible above plain dark leotards—were not only the center of attention, but the locus of most of the physical activity. A couple with masks made of dough gradually nibbled each other's face away. A couple with heads made of silicone putty engaged in a "dialogue" that brought them closer and closer until their heads blended into one inseparable mass. Two "gamblers" with racks of dice for heads played until one was left featureless. A funny and touching courting couple had rolls of toilet paper for features; their mouths unrolled to talk, their ears to listen, and when they wept, their eyes turned into streams of white paper tears.

The American public adored such engagingly articulated bits of wisdom.* The style of transformation made the social criticism less pungent than it would have been in agit-prop theater, but penetrating when remembered later, after the sugarcoating of performance whimsy had melted. In the *New Show,* which the trio brought to New York in 1986, the fantastic shapes onstage were less charming, more abstract, and often conspicuously headless. Interviewed by Glen Collins of *The New York Times* just before the *New Show* opened at the Joyce Theater, Bossard said, "Our culture encourages people to disconnect their heads from their bodies, to be distracted from their status as harmonious human beings. Headless, we run wherever they call us—to do, to buy, to be."

The acts in the *New Show*—predominantly colored black, white, and silver—suggested an aberrant technology spawning monsters. The basic leotard was built up to cover the performers' real heads, giving them bulky shoulders and a nub for heads to settle on or roll from. The performers, unlike classical mimes, used real objects, and they allowed these objects to display their own nature. Thus a human being could have an attaché case for a head, but he/she treated it like an attaché case, snapping it primly open and shut.

The disguises tended to be more complete in the *New Show:* a seductive shape that rippled like a sea anemone; a thrashing, phantasmagorical octopus; two dueling Mylar bags; a battleground of foam-rubber rings and cones, with dark human legs barely visible among them—like a Bauhaus construction. Creatures without heads, creatures in search of heads (why not this balloon?), creatures ignorant of their headlessness. A fellow with a long neck of fat vacuum-cleaner tubing could knot it to form the simulacrum of a head; a complex creation of silver tubing could untwist itself to stand for

---

*By 1978 the original three performers had trained new people to spell them in the show at the Bijou, and in 1979 they had three additional companies bringing the show to other cities.

Pilobolites in a classic pose (*Ciona,* 1973): (clockwise from left bottom) Moses Pendleton, Robby Barnett, Alison Chase, and Jonathan Wolken. Photograph by Lois Greenfield.

a second, becoming a stout, forbidding figure with cylindrical arms and legs and no head at all. Most frightening of all, a silver mass slowly inflated to giant size, threatening to spill off the stage and gobble the audience, finally making a meal of the suitcase-man. Seldom has a metaphor for the monster of technology and those who feed it been presented with such fairy-tale clarity.

Imagine this: two people lean toward each other in profile; two others stand on their backs and lean toward each other, also in profile. By bracing themselves with their arms and counterbalancing their weight while they slowly alter their positions, the four performers create a sequence of melting shapes—M's, a pair of eyeglasses, double arches, mandalas. The spectator's focus shifts from the created designs to the individual muscular bodies and back to the designs.

One member of Pilobolus, talking about their dance *Ciona* (1973), explained the process as one of ". . . multiplying simple forms and movements, making them bigger and more complex. For instance, we took an ordinary cartwheel and attached another person upside down. The motion itself was there, but it was no longer a single person doing a cartwheel; it was more like a coin oscillating on its edge. By joining together and confusing appendages, by making entities larger than the individual, you get a certain abstraction and at the same time an expansion of references."

Pilobolus, founded in 1971, can be thought of as a spawn of the seventies' counterculture. Neither the three students—Jonathan Wolken, Robb (later Moses) Pendleton, and Steve Johnson—who made a highly novel work, *Pilobolus,* in Alison Chase's dance-composition class at Dartmouth, nor the four who went on to build a company—Wolken, Pendleton, Robby Barnett, and Lee Harris—had formal dance training. They were athletic guys with Ivy League educations and brains to match, lively senses of fantasy, and lusty wits. They parlayed what Pendleton once called "college-days togetherness" into a highly successful dance company that was also a communal enterprise in country living, with collaborative dance-making as the chief activity.

The times and the limited dance training of the four original Pilobolites combined to produce an aesthetic as well as a creative process. Often in their complex shifting designs, with the men hanging off each other, every individual was off-balance, but together they created a stable structure. The "linkages" that were their trademark, said Pendleton in a 1976 interview, were ". . . the only way we could move together. We were afraid to move as independent movers. It was a very comfortable thing—where you never felt personally responsible." The analogy to the communes that proliferated in the late sixties and early seventies is inescapable.

Nowhere is the image of male bonding, which underlay almost all Pilobolus work until recently, clearer than in a film, *Pilobolus and Joan,* made in 1973 by Ed Emshwiller. The script, by Carol Emshwiller, asks us to accept that Pendleton, Barnett, Wolken, and Harris are a single insect who's trying to think and act like a man and understand the human world. Linked, feet on each other's shoulders, they walk down the street like a large pink caterpillar. (No one pays any attention.) Tiered, they explore a potter's studio. Sitting on one another's laps, they read a newspaper. Stacked on top of each other, they sleep beside their human sweetheart, folk singer Joan McDermott. Emshwiller's dazzling camera effects turn their dances into the splitting and rejoining of a single organism, into a slow-motion, erotic blossoming of flesh. ("To accept someone is to accept their transformations," says the foggily brainy narration.)

It was Murray Louis who "discovered" Pilobolus at a college dance festival, Louis and Alwin Nikolais who made it possible for the group to make its public debut at The Space in New York City, where Louis and Nikolais rehearsed their own companies and had offices. Their enthusiasm is understandable. The antic group designs, the lyrical, slow-motion Rorschachs and cantilevering play of opposing forces in early Pilobolus dances like *Pilobolus* and *Ocellus* have a lot in common with some of Nikolais's own illusion-games with limbs and heads and torsos. However, unlike Nikolais or any of the disguise artists previously discussed, Pilobolus only occasionally transforms the body by covering it; the performers' bodies are highly visible, sometimes naked, or nearly so.

The frankness about sexuality that had bloomed in the sixties, the healthy appreciation of the body (long overdue in American culture) seem crucial to the Pilobolus repertory. The ingenious ways in which the dancers manipulate their own and each other's bodies force you to consider those bodies closely. Used as props (Pendleton said in a 1975 interview, "You know, use a person instead of a bonnet or a sausage.") they become in the process more visible as bodies. A woman who is sitting on a man's head as a living helmet certainly makes you ponder her as a woman—both in terms of her structure and the meaning of her act.

At first, the advent of two trained women dancers in 1973 made little difference. In *Ciona,* Martha Clarke and Alison Chase, the men's former teacher, simply became lighter-weight components of Pilobolus's mandalas, whirligigs, interlocking structures. But gradually the nature of the transformations altered. Works like *Monkshood's Farewell* (1974) and *Untitled* (1975) exploited the inherent strangeness of the style and its natural suitability for suggesting the aberrant and the cruel as well as the beautiful and the funny. The ingenious acrobatics and group shapes acquired depth and resonance.*

Monkshood, a suite of medieval fantasies out of some bizarre bestiary or heraldic fantasy, does it all with bodies: a man slogs onstage, wearing two people as shoes; a birdlike creature stands on one leg and plays forlorn

*Both dances are still in the company's repertory, even though all but one (Michael Tracy) of the six who created them no longer perform with the company.

games with its other leg; three people together form horse, rider, and lance to attack a similar structure. Images of knights, monsters, priests metamorphose from six human beings.

In *Untitled,* perhaps Pilobolus's most famous work, costume and body linkage together produce a surreal tale that has both the ambiguity and the

Alison Chase and Martha Clarke in *Untitled.* Photograph by Lois Greenfield.

clarity of a dream—the kind of dream you rush to tell your analyst about. Two demurely playful ladies in straw hats and nineteenth-century gowns suddenly shoot up on long, hairy legs (they are, of course, sitting on the shoulders of two men). Like odd wading birds, they can fluff down and then rise disconcertingly, but their changes in size do not appear to discombobulate two men in summer suits with straw boaters who come to court them. Left alone, the women grow protuberant bellies and give birth to naked men—fantasy sons/lovers who dance with them dreamily. But the women can pull their skirts off a man and leave him lying on his back, knees bent, head held uncomfortably just off the floor—like a fly wrapped in cold storage waiting for a spider's meal.

With Philobolus, the erotic or suggestive is never far away. The shapes—beautiful or grotesque or whimsical or enigmatic—may be achieved by intimate contact; the heads against crotches, the bodies sliding together can emit fleeting images of idyllic group sex. A gimmick in *Molly's Not Dead:* men waddle on wearing women; the women, legs hooked over the men's shoulders, hang upside down, their buttocks under the men's chins, their backs forming the men's paunches, their heads dangling between the men's legs like scrotums. The comedy—and it *is* funny—arises not only from the kinetic humor, but from the raunchiness. When Arlene Croce referred to Pilobolus works of the 1970s as "satyr plays," it was, in part, this gleeful use of each other's bodies that she was referring to.

And always you get the dual image of a bunch of personable athletes *and* the fantasies that their physical skills engender. When in *The Empty Suitor,* a large, shabby gentlemen on a park bench turns out to have a greedy little female succubus hidden in his overcoat, or a pair of pink legs protrude from a ruffled pink skirt with no head and body in sight (Alison Chase's *Lost in Fauna*), or two lovers seem to be slowly growing together (the Chase-Pendleton *Alraune* or *Shizen*), the visible interaction between illusion and reality becomes as vivid, as mentally and physically provocative for contemporary viewers as Fuller's gigantic, luminous blossoms with a woman's body at their heart were for her turn-of-the-century fans.

Virtuosity—The Ultimate Disguise?

One might think, on the face of it, that the opposite extreme of complete disguise would be the full revelation of the dancer *as dancer:* the superbly coordinated and graceful athlete with lithe, well-muscled body, performing at a peak of physical skill. But as Jacques Rivière suggested back in 1913, and as the Judson Church revolutionaries believed in the sixties, virtuosity itself can be a kind of disguise. Of Nijinsky dancing in Fokine's *Le Spectre de la Rose,* Rivière wrote:

> From head to toe, the body in some way takes on fluidity and fullness. An added elegance casually descends and rests upon it. Like a heavily made-up actor, it is no longer recognizable. The *Specter of the Rose* offers the best example for this transfiguration. Nijinsky's body literally disappears in its own dance . . . The atmosphere in which he is submerged is dynamic rather than multicolored, but he is rendered as indistinct by it as Loïe Fuller by her luminous veils.

One difference between Nijinsky in *Spectre* and Loïe Fuller, however, is the object of the disguise. Fuller (Nikolais also) raised issues about our sensory perception—about the discrepancy between what we think we see and the world as we structure it in memory and dream. The virtuoso is a shaman too, but the illusion he creates is fundamentally always the same: the ideal physical being, the body we'd all like to be. Pilobolus, using highly visible muscular technique as a component of illusion, might be said to straddle the line between this sort of disguising and magic with veils and masks.

The virtuosic dancer goes beyond the great athlete, who develops speed and coordination and strength in service of a complex of goals, and beyond the circus acrobat, who perfects variations on a single superb trick—like my childhood favorite, Unus, who, wearing a suit and a top hat, stood on one finger on top of a dizzying pile of precariously balanced objects. The dancer, like athlete and acrobat, presents speed, equilibrium, coordination, limberness, strength—human attributes that, although within the reach of almost everybody, rarely combine as excellently in life as we'd like them to. But because the dancer is not limited by the athlete's concrete purposes and the acrobat's single-minded focus, he or she more freely presents us with an ideal vision of ourselves.

The high degree of technical mastery that we call virtuosity is a tricky subject. Superb performers are always in some sense virtuosi, even when, like Isadora Duncan, they seek to look natural, untrained. On the other hand, great dancers, such as Sara Rudner of Twyla Tharp's company or Suzanne Farrell of the New York City Ballet may operate at a high level of physical daring, and yet not be perceived primarily as virtuosi; their spiritual transparency, their air of being divinely inspired colors them differently. Too, many dancers with a virtuosic command of technique can lack the indefinable quality that moves audiences deeply. We often speak of "empty virtuosity," when faulting dancers who call our attention to their feats *as feats* and admire choreographers and dancers who put their technical expertise in the service of expression or musicality. Even *fouettés,* those whipping turns on one leg that became a popular ballerina stunt in the Russia of Petipa's heyday, can be significant when beautifully performed or effectively situated in a ballet. Selma Jeanne Cohen, noting Mikhail Fokine's aversion to sequences of *fouettés,* points out that he justified their inclusion in *Bluebeard* in 1945: "As Bluebeard's sixth wife—and, she hoped, his last—[Irina] Baronova cornered Alicia Markova, who seemed headed for the status of number seven, whipping her across the stage in a series of thirty-two furious fouettés. They were not about skill, they were about jealousy."

Although it is pleasant to see difficult steps and dancerly expertise submerged into the texture of the whole work by a choreographer's subtlety and a dancer's good taste, virtuosic steps have an expressive power of their own, and they define the dancer-as-superperson whether the performance in question is a great work of art or a flash act. The extraordinary poetic resonance of Mikhail Baryshnikov's leaps certainly distinguishes them from those of the talented aspirant at a ballet competition, but their ethos is fundamentally the same.

The steps that make audiences sit enraptured, or gasp, or scream approval seem to involve the illusion that the dancer has briefly transcended some basic condition of the body and of the forces that act on it. To the nineteenth-century dance public, the ethereal hovering of sylphs had the potential to suggest spiritual flight through physical action. This notion could exist only in a Judaeo-Christian society, which is apt to identify an upward direction in space with nearness to God and the afterlife. But "lightness" in the form of leaps figures prominently in the dances of cultures that worship many gods immanent in nature, not a single power above.

Surely it's simply the momentary escape from gravity that the leap reveals which makes it potent—regardless of any spiritual significance we care to attribute to it. Gravity binds us all to the earth. Falls, and a subsequent rise, can articulate the struggle against it (as did the falls that absorbed Martha Graham and Doris Humphrey), or show us the delight of a voluntary yielding: "I choose to fall," was in effect the message of Trisha Brown's *Falling Duet,* and the daring and exuberance with which she and her partner plummeted to the floor turned them into heroes, even though their relaxed behavior minimalized the dehumanizing effect of virtuosity.

Equilibrium is something that healthy, unintoxicated humans don't de-

Mikhail Baryshnikov in Eliot Feld's *Santa Fe Saga* (1978). Photograph by Lois Greenfield.

vote much conscious thought to. When we're knocked off-balance or slip on the ice or walk a narrow ledge, then we realize how precarious a thing it is. The dancer balancing on one foot or the tip of one toe affirms a harmonious relation with the physical or emotional forces that threaten stability. The amazing supports that an experienced pair of Contact Improvisers can come up with (not conceived as virtuosic, but often unavoidably so) epitomize the risks and the intricate, constantly changing balance inherent in an intimate human relationship. Spinning—whether by the ballet dancer or the dervish or Laura Dean's contemporary company—also tests equilibrium. It might be related as well to the fact of the earth's turning. I don't mean our knowledge of this, but some deep, unconscious sense that the world is turning, that we are all turning all the time.

Cultural factors that affect dance style inevitably affect virtuosity too. Choreographers Alvin Ailey and Talley Beatty came out of modern dance and admired ballet, yet the specifically black subject matter that often interests them and the predominantly black dancers they work with have set certain parameters (which was true also for pioneers Pearl Primus and Katherine Dunham, who drew from African and Caribbean sources). In his dramatic works, or strongly emotional ones like the enduringly popular *Revelations,* Ailey has created black archetypes that answer to the Siegfrieds and Odettes of ballet or the "Greek" heroines of Martha Graham: the desperate man in flight (from slavery, from drugs); the suave, sharp-footed dude, who can turn-on-a-dime, who's never at a loss; the laborer; the prisoner (whore or jailed man or sinner) yearning upward for freedom, for release; the provocative and dangerous female; the damaged innocent; the luscious goddess out of Africa (like the great Judith Jamison, transfigured in the last movement of the solo *Cry*). These roles make certain steps come to the fore: high leaps and plunging falls and rolls, rapid multiple turns, legs not simply lifted but flung triumphantly high, backs deeply arched in abandon or despair, and, often, a fluent swaying in the hips as well.

There is an aspect of virtuosity that doesn't figure in disguising via fabric and mask: it tends to be progressive. Early in the nineteenth century, a dancer rose for a second onto the tips of her toes; before many decades had elapsed, women were hopping on pointe. Once a double pirouette was thought a startling feat; now dancers aim for ten. This is not to say that *dancing* gets better, only that the technical ante has been upped.

Tanaquil LeClercq, comparing her New York City Ballet colleagues of the 1950's to today's NYCB dancers, said, "Now it's like a master race." And it's not only ballet dancers whose legs swing up beside their ears,

Judith Jamison in *Cry*. Photograph by Max Waldman.

whose *arabesques* point to twelve noon. Today dancers in Martha Graham's company, in Alvin Ailey's, in Merce Cunningham's display a purely physical range of accomplishment beyond that of dancers in those same companies twenty or more years ago. The standards for this new virtuosity are determined, to a considerable extent, by ballet: height of extensions and long, clean lines have become more important than nuances in dynamics or subtleties of expression or power in the center of the body.

Lifts, used sparingly by Balanchine and the first modern dancers alike, have acquired new importance in ballet and modern dance in recent decades. They have a history as stunts, but they've also been used by choreographers to suggest the heights of elation and abandon, or, in dramatic ballets, to stand for passion. In the hands of such choreographers as Gerald Arpino of the Joffrey Ballet or John Neumeier of the Hamburg Ballet, the constant manipulation of women by men into beautiful and erotic designs acquires a new significance: eroticism *as* virtuosity. (In a time when women are increasing their power and independence, the glut of lifts glamorizes a curiously reactionary view.)

Virtuosity is, of course, conditioned by the dancerly images of the day (images that sometimes—as in modern dance of the spartan thirties and antielitist sixties, for instance—made any display of prowess completely taboo). A predatory female in nineteenth-century ballet—Odile in *Swan Lake,* say—might stun a man with her *fouettés;* she would not be likely to hurl herself on him from ten feet away and end with her legs tightening around his neck, like some contemporary femmes fatales.

During the early seventies, when Eastern mysticism and states of altered consciousness were of interest to many, Laura Dean and two colleagues spun in silence and in perfect synchrony for a solid hour (*Spinning Dance*). The spinning wasn't intended as exhibitionistic virtuosity, but the somewhat adventurous audience that Dean drew responded to the women's stamina, inner equilibrium, and sensitivity to each other as if they had witnessed virtuosic display of the highest order. Now in the 1980's, the emphasis on technical range and expertise reflects society's current preoccupation with success and physical prowess.

Today, dancers concerned with innovation, as Dean was, offer alternatives to standard ballet pyrotechnics. Molissa Fenley rejects ballet technique and works out by running miles and pumping iron. Her body, sleek and trim and powerfully muscled, is not elegant in the sense that the contemporary ballet dancer's body is. She looks more like a triathlete, the aerobic ideal of health-conscious America; and in the patterns of nonstop, high-energy traveling steps that make up her dances, she creates her own form of ordeal. The lifts that stud the dances of Bill T. Jones and Arnie Zane are designed to show athletic skill, courage, and split-second timing rather than sculptural beauty. Dancers of both genders run up each other's backs, vault over each other, dive onto one another. These days, it's almost a tradition in certain postmodern circles for a dancer to come onstage looking and acting like nothing special, and then stun the audience with some brilliant move.

Real physical ordeal, even as a component of such dramatic and ostensibly nonvirtuosic dances as Kei Takei's ongoing *Light* or Pina Bausch's *Tanztheater* pieces, can be formalized into a kind of transformative virtuosity. Bausch's performers often surrender their individuality in the extremity of their struggles—becoming "runners" or "screamers" or "she who gets whipped." When they stumble out to take their curtain call, glazed and exhausted, the audience cheers wildly—as if these actor-dancers, in giving their emotional and physical all, have won some kind of Olympic event.

Toward the end of the 1960's, intelligence and rhythmic acuity also entered the domain of virtuosity. When in Twyla Tharp's brilliant *The Fugue* (1970), Tharp, Sara Rudner, and Rose Marie Wright, unaccompanied by music, hammering out a variegated and demanding set of canonic variations with boots on a miked floor, ended each sequence in perfect unison, spectators were invariably awed. Trisha Brown's dancers could go forward or backward through three separate dance sequences in *Line Up* (1977), or shunt from one to another on the verbal commands of a "caller" who, in a virtuosic display of her own, brought them all into unison at the end. At a concert by Lucinda Childs's dance company in Leuven, Belgium, in 1985, something went wrong with the computerized power supply, halting the performance in the middle of a highly repetitive, patterned, nonstop score by Philip Glass and a highly repetitive, patterned, nonstop dance by Childs. The dancers waited calmly, saying little to each other, while the stage manager and crew fiddled with switches and dials. When the music came back on, the dancers instantly began to dance. To the thrilled audience, for whom the dance and music had few landmarks, it seemed as if these people had memories and instincts the rest of us had yet to evolve.

Whether current virtuosic dancing presents the dancers as superathletes or sleek projectiles hurtling through the air or quick-witted components of a complicated pattern, much of it seems to draw its imagery from modern technology—from rockets and missiles and computers. Skip Blumberg, making a video treatment for the series *Alive from Off-Center* of Charles Moulton's *Nine-Person Ball Passing* (just that—nine people in three tiers, nine balls), kept multiplying the image, getting it smaller and smaller until what at first had been a bunch of charming individuals playing a neat rhythmic game became a symmetrical pattern of winking dots. It was an acknowledgment both beguiling and chilling of our fascination with and our uneasiness about computers—feelings analogous to the ambivalence about machines that colored the first part of the century.

Martha Graham has likened superb dancers to the medieval *athletae dei,* "acrobats of god," who sought transfiguration by purifying the flesh of all laziness, all excess, as if they hoped that in some white-hot moment, flesh might almost become spirit. Today, virtuosity as valor emphasizes the power of spirit and will with a different accent. Not only does it sometimes propose the human as alternative to, and rival of, the computer, it can suggest a hope that human virtues will survive in a universe at the edge of destruction. The dancers in the "Golden Section" of Twyla Tharp's *The Catherine Wheel* or Trisha Brown's *Set and Reset* do not show you stunts or feats, even as they are executing the most amazing feats you're likely to have seen. Their lack of theatrical pretense, their variegated timing and attack, the complexity of their activity make their transformation by skill and daring all the more stirring. They aren't rendered "indistinct" as humans by their dancing—as Rivière thought Nijinsky was: they stand out with a hyperclarity, as if a nimbus of light surrounded their bodies.

Christine Uchida, John Carrafa, Shelley Washington, and William Whitener in Twyla Tharp's *Baker's Dozen.* Photograph by Lois Greenfield.

Notes

The material for this book has been drawn not only from books, articles, and live performances that I have seen, but from letters, films, photographs, programs, route sheets, diary entries, notes for dances, and interviews. Most of these sources are housed in the Dance Collection of the Library and Museum of the Performing Arts at Lincoln Center (a branch of the New York Public Library). References to NYPL in the Notes refer to the Dance Collection unless it is stipulated otherwise.

The principal collections that I have studied are the Denishawn Scrapbooks, the Doris Humphrey Collection, the Doris Humphrey Letters, the Loïe Fuller Collections, the Irma Duncan Collection, the Louis Horst Scrapbooks, the Humphrey-Weidman Scrapbooks, the Hanya Holm Scrapbooks.

The Library's Jerome Robbins Archive has supplied me with visual documentation of dancers and dances. These sources range from film clips dating from the first half of the twentieth century to videotapes made in the last twenty-five years.

In addition, I have made use of tapes and transcripts in the Library's Oral History Archives, of the Bennington School of Dance Oral History Tapes in the Oral History Archives at Columbia University, of material in the Regional Oral History Office at the Bancroft Library, University of California at Berkeley (see Mitchell and Riess in the Sources), and of tapes kindly lent me by Selma Jeanne Cohen.

The principal interviewees on the tapes were: Ethel Butler, Jane Dudley, Don Oscar Becque, Raymond Duncan, and Isamu Noguchi (NYPL); Claudia Moore Read, William Bales, Helen Priest Rogers, Dorothy Bird, and Otto Luening (Bennington tapes); Sulgwynn and Charles Quitzow (Bancroft Library); Bonnie Bird, Jane Dudley, Robert Cohan, Mary Hinkson, and Bertram Ross (Cohen tapes).

The NYPL Dance Collection's iconography and clipping files for choreographers and companies are a treasure trove for the researcher. I have, in addition, consulted the following major periodicals: *Dance Magazine* (formerly *American Dancer*),

Dance Index, The Dancing Times, The Dance Magazine, Dance Observer, Dance News, Dance Chronicle, Dance Scope, Ballet Review, The Drama Review, The Director, Dance and Dancers, Ballet, Eddy, Ballet News, The New York Times, and *The Village Voice.*

Additional material has been found in the New York Public Library's Theater Collection and in the Bender Library at Mills College, Oakland, California.

The books referred to in the Notes, as well as other sources, are listed in full in the Sources.

FLESH AND SPIRIT: THE UNEASY BALANCE

In Pursuit of the Sylph

30 "Qui me visitent . . . ," "A Trilby, le lutin d'Argail," Hugo, 118.

32 Several scholars . . . , notably Chapman in "An Unromantic View of Nineteenth-Century Romanticism" and Aschengreen.

"Dancing is little adapted . . . ," Gautier, "Fanny Elssler in *La Tempête," La Presse,* 9/11/37, in *The Romantic Ballet as Seen by Théophile Gautier,* 17.

". . . a perspective upon . . . ," cited in Rosen and Zerner, 17.

33 "Is Europe saying . . . ," *Petit courrier des dames, journal des modes,* January 1839, 31.

"THE FIRST NIGHT OF THE REAL WATER!," playbill for the Sadler's Wells Theatre, week beginning August 23, 1841.

"And the rising moon . . . ," Gautier, *Les Beautés de l'Opera,* Book 1, 15.

"enchantresses of German ballads," *Petit courrier des dames,* July 1839.

. . . for remarking that Giselle died . . . , Chapman, "An Unromantic View of Nineteenth-Century Romanticism," 37.

34 "dance their mazy fascinations . . . ," Heath, 80.

35 "intimacy, spirituality . . . ," cited in Rosen and Zerner, 22.

". . . painted the foreground . . . ," Honour, 78.

37 "Si tu me suis . . . ," "La Fée et la Péri," Hugo, 196.

"By the hundred they rise . . . ," Aschengreen, 15.

". . . drop in like flakes . . . ," cited in Guest, *The Romantic Ballet in Paris,* 112.

38 As historian Marian Hannah Winter has remarked . . . , Winter, 3.

When Taglioni saw . . . , Vaillat, 85–86.

39 "so ethereal . . . ," *The Times,* 3/13/43, cited in Guest, *Jules Perrot: Master of the Romantic Ballet,* 92.

40 "Eoline is only the envelope . . . ," *Le Journal de Saint-Petersbourg,* 11/23/58, cited in Guest, ibid., 311.

"the aerial Perrot . . . ," Gautier, "Perrot and Carlotta Grisi in *Le Zingaro," La Presse,* 3/2/40, cited in Guest, ibid., 57.

"dancing excels . . . ," *Morning Post,* 3/4/43, cited in Cohen, "The English Critic and the Romantic Ballet," 87.

"vanish at the edges," Rosen and Zerner, 79.

42 "She is supported . . . ," *The Times,* 9/30/43, cited in Cohen, op. cit., 85.
"with her tall sister . . . ," Bournonville, *My Theater Life,* 49.
". . . too heavy, too stupid . . . ," Gautier, *Les Beautés de l'Opéra,* Book 1, 17.
"Laughing terribly . . . ," Heine, *Les Nuits florentines,* cited in Guest, *Jules Perrot: Master of the Romantic Ballet,* 67.
"Fear to fix . . . ," Chasles, *Les Beautés de l'Opéra,* Book 7, 4–5.

43 "C'était une question . . . ," *Petit courrier des dames,* December 1839, 207.
"supple and sinewy foot . . . ," *Le Théâtre,* 8/11/56, cited in Beaumont, *The Complete Book of Ballets,* 225.
". . . hops on the tip of the toe . . . ," Gautier, "Paquita," *La Presse,* 4/6/46, cited in Guest, *The Romantic Ballet in Paris,* 254.
Adice's syllabus . . . , *Théorie de la gymnastique de la danse théâtrale,* excerpted in Cohen, *Dance as a Theatre Art,* 71–77.
Bournonville dancers hoisted . . . Allan Fridericia, "The Bournonville School," in Mørk, 97.

44 "knows how to soften . . . ," Gautier, review of *Orfa,* in *La Presse,* 1/11/55, cited in Chapman, op. cit., 33.
A 1984 exhibit . . . , *Fanny Elssler: Materialen.*

46 Benjamin Lumley . . . , cited in Guest, "Dandies and Dancers," 46.
"Spanish shape," Second, 188.
"Well, Fanny . . . ," Grote, cited in Guest, *The Romantic Ballet in England,* 67.

47 "two bony triangles . . . ," Gautier, "Fanny Elssler," *La Presse,* 10/19/37, in *The Romantic Ballet as Seen by Théophile Gautier,* 21.
"It was . . . ," N. P. Willis, *Famous Persons and Famous Places,* 1854, cited in Beaumont, op. cit., 112.

Heroism in the Harem

50 "Europe's collective daydream . . . ," V. G. Kiernan, *Lords of Humankind,* cited in Said, 52.
"Never . . . have so many intellects . . . ," Hugo, 210.
"The more superficial . . . ," Schwab, 412.

51 "For me . . . a lotus . . . ," de Nerval, cited in Said, 101.
Byron's epic poem . . . , Byron (Introduction to *Le Corsaire*).
"What matter if . . . ," Howard Mumford Jones, cited in Sticklor, 25.
"Le sérail . . . ," "Les Têtes de Sérail," Hugo, 243.

52 "It is the fashion now . . . ," Gautier, *Mademoiselle de Maupin,* Preface, VII.

54 "It is not easy . . . ," Gautier, "La Gipsy," 2/4/39, in *The Romantic Ballet as Seen by Théophile Gautier,* 29.
"In the interior . . ." Gautier, letter to de Nerval, 7/25/43, in ibid., 62.
". . . a long psychological study . . . ," Binney, 331.

55 "The fierce, expressive eyes . . . ," *Il Caricaturista,* 1/22/82, cited in Guest, *The Divine Virginia,* 34.

57 "The captain orders . . . ," Beaumont, *The Complete Book of Ballets,* 76.

59 "Yet are Spain's maids . . . ," Byron, *Childe Harold,* I, 57. Both citation and anecdote in Borst, 28.

"This son of Hermes . . . ," Gautier, *Mademoiselle de Maupin,* 43–44.

60 ". . . I looked at her again . . . ," Guest, *The Divine Virginia,* 50.

61 ". . . Never did foot . . . ," Gautier, "Paquita," *La Presse,* 4/6/46, in *The Romantic Ballet as Seen by Théophile Gautier,* 77.

62 If recent reconstructions . . . , notably by Ann Hutchinson Guest, from Friedrich Zorn's notation (filmed in 1980 with Margaret Barbieri performing).

"What fire! . . . ," Gautier, *Les Beautés de l'Opéra,* (trans., Guest), cited in Guest, *The Romantic Ballet in Paris,* 152.

"Those contortions . . . ," de Boigne, *Petits mémoires de l'Opéra,* 1857, cited in Guest, *The Romantic Ballet in Paris,* 152.

64 "The pink scarves . . . ," Guest, ibid., 103.

Hermes, The Shell . . . , names as given by Guest, *Jules Perrot: Master of the Romantic Ballet,* 168.

65 "our imagination . . . ," Gautier, cited in Binney, 129.

THE SEARCH FOR MOTION

69 "It is far back . . . ," Mary Fanton Roberts, cited in Duncan, *My Life,* 222.

70 ". . . still trying to give . . . ," George Bernard Shaw, *World,* 1/24/94, in *Music in London, 1890–1894,* Vol. III, 145.

72 "she threw a ball . . . ," Rambert, 35.

73 "could leap to the ceiling . . . ," Henry Boynton, quoted by Sulgwynn Quitzow, in Mitchell and Riess, Vol. 1, 88.

"We watched her as if . . . ," Yeats, *Letters to His Son, W. B. Yeats and Others (1869–1922),* cited in Steegmuller, 396.

". . . her performance . . . ," Nijinsky, quoted in Nijinska, 224.

". . . as there are probably . . . ," Van Vechten, 20.

"The little figure . . . ," Sibmacher Zynen, *Algemeen Handelsblad,* 11/1/07, cited in Loewenthal, 250.

74 "See the centuried mist . . . ," Joseph Duncan, "Intaglio: Lines on a Beautiful Antique," in Harte, 140–141.

76 ". . . even in nature . . . ," Duncan, "Movement is Life," (ca. 1908) in *The Art of the Dance,* 79.

"Dear . . . it's rather discouraging . . . ," Duncan, letter to Edward Gordon Craig, Warsaw, December 1906 (No. 111, Craig-Duncan Collection, NYPL), in Steegmuller, 170.

"I had found my dance . . . ," Duncan, "The Parthenon" (1903 or 1904), in *The Art of the Dance,* 65.

78 "I go each morning . . . ," Duncan, letter to Edward Gordon Craig, St. Petersburg, 1908 (No. 216, Craig-Duncan Collection, NYPL), in Steegmuller, 284.

"Get your slouching John . . . ," Chadwick, Introduction (unpaginated).

"Delsarte, the master . . . ," Duncan, interviewed in the *New York Herald Sun,* 2/20/98, reprinted in *The Director,* Vol. I, No. 4, March 1898.

80 "Swedenborg geometrized," Professor Monroe, cited in Stebbins, *Delsarte System of Expression,* 114.

"quack religion," Shaw, "A Musical Farce," *The Saturday Review,* 1/9/97, in *Our Theatres in the Nineties,* 13.

"(a) Assume attitude of explosion . . . ," Stebbins, op. cit., 459.

81 "no art of Delsartean formulas . . . ," Mason, *San Francisco Chronicle,* 1/4/18, cited in Rather, 77.

"Strength at the centre . . . ," Chadwick, Introduction (unpaginated).

"I use my body . . . ," Duncan interview, ca. 1922–23, cited in Irma Duncan and MacDougall, 168.

82 "scattering seeds as she goes . . . ," *The Director,* Vol. I, No. 4, March 1898.

"Dressed in a simple white frock . . . ," "Isadora Duncan," *The Lady,* 8/2/1900 (signed C.M.).

84 "I see America dancing . . . ," essay of that title (1927), in Duncan, *The Art of the Dance,* 49.

86 "a state of mind," St. Denis, cited in Terry, "The Legacy of Isadora Duncan and Ruth St. Denis," *Dance Perspectives 5,* Winter 1960, 26.

Gilbert Murray marveled . . . , Murray, introductory essay to Vol. III (Euripides) of *The Athenian Drama for English Readers* (Irma Duncan Collection, NYPL), xxi.

"Greek sensuousness, therefore . . . ," Pater, *The Renaissance,* 149.

87 "It was with a gesture . . . ," Alexander Rumnev, quoted in Roslaveva, "Prechistenka 20: The Isadora Duncan School in Moscow," 11.

"Everything was touched . . . ," *Nieuwe Rotterdamer Courant,* 4/13/03, cited (and translated) in Loewenthal, 252.

"the seat of the appetites," Stebbins, op. cit., 207.

88 "Oh, they like a colt . . . ," Euripides, *The Bacchae,* in *The Athenian Drama for English Readers,* Vol. III, 86.

89 ". . . the scene, together with . . . ," Nietzsche, 217.

". . . When I have danced . . . ," Duncan, fragments collected as "The Dance of the Greeks," in *The Art of the Dance,* 96.

". . . bound endlessly . . . ," Duncan, "Depth," in ibid., 99.

90 ". . . the factory walls crumbled . . . ," W. R. Tiverton, quoted in Flitch, 110.

"The true dance must be . . . ," Duncan, notebooks 1900–1903 (Folder 141, Irma Duncan Collection, NYPL).

"rays and vibrations," "crater of motor power," Duncan, *My Life,* 75.

"She had a wonderful way . . . ," Ashton, quoted in Vaughan, 5.

91 "gift of Isis" or "child of Isis," Duncan, letter to Edward Gordon Craig, Brussels, March 1905 [?] (No. 40 in Craig-Duncan Collection, NYPL), in Steegmuller, 91.

"Osiris, the all-powerful . . . ," Stebbins, op. cit., 329.

"Waves—love waves . . . ," Duncan, notes in Craig's sketchbook (No. 23 in Craig-Duncan Collection, NYPL), in Steegmuller, 56.

95 "I would make him the rival . . . ," Rousseau, 101.

"Let him learn . . . ," ibid., 24.

97 ". . . too obviously . . . ," H. T. Parker, "Isadora Incontinent," *Boston Evening Transcript,* 10/23/22, in Holmes, 72.

"schools of life," Duncan, fragment in *The Art of the Dance,* 141 (where it appears as "a school . . .").

"They appear . . . ," Austin Harrison, *Observer,* 7/12/08, cited in Bardsley, 241.

99 In 1979 Therese . . . said, Maria-Theresa, quoted in Jowitt, "Maria-Theresa, Isadora's Last Dancing Daughter," *The Village Voice,* 2/26/79. In Jowitt, *The Dance in Mind,* 133–138 (138).

"child of nature," Claire de Morinni, "'Loïe Fuller—the Fairy of Light," in Magriel, 215.

"the first to try it . . . ," Craig, marginal notataion on page 112 of his private copy of Flitch, in Bender Library at Mills College.

100 "Isadora was building . . . ," Mitchell and Riess, 65.

In photos, the girls . . . , Hinman, Vol. IV, plate between 8 and 9 and frontispiece.

101 "Duncan proved . . . ," Fokine, 256.

"awful . . . unbelievable . . . ," Balanchine, quoted in Lydia Joel, "Finding Isadora—Notes from the Editor's Desk," *Dance Magazine,* June 1969, 51.

"The way she used . . . ," Ashton, cited in ibid.

SPHINXES, SLAVES AND GODDESSES

The Veils of Salome

107 ". . . I surmised . . . ," Petipa, 50.

108 "unbroken divertissement," Levinson, 40.

". . . a special proud stance . . . ," Fokine, 49.

"The bearers set down . . . ," Cocteau (and Alexandre), *The Decorative Art of Léon Bakst.* (This translation, by Adrienne Foulke, is excerpted in Kochno, 37.)

110 "With your eastern dance . . . ," O. V. Milosz, cited in Jullian, *Dreamers of Decadence,* 40.

111 "Her naked feet . . . ," *Truth,* 3/18/08, cited in Cherniavsky, "Maud Allan, Part III: Two Years of Triumph, 1908–1909," 122.

114 . . . it was Benois (or Bakst) . . . The issue of inspiration for details of productions depends on whose memoirs one reads and which the major biographers find most credible.

116 "whatever that may be . . . ," Shaw, "The Shooting Star Season," *The Saturday Review,* 7/10/97, in *Our Theatres in the Nineties,* III, 188.
André Levinson complained . . . , Levinson, 41.

118 "spirals of sound . . . ," Cocteau, in Kochno, 37.
"Orientals do not live . . . ," Fokine, 154.
"The Orient . . . ," ibid., 155.
". . . swollen with love . . . ," Samain, "Cléopâtre," in *The Oxford Book of French Poetry,* 501.

119 "To say that its story . . . ," Brussel, cited in Beaumont, *Mikhail Fokine and His Ballets,* 90.

The Veil of Isis

125 "as a flower of poison . . . ," letter from Kessler to Hofmannsthal, 11/24/06, cited in Shelton, *Divine Dancer,* 77 (trans., Susan Pustejovsky).
"intense artistic egoism," St. Denis, *An Unfinished Life,* 110.
"In front of an idol . . . ," Hofmannsthal's sketch for *Salome,* cited in Sorell, *Dance in Its Time,* 337.
"soulless splendor," Jullian, *Dreamers of Decadence,* 152.

127 ". . . stately, rich in color . . . ," "St. Denis Adds Peacock Dance," *San Francisco Bulletin,* 6/25/19.
"Passage indeed O soul . . . ," Whitman, 193, 196.

128 "I demand of the dance . . . ," St. Denis, draft of article, 1925 (Folder 216, Denishawn Collection, NYPL).
"I am Kuan Yin . . . ," St. Denis, "Kuan Yin," cited in Shelton, op. cit., 209.

130 "What can he transmit . . . ," Vivekânanda, 275.
"I send out . . . ," St. Denis, interviewed in "Ruth St. Denis, A Soul Dancer," *Kansas City Times,* 11/24/13.
It is a poster . . . The cigarette poster is in the Denishawn Collection of the NYPL and is reproduced opposite page 51 in St. Denis, *An Unfinished Life.*
"I knew that my destiny . . . ," St. Denis, *An Unfinished Life,* 52.

131 "I did not go to India . . . ," quoted in "Truths of Buddhism Taught in Dance: A New Jersey Girl's Interpretation of Hindu Mysteries," *The Sun,* 2/4/06.
"bridged the gap . . . ," ibid.

133 ". . . a Little Jacket . . . ," "Not a Hindoo, Only a Jersey Buddhist, Ruth St. Denis by Name, and She's a Dancer: Her Costume? Ah, well. She Wears a Little Jacket and a Skirt of Gauze and Maybe an Anklet or Two," *Morning Telegraph,* 1/27/06.

a variety bill that included . . . , playbill in Denishawn Scrapbooks (NYPL).

"BAREFOOTED WRIGGLER . . . ," Alice Rohe, *The World,* 3/23/06.

". . . DANCES THAT HAVE . . . ," "Alan Dale Describes, etc.," Alan Dale, *Journal,* 3/23/06.

134 a critic wrote approvingly . . . , "Ruth St. Denis in Hindoo Dances," *New York Globe,* 11/17/[09].

and another claimed . . . , Alice Martin, "Public Forced to Accept Ideals of Hindoo Faith in Ruth St. Denis's Dances," *St. Louis Republic,* 3/21/10.

136 ". . . glide around her neck . . . ," "Bringing Temple Dances from the Orient to Broadway," *New York Times,* 3/25/06.

". . . an embodiment . . . ," Ruth St. Denis, *An Unfinished Life,* 140.

"I'm not a music dancer . . . ," public interview with Walter Terry on his series at the 92nd Street YMHA, 1950–1951. Quoted in Sherman, *The Drama of Denishawn Dance,* 45.

". . . to wear the feathered . . . ," Parker, "Miss St. Denis Dances," *Boston Evening Transcript,* 2/15/16, in Holmes, 88.

". . . the light falling . . . ," Parker, "An Exotic Evening," *Boston Evening Transcript,* 12/20/10, in Holmes, 84.

137 ". . . I had to be . . . ," cited in Shelton, op. cit., 99.

". . . her dance . . . ," Shawn, *Ruth St. Denis: Pioneer and Prophet,* Vol. 1, 51.

138 "look like last year's raincoat," Charles Darnton, "Ruth St. Denis Beautiful to See in Dance Plays," *Evening World,* 3/13/13.

139 "They followed . . . ," Shelton, op. cit., 97.

"She is mistress still . . . ," Parker, "Miss St. Denis, Mr. Shawn and Their Youthful Company," *Boston Evening Transcript,* 1/18/23, in Holmes, 157.

141 "a clinging gauze . . . ," "Davis [theatre]—Vaudeville," *Post* (Pittsburgh), 1/15/17.

"a succession of . . . ," J.A.S., "Dancers a Big Surprise," *Kansas City Times,* 11/16/22.

"The curtain rose . . . ," Shawn, *One Thousand and One Night Stands,* 34–35.

142 "Dancing is a manly sport . . . ," quoted in "Manly Sport, Not Fine Art, Is Dancing—Shawn," *San Diego Union,* 7/18/13.

143 "orchidaceous," "Ruth St. Denis Back in Dance of Sea Nymph," Redfern Mason, *San Francisco Examiner,* 10/13/15.

144 "Dancing for Powerful Muscles," *Physical Culture,* July [1914?].

"How Dancing Develops a Beautiful Figure," *Denver Post,* 1913.

"Simplicity," he said . . . , Henry Christeen Warnack, *Los Angeles Times,* 1/12/15.
A 1918 brochure . . . , Louis Horst Scrapbooks, Vol. 1 (NYPL).
145 "The barriers . . . ," "A Unique Dance Theatre at Denishawn," *Graphic* (Los Angeles), 8/20/17.
147 ". . . One may well imagine . . . ," Thiess, 26ff. (trans., Linda Roesner).

MODERN MOVERS

The Created Self

151 ". . . the difficulties of eating . . . ," Helen Eager, review in the *Boston Traveller,* 12/31/35 (Ziegfeld Follies clipping files in the Theater Collection of the NYPL).
152 "Ugly girl makes . . . ," quoted in Terry, *Frontiers of Dance: The Life of Martha Graham,* 71. Terry has taken some liberties with his own 1937 interview with Fokine for the *Boston Herald* ("Coq d'Or Created in 16 Days, Says Michel Fokine on Boston Visit," 11/4/37). The original appears in his *I Was There,* 13.
"Generally speaking . . . ," Apollinaire, "The Cubist Painters" (1913), excerpted in Chipp, 222.
"By common consent . . . ," Parker, "The Dance as Never Before in Boston Eye," *Boston Evening Transcript,* 12/4/29, in Holmes, 203.
154 ". . . riveted to the floor . . . ," Selden, 97.
155 ". . . it had to be harder . . . ," Jowitt, "A Conversation with Bessie Schönberg," 46.
157 "There were no pretty bodies . . . ," Dudley and Robert Cohan, interviewed by Selma Jeanne Cohen in 1973.
"heroic and maenadic . . . ," Schlundt, *Tamiris: A Chronicle of Her Dance Career, 1927–1955,* 7.
"spirit which is deep . . . ," Blitzstein, *Modern Music,* March–April 1931.
"I had to keep . . . ," Karsavina, *Theatre Street,* 291.
158 . . . a clever fictitious interview . . . , Marcellus Schiffer, "Besuch bei Mary Wigman," *Die Weltbühne,* Vol. 22, No. 1, 1/5/26. I am grateful to Wigman scholar Susan Allene Manning for information concerning this interview.
161 "the weakling exoticism . . . ," Graham, "Seeking an American Art of the Dance," in Sayler, 257.
"The dancers become . . . ," Bridle, "Visualized Music Is Denishawan Programs," *Toronto Star,* 4/26/24.
(footnote) "takeoff on the Madonna interpretations," Tamiris reviewed, *The Dance Magazine,* April 1928.
163 "moldy green," Humphrey, letter to her parents, 2/13/28, Doris Humphrey Letters (NYPL).

"strong, free, joyous . . . ," program note, Louis Horst Scrapbooks (NYPL).

164 ". . . the form is . . . ," Kandinsky, "On the Problem of Form" (*Der Blaue Reiter,* 1912), in Chipp, 157 (trans., Kenneth Lindsay).

"Out of emotion . . . ," Graham, cited in Armitage, 97.

"I'm putting some of my ideas . . . ," Humphrey, letter to her parents, 8/8/27, Doris Humphrey Letters (NYPL).

according to John Martin . . . , a *New York Times* review, 11/26/33, excerpted in Armitage, 16, 17.

165 "barbaric crescendo of falls," John Martin, "Martha Graham, Dancer, Is Cheered," *New York Times,* 12/7/31.

167 ". . . of the dramatic implications . . . ," Holm, "The German Dance in the American Scene," in Stewart and Armitage, 84.

A program . . . , lecture-demonstration given at the Wadsworth Atheneum, Hartford, Conn., January 1937, and elsewhere during that year, Holm Scrapbooks, NYPL (reproduced in Sorell, *Hanya Holm: The Biography of an Artist,* 49).

168 "thoroughly modern dance creations," *The Dance Magazine,* October 1926.

"The Dancer of Dresden," Diane Hubert, *The Dance Magazine,* July 1927.

"before she goes over . . . ," unsigned article, *New York Times,* 2/13/28.

"the recent German methods," publicity material in "Dorsha's" clipping file (NYPL).

169 "obtrusive beauty," Kandinsky, op. cit., in Chipp, 161.

171 . . . her angular gestures . . . , publicity material in Louis Horst Scrapbooks, ca. 1931 (NYPL).

". . . unbroken in effect . . . ," Hitchcock and Johnson, 45.

"To create work . . . ," Frankl, 25.

"The things we make . . . ," Teague, "Will It Last?," *Advertising Arts,* March 1931, cited in Meikle, 137.

As painter Fernand Léger pointed out . . . , Chipp, 278–279.

175 "We must imitate . . . ," "The Futurist Dance," trans., Elizabeth Delza, *Dance Observer,* Vol. II, No. 7, October 1935, 75–76 (75).

176 "machine-like accuracy . . . ," Lillian Ray, review in *The Dance Magazine,* December 1927.

"The dance today . . . ," Graham (1934), cited in Armitage, 101.

An American dancer who studied . . . , Claudia Moore Read, Bennington School of Dance Oral History Tapes, Columbia University Library.

"moves across the stage . . . ," "Streamlined Dancing," *Rocky Mountain News,* 11/18/36, Holm Scrapbooks (NYPL).

"Five years ago . . . ," Edwin Denby, "Balanchine and Tchaikovsky; 'Ballet Imperial'; Carmen Amaya; Doris Humphrey," *Modern Music,* January–February 1943, in Denby, 105.

177 "Art is international . . . ," printed in a program for Tamiris's concert at the Little Theatre, New York, 1/29/28, in Tamiris clipping file (NYPL).

Group Spirits

178 "Now individualism must . . . ," Graham, in Sayler, 254.
fervently likened . . . , Humphrey, letter to her parents, 9/28/28, Doris Humphrey Letters (NYPL).
". . . the group can express," Humphrey, letter to her parents, 11/24/29, Doris Humphrey Letters (NYPL).

180 "The artist goes forth . . . ," program, 1937, Humphrey-Weidman Scrapbooks (NYPL).

181 It was in those terms . . . , Holm, March 1938, cited in Kriegsman, 160–161.
". . . the breakup of mass . . . ," John Martin, "A Major Work," *New York Times,* 8/22/37, cited in ibid., 162.

182 "All the major forms . . . ," Siegel, *The Shapes of Change: Images of American Dance,* 92.

184 "I want to visualize . . . ," draft of a letter to Letitia Ide (also sent to Ernestine Henoch and other prospective company members), 1929, Doris Humphrey Letters (NYPL).
"Then there is the group . . ." Humphrey to her parents, 11/24/29, Doris Humphrey Letters (NYPL).

185 Alwin Nikolais says . . . , Jowitt, "Man the Marvelous Mechanism," *The Village Voice,* 10/19/82. In Jowitt, *The Dance in Mind,* 199–203.
According to one of them . . . , King, 184.

187 Eleanor King recalls . . . , King, 22.
According to Helen Priest Rogers . . . , interview for the Bennington School of Dance Oral History Tapes (1978–1979), Columbia University Library.

188 ". . . her Spartan band . . . ," Kirstein essay in Armitage, 26.
"artistic militarism," Selden, 120.
"I still remember . . . ," Terry, quoted by Sorell, *Hanya Holm: The Biography of an Artist,* 77.

189 "MARTHA GRAHAM . . . ," ad under John Martin's dance column, *New York Times,* 11/11/34.

191 "The people . . . ," Burt Supree, "American Originals," *The Village Voice,* 8/30/83.

193 "Four abstract themes . . . ," Humphrey, "What Shall We Dance About?," *Trend: A Quarterly of the Seven Arts,* June–July–August 1932. Reprinted in full in Cohen, *Doris Humphrey: An Artist First,* 252.

195 . . . to "express" the triangle . . . , Humphrey, letter to her parents, 8/27/27, Doris Humphrey Letters (NYPL).
"form rhythms," Hambidge, 5.
"rectangle of the whirling squares," ibid., 29 and elsewhere.

196 ("It was here . . ."), Humphrey, draft article (1936), Doris Humphrey Collection (NYPL). Reprinted in full in Cohen, *Doris Humphrey: An Artist First,* 240.

THE HEROINES WITHIN

201 "If people would only . . . ," Graham, interviewed by Margaret Lloyd, *Christian Science Monitor,* 3/9/35.

202 "stark impersonality . . . ," Martin, "Graham Again," *New York Times,* 11/25/34.

"The theory of relativity . . . ," Sypher, 266.

"She has a quality . . . ," "Ted Shawn Delights with Artistic Dances," *Omaha World Herald,* 10/23/21.

203 "Dark, swiftly moving . . . ," George Beal, review of Blanche Yurka's *Electra, Boston Post,* 5/19/31.

204 ". . . a kind of fever chart . . . ," cited in Ernestine Stodelle, "Midstream, The Second Decade of Modern Dance: Martha Graham," *Dance Observer,* May 1962, 69–71 (70).

Dance is action . . . , Graham (1935), cited in Armitage, 103. The exact quote is "This is a time of action, not re-action. The dance is action, not attitude, not an interpretation."

206 "the Woman-Soul . . . ," Graham, *The Notebooks of Martha Graham,* 169.

"The interest in myth . . . ," Gottlieb, quoted in Clearwater, 23.

". . . are the eternal symbols . . . ," Rothko, quoted in ibid., 23–24.

"The man who speaks . . . ," Jung, cited in Wickes, prefatory note.

. . . a colleague recalls . . . , interview with William Bales, Bennington School of Dance Oral History Tapes, 1978–1979.

207 "The more clearly . . . ," Wickes, *The Inner World of Choice,* cited by Sears, 30.

209 "The one in red . . . ," program, Bennington Festival, 8/11/40, Louis Horst Scrapbooks (NYPL).

"one like Medea . . ." continues to be the program credit.

". . . it was not the relation . . . ," Denby, "Graham's 'El Penitente' and 'Letter to the World'; Balanchine's 'Balustrade'," *Modern Music,* March–April 1941, in Denby, 72.

"the meadow of choice . . . ," Graham, cited in Noguchi, 126.

"moment of choice . . . ," Wickes, 77–78.

"When a woman . . . ," Graham, quoted in Friedman, 30.

"the fear of the artist . . . ," Graham, in a lecture to a student composers' forum at Juilliard, March 25, 1952 (tape in NYPL).

211 Elsewhere she noted . . . , Graham, *The Notebooks of Martha Graham,* 255, 258.

". . . Where does a parallel come . . . ," ibid., 190.

"call to adventure," Campbell, 51 and elsewhere.

"willed introversion," ibid., 64.

". . . to know the Self . . . ," Wickes, 276.

212 "In these dances . . . ," Denby, " 'Appalachian Spring' and 'Herodiade' a Second Time," *New York Herald Tribune,* 5/20/45, in Denby, 318.

214 "The thing to keep . . . ," W. E. Oliver, "Graham Has New Dances," *Los Angeles Herald Express,* 3/30/46.

215 "In *The Happy Journey* . . . ," Wilder, Preface, XI.

218 "The primary point . . . ," Bowers, 17.

221 "I have come so fast . . . ," Waley, 36.

"She used the opening . . . ," Ross, quoted in Fellom, 5. Revised and expanded by Ross and Jowitt, 1987.

223 . . . he [Noguchi] once said . . . , Noguchi, 266.

225 ". . . the possible condition of ecstasy . . . ," cited in Stodelle, 172.

226 "Rest comes through change . . . ," Graham quoted by Bonnie Bird in technique classes given for Senta Driver's company, Harry, November 1981.

"There is no facility . . . ," Jane Dudley and Robert Cohan, interviewed by Selma Jeanne Cohen, 1973 (tapes in archives of Dance Perspectives Foundation).

"She wanted you . . . ," ibid.

227 ". . . there is really . . . ," Robert Horan, "The Recent Theater of Martha Graham" (1947), in Magriel, 251.

"Her compositions . . . ," John Martin, "Graham in Retrospect," *New York Times,* 5/21/44.

230 "doom eager," Graham, program note for *Deaths and Entrances,* 1943, Louis Horst Scrapbooks (NYPL).

("We're usually stiff foils . . ."), Taylor, 85.

BY THEIR STEPS YOU SHALL KNOW THEM

In the Royal Image

237 "What he wants (as usual) . . . ," cited in Garis, 17.

"musique dansante," quoted in Goldner, 13.

". . . would seem to have found . . . ," John Martin, "Balanchine," *New York Times,* 7/17/38.

238 ". . . by the time we saw . . . ," P. W. Manchester, "Symphonie Concertante." (This was one of many reviews by various British critics, titled "14 Works Presented by New York City Ballet," *Ballet,* Vol. 10, No. 2, September–October 1950, 27.)

"They are mere . . . ," Buckle, "Critics' Sabbath," *Ballet,* ibid., 5.

"We want kings . . . ," ibid., 6.

"He has made our dancers . . . ," Denby, "Ballet: The American Position," *Town & Country,* April 1947, in Denby, 511.

240 "A woman appears . . . ," Balanchine, "Balanchine Defines Dance as a Visual Art," *Musical America,* 3/5/44.
"unconscious images," Denby, "A Letter on New York City's Ballet," *Ballet,* August 1952, in Denby, 418.
241 "Forward to Petipa," Volkov, 162.
"childlike interludes . . . ," Billington, 437.
243 According to Russian ballet historian . . . , Chujoy, 49.
"children, hussars, and . . . ," cited in Wiley, "Alexandre Benois Comments on the First *Saisons Russes," The Dancing Times,* October 1980, 29.
244 "the fountain girls," Maiorano and Brooks, 143. Levien (67) calls them "les ballerines près de l'eau," and says that they simply did classical arm movements.
Petipa's notes contain . . . , Krassovskaya, 12.
His original plans . . . , Wiley, *Tchaikovsky's Ballets,* 258.
"There are everywhere . . . ," Volynsky, cited in Slonimsky, "Marius Petipa," 121.
245 "Ogresses of the waltz," Gautier, *Les Beautés de l'Opéra,* Book 1, 20.
246 He had been struck . . . , Roslaveva, *Era of the Russian Ballet,* 100.
247 "My favorite moment . . . ," Croce, "Makarova's Miracle," *The New Yorker,* 7/22/74, in *Afterimages,* 72.
We can infer from a review . . . , a Russian review of 1858, cited in Guest, *Jules Perrot: Master of the Romantic Ballet,* 314.
248 "lacked lightness," cited in Wiley, *Tchaikovsky's Ballets,* 147.
"grace, art, precision . . . ," cited in ibid., 270.
249 Virginia Zucchi was rebuked . . . , Racster, 135.
251 . . . as one had in 1879 . . . in *Souffleur,* cited in Slonimsky, "Marius Petipa," 112.

Forward to Petipa

253 "You discover . . . ," Balanchine, *By George Balanchine,* 26.
"The basic gesture of *Agon. . . ,"* Denby, "Three Sides of 'Agon'," *Evergreen Review,* Winter 1959, in Denby, 460.
"Dissonance makes us aware . . . ," Balanchine, "The Dance Element in Stravinsky's Music," from "Stravinsky in the Theater," *Dance Index,* 1947, excerpted in Goldner, 25.
255 ". . . the most interesting . . . ," Balanchine, *By George Balanchine,* 10.
"There is that love of bigness . . . ," Balanchine," quoted by Claudia Cassidy, "On the Aisle," *Chicago Journal of Commerce,* 8/17/35.
"To dance his ballets . . . ," Bonnefous, interviewed in Newman, 342.
"cold, luminous . . . ," Balanchine, "The American Dancer," *Dance News,* April 1944.
256 (Petipa, too, it is said . . .), Guest, *The Divine Virginia,* 76.
"Gisellititis," Bentley, *Winter Season: A Dancer's Journal,* 61.

"It should look maximum . . . ," LeClercq, interviewed in Newman, 153.
"From 16–24 bars . . . ," " 'La Belle au Bois Dormant' ('The Sleeping Princess'): Marius Petipa's Original Programme for this Famous ballet (Act 1)," trans., Joan Lawson, *The Dancing Times,* January 1943, 168–170 (169).
259 ". . . time that shows us . . . ," Balanchine, quoted in Goldner, 14.
260 Balanchine remarked . . . , Balanchine in conversation with Walter Terry on Terry's "Open Interview" series at the 92nd Street YMHA, 11/5/50.
"Evocations of the great Petipa role . . . ," Croce, "From a Far Country," *The New Yorker,* 5/22/78, in Croce, *Going to the Dance,* 89.
". . . Aurora rewritten in lightning," ibid.
261 "space-age wilis," Goldner, 75.
263 "The dramatic elements . . . ," "Balanchine Defines Dance as a Visual Art."
264 "You make yourself a Balanchine ballerina . . . ," Hayden, interviewed in Tracy, 112.
265 "pinheads," Blackmur, "The Swan in Zurich," *A Primer of Ignorance* (1956), in Copeland and Cohen, 357.
Balanchine had said . . . , Balanchine, *By George Balanchine,* 17.
"big girls with long legs . . . ," quoted in Maiorano and Brooks, 71.
As Balanchine has commented . . . The story of the Barnes review and Balanchine's response are in Maiorano and Brooks, 163.
He told Lincoln Kirstein . . . , Taper, 152.
266 "Which leg . . ." This is not a specific quote. Variants of it appear in many memoirs and accounts. See Ashley, 161, for example.
Mary Ellen Moylan . . . says . . . , Moylan, interviewed in Tracy, 86.
"rubber orchid," LeClercq, interviewed in Newman, 156.
As Arlene Croce has pointed out . . . , "The Children of Sixty-sixth Street," *The New Yorker,* 6/5/78, in *Going to the Dance,* 93–94.
268 . . . in an interview with Simon Volkov . . . , Volkov, 152, 153.
"These days . . . ," Verdy, interviewed in Tracy, 137.
. . . according to a recent study . . . , Vincent, 85–92.
269 "clean emotion angelically . . . ," Balanchine, "The American Dancer."
270 "The ballet is . . . a woman . . . ," Balanchine, *By George Balanchine,* 16.
"You are ladies . . . ," Balanchine, quoted in Maiorano and Brooks, 119.
". . . the manipulated ballerina . . . ," Daly, 14.
"In his pattern of possession . . . ," Pierpont, 358.
271 "I didn't think it was demeaning . . . ," Martins, interviewed in Newman, 358.
272 "At the end . . . ," Goldner, 80.
273 "pig snouting truffles," Stravinsky, quoted in Goldner, 9.
"as if you're asking . . . ," Martins, *Far from Denmark,* 61.
"The self is defined . . . ," Bentley, "Balanchine and the Kirov," 40.
"echoless technique," Blackmur, in Copeland and Cohen, 358.

". . . selfless bravura . . . ," Crisp, Foreword to Ashley, xi.

". . . Grace has . . . ," Kleist, "Puppet Theatre" (trans., Arthur Waley), in Copeland and Cohen, 184. (The emphasis is mine.)

274 "ballerinas are kin . . . ," Kirstein, *Portrait of Mr. B.,* 23.

"a clear if complex blending . . . ," Kirstein, "What Ballet Is About; An American Glossary," 6–7.

ILLUSION OF CHOICE—ACCEPTANCE OF CHANCE

278 ". . . being avant-garde . . . ," Louis Horst, "Three Concerts are Hailed," *New London Evening Day,* 8/18/58. (During summers at the American Dance Festival, Horst occasionally served as guest critic for the local paper.)

". . . a little sparse . . . ," Hering, "American Dance Festival: #11," *Dance Magazine,* October 1958, 33–34 (34).

"sensuously lovely," ibid., 33.

279 ". . . Like the passing of birds . . . ," Cunningham, "Summerspace Story," *Dance Magazine,* June 1966, 52–54 (52).

"They are shapes . . . ," Hering, op. cit., 33.

280 ". . . five minutes of watching . . . ," Terry, "Four Solo Recitals Here Called Indicative of Future of Dance," *New York Herald Tribune.* In Terry, *I Was There: Selected Dance Reviews and Articles—1936–1976,* 168.

"His build resembles . . . ," Denby, "Merce Cunningham," *New York Herald Tribune,* 4/6/44, in Denby, 207.

281 "Merce . . . is dedicated . . . ," Dunn, interviewed by Jowitt, 1976.

282 He confesses . . . , Cunningham, interviewed on "Merce Cunningham," *South Bank Show* (London Weekend Television), August 1979, filmed by Geoff Dunlop.

"The function of the artist . . . ," Coomaraswamy, cited in Cage, 194.

"I think of dance . . . ," Cunningham interviewed by Lesschaeve, *Merce Cunningham: The Dancer and the Dance,* 84.

283 "Let everything be allowed . . . ," Chuang Tzu, cited in Capra, 105.

"Art . . . must be part of life . . . ," Tompkins, *Off the Wall,* 274.

"Buddha . . . reduces substances . . . ," S. Radhakrishnan, cited in Capra, 177.

"The world thus appears . . . ," Heisenberg, *Physics and Philosophy,* cited in Capra, 125.

A former Cunningham dancer remembers . . . , Carolyn Brown, in "Time to Walk in Space," *Dance Perspectives 34,* Summer 1968, 38.

284 ". . . can only be compared . . . ," Jeans, 176–177.

"unbroken wholeness," Zukav, 309.

"chaos has come again," Cunningham, "Two Questions and Five Dances," in "Time to Walk in Space," op. cit.

286 "The feeling I have . . . ," Cunningham, "The Impermanent Art," 71.

287 ". . . their bursts of dynamic force . . . ," Terry, "A Young Dancer Lost in the Maze of Choreographic Experimentation," *New York Herald Tribune,* 2/4/51.

"a dancer may be . . . ," Frances A. Klein, "Cunningham's Dance Company Shows Artistry," *St. Louis Globe-Democrat,* 11/10/56.

"privileged moments," Cunningham, interviewed by Lesschaeve, *Merce Cunningham: Le Danseur et la danse,* 193. (The French edition of this book, published in 1980, differs in parts from the English version published five years later. These notes include references to both.)

289 "He didn't have his head . . . ," Johnston, "Time Tunnel," *The Village Voice,* 2/8/68, in Johnston, 126.

"At this moment I read . . . ," Lesschaeve, *Merce Cunningham: Le Danseur et la danse,* 16.

"Being and fading . . . ," W. Thirring, "Urbausteine der Materie," cited in Capra, 208.

290 It is on record . . . , Cunningham, *Changes: Notes on Choreography* (unpaginated).

291 Cunningham has spoken . . . , Lesschaeve, *Merce Cunningham: The Dancer and the Dance,* 140.

292 "Experience of our age . . . ," cited in Bernice Rose, 12.

He has remembered . . . , Lesschaeve, *Merce Cunningham: The Dancer and the Dance,* 34.

293 "Enlightenment . . . consists merely . . . ," Capra, 111.

294 "an image of the same thing," Cunningham, interviewed on *Place, Scramble, How to Pass, Kick, Fall and Run,* WNED/TV, Buffalo, 1968, filmed by Roger Englander.

"Perfection is something . . . ," Cunningham, quoted by Stephanie Jordan, "Cunningham and Cage at the Laban Centre," *Dancing Times,* Vol. LXXXI, No. 841, October 1980, 38–39 (39).

"nature puppet . . . ," Cunningham, "The Impermanent Art," 73.

"cultivate nonexpressiveness," Lesschaeve, *Merce Cunningham: Le Danseur et la danse,* 129.

"So if you really dance . . . ," Cunningham, "The Impermanent Art," 73.

297 "a little community," Lesschaeve, *Merce Cunningham: The Dancer and the Dance,* 14.

"a dramatic dance . . . ," ibid., 120.

299 "an intimate connection," Lesschaeve, ibid., 107.

"What's this exotic fall . . . ," Neels, in *498 Third Avenue,* Norddeutscher Rundfunk, Hamburg, 1967, filmed by Klaus Wildenhahn.

300 Fixing a tricky spot . . . This passage is drawn from "Merce Cunningham," *South Bank Show* (London Weekend Television), August 1979, filmed by Geoff Dunlop.

302 "It's much like riding . . . ," Capra, 111.
"Dancing is of divine origin . . . ," Cunningham, in "The Non-Objective Choreographers' Symposium: Four Dancers Speak and Dance," *Dance Magazine,* November 1957. (This is a transcript of a symposium led by David Vaughan, 4/27/57, at the Henry Street Playhouse.)

EVERYDAY BODIES

305 . . . revival of *Flat* . . . The account of the dance is based on a performance videotaped by Michael Rowe for the Bennington College Judson Project. See *Judson Dance Theater: 1962–1966.*
309 ". . . it's like bubble gum . . . ," Tharp, interviewed in "Twyla Tharp: Questions and Answers," *Ballet Review,* Vol. 4, No. 1, 1971, 47.
310 "screaming fit downstage right . . . ," Rainer, *Work 1961–73,* 286.
"I envisioned myself . . . ," ibid., 77.
"Their eyes . . . ," Schneemann, 18.
312 "Does nothing . . . ," Hering, "Seven New Dances by Paul Taylor," *Dance Magazine,* December 1957, 84.
313 "If it was an explicit assault . . . ," Rainer, in *Judson Dance Theater: 1962–1966,* 54.
314 "It is often necessary . . . ," Young, 79.
315 "violin, drum, trombone . . . ," Schwitters, cited in Kostelanetz, *The Theater of Mixed Means,* 12.
316 "Bring a bale of hay . . . ," Young, in *An Anthology* (eds., Young and Jackson Mac Low, 1970), included in Haskell, 54.
"When I began performing . . . ," Gordon, 43.
"I *would not* hold . . . ," Forti, 34.
"We must have been . . . ," ibid., 29.
317 . . . Remy Charlip recalled . . . Banes, *Democracy's Body: Judson Dance Theater 1962–1964,* 33.
321 "enslaving sounds," La Monte Young, 81.
". . . give a body definition . . . ," Morris, "Notes on Dance," 179.
"unenhanced physicality," statement in program for *The Mind Is a Muscle* performances at the Anderson Theater, April 11, 14, 15, 1968, reprinted in Rainer, *Work 1961–73,* 71.
322 "There's a kinaesthetic reality . . . , "Brown, interviewed by Jowitt, May 1982.
". . . wouldn't just stand around . . . ," Rauschenberg, in videotaped interview with Trisha Brown and Alex Hay (interviewer, Sally Banes; camera, Meg Cottam/Amanda Degener), Bennington College Judson Project. (See *Judson Dance Theater 1962–1966* in Sources.)

323 "not buttering it up," Brown, interviewed by Jowitt, May 1982.
324 ". . . as diverse as . . . ," Rainer, unpublished notebooks, cited in Banes, "Icon and Image in New Dance," *New Dance USA,* 13.
". . . thirty two any old . . . ," Johnston, "Paxton's People," *The Village Voice,* 4/14/68, in Johnston, *Marmalade Me,* 137. (The line "There is a way of looking at things that renders them performance" is a statement of Paxton's which Johnston introduced early in the review.)
325 "on tiptoe . . . ," Brown, interviewed by Jowitt, May 1982.
326 "to act without forcing . . . ," Huang, 2.
327 ". . . pitched forward with such speed . . . ," Johnston," Judson Collaboration," *The Village Voice,* 11/28/63.
328 "expressive in the broadest sense . . . ," Carroll, "Post-Modern Dance and Expression," in Fanchon and Myers, 101.
". . . did not have much physical magnetism," Hering, "The Engineers Had All the Fun," *Dance Magazine,* December 1966, 38. (Review of Hay's group piece, *Solo,* in Nine Evenings of Theatre and Engineering, October 13–23, 1966.)
330 ". . . a new kind of subjectivity . . . ," Barbara Haskell, 83.
"stand-ins for personae," Rainer, letter to the author, 4/13/73, excerpted in Jowitt, *Dance Beat,* 150.
333 as Paxton discovered . . . , *Judson Dance Theater,* 50.
336 "a grown-up person . . . ," Tharp, program for *Dancing in the Streets of London and Paris, Continued in Stockholm and Sometimes Madrid,* Metropolitan Museum of Art, 1969.

THE ALLURE OF METAMORPHOSIS

342 ". . . Human beings will always love . . . ," Schlemmer, diary entry, 7/5/26, in *The Letters and Diaries of Oskar Schlemmer,* 196.
"I have never created . . . ," Schlemmer, diary entry, 9/7/31, *ibid.,* 283.
344 "fantastic and passional vegetation . . . ," Marx, "Loïe Fuller" (title of publication missing, Loïe Fuller clipping files, NYPL), January–June 1905, 265–357 (354).
"the visual embodiment of the idea," cited in Kermode, 19.
"J'adore les indécis . . . ," Samain, "Dilection," cited in Sypher, 140.
345 ". . . an artistic intoxication . . . ," Kermode, 18.
One patent application . . . ," Loïe Fuller clipping files (NYPL).
"It's a butterfly . . . ," Fuller, 31.
346 "It is not a skirt dance . . . ," *The New York Spirit of the Times,* 1/16/92, cited in Sommer, "Loïe Fuller," 57.
". . . evoke the dream . . . ," Jules Claretie in *Le Temps,* quoted in pro-

gram for *La Tragedie de Salomé,* 1907, Loïe Fuller clipping files (NYPL).
"... half element, half human . . . ," Marx, 273.
"Distinctly visible . . . ," Hugh Morton, "Loïe Fuller and Her Strange Art," *Metropolitan Magazine* (no date in Loïe Fuller clipping files, NYPL), 277–284.

348 In an 1896 critic's scenario . . . , review cited in Sommer, "Loïe Fuller," 60.
"... exotic visions . . . ," Battersby, Introduction to Harris, 8.
"... the history of the transfiguration . . . ," Schlemmer et al., "Man and Art Figure," *Theatre of the Bauhaus,* 17.
"Life has become . . . ," Schlemmer, diary entry, September 1922, *The Letters and Diaries of Oskar Schlemmer,* 126.
("the egg shape . . ."), Schlemmer et al., "Man and Art Figure," *Theatre of the Bauhaus,* 27.

349 "the Measure of All Things," Schlemmer, diary entry, November 1922, *The Letters and Diaries of Oskar Schlemmer,* 133.
"in the past. . . ," Schlemmer, article in *Stuttgarter Neues Tagblatt,* 9/29/22, cited in *Oskar Schlemmer: The Triadic Ballet,* 24.

351 "What is this . . . ," Michel, cited in Schlemmer diary entry, May 1929, *The Letters and Diaries of Oskar Schlemmer,* 242.
"abstraction, mechanization . . . ," Schlemmer et al., "Man and Art Figure," *Theatre of the Bauhaus,* 17.

352 "... magic of transforming . . . ," Gropius, Introduction to *Theatre of the Bauhaus,* 9.
"We must be content . . . ," Schlemmer, letter to Otto Meyer, 6/14/21, *The Letters and Diaries of Oskar Schlemmer,* 107.
"with its static . . . ," A. Ho, *Thuringer Allgemeine Zeitung,* 8/18/23, cited in *Oskar Schlemmer: The Triadic Ballet,* 25.
"bright games, disguises . . . ," Schlemmer, diary entry, 7/5/26, *The Letters and Diaries of Oskar Schlemmer,* 196.

353 "fairly bold . . . ," Martin, "Bright Augury," *New York Times,* 2/26/56
"In these days . . . ," Martin, "New Life," *New York Times,* 2/19/56.

354 "I began to establish . . . ," Nikolais in Siegel, "Nik: A Documentary," 11.
"Shaman dances . . . ," Kathy Duncan, "Nikolais: What's at the Bottom of His Bag of Tricks?," *The Village Voice,* 2/28/77.

357 "symmetry, flexibility, lightness," Kleist, "Puppet Theatre" (trans., Arthur Waley), in Copeland and Cohen, 181.

360 "Our culture . . . ," Bossard, interviewed by Glenn Collins, "Mummenschanz Is Back, with a New Bag of Tricks," *New York Times,* 4/20/86.

362 "... multiplying simple forms . . . ," unidentified Pilobolite in Matson, 72.
"college-days togetherness," Pendleton, interviewed by Anna Kisselgoff, "Pilobolus Dancing Its Way to Togetherness," *New York Times,* 3/5/76.

". . . the only way we . . . ," ibid.

363 ("To accept someone . . ."), *Pilobolus and Joan,* a film by Ed Emshwiller, script by Carol Emshwiller, 1973.

"You know, use a person . . . ," Pendleton, interviewed by Linda Winer, "A Fun(gus) Troupe That'll Grow on You," *Chicago Tribune,* 1/24/75.

365 "satyr plays," Croce, "Keeping It All Together," *The New Yorker,* 3/18/85.

366 "From head to toe . . . ," Rivière, "Le Sacré du Printemps," November 1913. Reprinted in Kirstein, *Nijinsky Dancing* (in a translation by Miriam Lassman), 165.

367 "As Bluebeard's sixth wife . . . ," Cohen, *Next Week Swan Lake: Reflections on Dance and Dancers,* 74.

369 "Now it's like . . . ," LeClercq, interviewed in Newman, 149.

Sources

Adams, Henry. *The Education of Henry Adams* (1918). Boston: Houghton Mifflin Company, 1961.

Agniel, Margaret. *The Art of the Body.* New York: Harcourt, Brace & Company, 1931.

Alexandre, Arsène, and Jean Cocteau. *The Decorative Art of Leon Bakst,* trans. Harry Melvill (1913). New York: Dover Publications, 1972.

Allevy, Marie-Antoinette. *La Mise-en-scène en France dans la première moitié du dix-neuvième siècle.* Paris: E. Droz, 1938.

Anscombe, Isabelle, and Charlotte Gere. *Arts and Crafts in Britain and America.* New York: Rizzoli, 1978.

Armitage, Merle (ed.). *Martha Graham: The Early Years* (1937). New York: Da Capo Press, Inc., 1978.

Arnold, Edwin. *Poetical Works of Edwin Arnold.* New York: Thomas Y. Crowell & Company, 189–.

Aschengreen, Erik. "The Beautiful Danger: Facets of Romantic Ballet," trans. Patricia N. McAndrew. *Dance Perspectives 58,* Summer 1974.

Ashley, Merrill. *Dancing for Balanchine.* New York: E. P. Dutton, Inc., 1984.

Balanchine, George. "The American Dancer." *Dance News,* April 1944, pp. 3, 6.

———. "Balanchine Defines Dance as a Visual Art" (as told to Robert Sabin), *Musical America,* 3/25/44, p. 27.

———. *By George Balanchine* (quotes drawn from various unattributed sources). New York: The Eakins Press, 1984.

———. "Notes on Choreography." *Dance Index,* Vol. 4, Nos. 2 and 3, February–March 1945.

Bancroft, Hubert Howe. *The Book of the Fair: an Historical and Descriptive Presentation of the World's Science, Art, and Industry, as viewed through the Columbian Exposition at Chicago in 1893,* Vol. 1. New York: Bounty Books, 1894.

Banes, Sally. *Democracy's Body: Judson Dance Theater 1962–1964.* Studies in the Fine Arts. Ann Arbor: UMI Research Press, 1983.

———. *Terpsichore in Sneakers: Post-Modern Dance.* Boston: Houghton Mifflin Company, 1980.

Bardsley, Kay. "Isadora Duncan's First School: The First Generation Founders of the Tradition." *CORD Dance Research Annual X.* New York: CORD, 1979, pp. 219–250.

Battcock, Gregory (ed.). *The New Art: A Critical Anthology* (1966), new revised ed. New York: E. P. Dutton & Company, Inc., 1973.

Beaumont, Cyril. *The Complete Book of Ballets* (1937). New York: G. P. Putnam's Sons, 1938.

———. *The Ballet Called Giselle,* 2nd revised ed., 1945. Brooklyn, N.Y.: Dance Horizons, 1969.

———. *The Ballet Called Swan Lake.* London: C. W. Beaumont, 1952.

———. *Mikhail Fokine and His Ballets* (1935). Brooklyn, N.Y.: Dance Horizons, 1981.

Benois, Alexander. *Reminiscences of the Russian Ballet,* trans. Mary Britnieva. London: Putnam, 1947.

Bentley, Toni. "Balanchine and the Kirov." *Ballet Review,* Vol. 14, No. 3, Fall 1986, pp. 38–43.

———. *Winter Season: A Dancer's Journal.* New York: Random House, 1982.

Billington, James. *Icon and Axe* (1966). New York: Vintage Books, 1970.

Binney, Edwin, III. *Les Ballets de Théophile Gautier.* Paris: Librairie Nizet, 1965.

Blair, Frederika. *Isadora Duncan: Portrait of the Artist As a Woman.* New York: McGraw-Hill Book Company, 1986.

Blasis, Carlo. *The Code of Terpsichore: A Practical and Historical Treatise of the Ballet, Dancing, and Pantomime with Complete Theory of the Art of Dancing,* trans. R. Barton (1828). Brooklyn, N.Y.: Dance Horizons, n.d.

Bolitho, William. *Twelve Against the Gods: The Story of Adventure.* New York: Simon & Schuster, 1929.

Borst, William A. *Lord Byron's First Pilgrimage.* New Haven, Conn.: Yale University Press, 1948.

Bournonville, August. *My Theater Life,* trans. Patricia N. McAndrew. Middletown, Conn.: Wesleyan University Press, 1979.

Bowers, Faubion. *Japanese Theater.* Rutland, Vt., and Tokyo: Charles E. Tuttle Company, 1974.

Bridges, Hal. *American Mysticism: From William James to Zen.* New York: Harper & Row, Publishers, 1970.

Brown, Carolyn. "On Chance." *Ballet Review,* Vol. 2, No. 2., pp. 7–25.

Buckle, Richard. *Diaghilev.* New York: Atheneum, 1979.

———. *Nijinsky.* New York: Simon & Schuster, 1971.

Bührer, Michel. *Mummenschanz.* New York: Rizzoli International Publications, Inc., 1986.

Byron, George Gordon, Lord. *Works of Lord Byron,* Vol. III, *Poetry.* London: John Murray/New York: Scribner's Sons, 1922.
Cage, John. *Silence: Lectures and Writings* (1961). Middletown, Conn.: Wesleyan University Press, 1973.
Caldwell, Helen. *Michio Ito: The Dancer and His Dances.* Berkeley: University of California Press, 1977.
Campbell, Joseph. *The Hero with a Thousand Faces.* Bollingen Series XVII. New York: Pantheon, 1949.
Capra, Fritjof. *The Tao of Physics: An Exploration of the Parallels Between Modern Physics and Eastern Mysticism* (1976). New York: Bantam Books, 1977.
Chadwick, Mara L. [Pratt]. *The New Calisthenics: A Manual of Health and Beauty.* Boston and New York: Educational Publishing Company, 1889.
Chapman, John. "The Aesthetic Interpretation of Dance History." *Dance Chronicle,* Vol. 3, No. 3, 1979–1980, pp. 254–274.
———. "An Unromantic View of Nineteenth-Century Romanticism." *York Dance Review,* No. 7, Spring 1978, pp. 28–40.
Chasles, Philarète (see under Gautier).
Cheney, Sheldon and Martha Candler. *Art and the Machine.* New York: McGraw-Hill Book Company, Inc., Whittlesey House, 1936.
Cherniavsky, Felix. "Maud Allan, Part I: The Early Years, 1873–1903." *Dance Chronicle,* Vol. 6, No. 1, 1983, pp. 1–36.
———. "Maud Allan, Part II. First Steps to a Dancing Career, 1904–1906." *Dance Chronicle,* Vol. 6, No. 3, 1983, pp. 189–227.
———. "Maud Allan, Part III: Two Years of Triumph, 1908–1909." *Dance Chronicle,* Vol. 7, No. 2, 1984, pp. 119–158.
Chipp, Hershel B., with contributions by Peter Selz and Joshua C. Taylor. *Theories of Modern Art, A Source Book by Artists and Critics.* Berkeley: University of California Press, 1968.
Chisholm, Lawrence. *Fenellosa: The Far East and American Culture.* New Haven: Yale University Press, 1963.
Chujoy, Anatole. "Russian Balletomania." *Dance Index,* Vol. 7, No. 3, March 1948.
Clearwater, Bonnie. *Mark Rothko: Works on Paper.* New York: Hudson Hills Press, 1984.
Cocuzza, Ginnine. "The Theatre of Angna Enters, American Dance-Mime." Doctoral dissertation, New York University, 1987.
Cohen, Barbara Naomi. "The Borrowed Art of Gertrude Hoffmann." *Dance Data,* No. 2, 1977. Brooklyn, N.Y.: Dance Horizons, 1977.
Cohen, Selma Jeanne (ed.). *Dance As a Theater Art: Source Readings in Dance History from 1581 to the Present.* New York: Dodd, Mead & Company, 1974.
———. *Doris Humphrey: An Artist First, An Autobiography,* ed. and completed by Cohen. Middletown, Conn.: Wesleyan University Press, 1972.
———. "The English Critic and the Romantic Ballet." *Theatre Survey,* Vol. XVII, No. 1, May 1976, pp. 82–91.

———. *Next Week Swan Lake: Reflections on Dance and Dancers.* Middletown, Conn.; Wesleyan University Press, 1982.
Copeland, Roger, and Marshall Cohen. *What Is Dance? Readings in Theory and Criticism.* New York: Oxford University Press, Inc., 1983.
Craig, Edward Gordon. *On the Art of the Theatre* (essays written between 1904 and 1910). New York: Theatre Arts Books, 1956.
Croce, Arlene. *Afterimages.* New York: Alfred A. Knopf, 1977.
———. *Going to the Dance.* New York: Alfred A. Knopf, 1982.
Crunden, Robert. *From Self to Society, 1919–1941.* Transitions in American Thought Series. Englewood Cliffs, N.J.: Prentice-Hall, Inc., 1972.
Cunningham, Merce. *Changes; Notes on Choreography,* ed. Frances Starr. New York: Something Else Press, Inc., 1968.
———. "The Impermanent Art." 7 *Arts,* No. 3, ed. Fernando Puma. Indian Hills, Colo.: The Falcon's Wing Press, 1955, pp. 69–77.
Daly, Ann. "The Balanchine Woman: Of Hummingbirds and Channel Swimmers." *The Drama Review,* Vol. 31, No. 1 (T-113), Spring 1987.
Daniel, David. "A Conversation with Suzanne Farrell." *Ballet Review,* Vol. 7, No. 1, pp. 1–15.
Delarue, Alison (ed.). *Fanny Elssler in America.* Brooklyn, N.Y.: Dance Horizons, 1976.
Denby, Edwin. *Dance Writings,* ed. Robert Cornfield and William Mackay. New York: Alfred A. Knopf, 1986.
Dijkstra, Bram. *Idols of Perversity: Fantasies of Feminine Evil in Fin-de-Siècle Culture.* New York: Oxford University Press, 1986.
Drewal, Margaret Thompson. "Isis and Isadora." Paper delivered at the 10th Annual Conference of the Dance History Scholars, University of California, Irvine, 2/14/87.
Duncan, Irma. *Duncan Dancer: An Autobiography.* Middletown, Conn.: Wesleyan University Press, 1965.
———, and Allan Ross MacDougall. *Isadora Duncan's Russian Days.* New York: Covici-Friede, Publishers, 1929.
Duncan, Isadora. *The Art of the Dance,* ed. Sheldon Cheney (1928). New York: Theatre Arts Books, 1977.
———. *My Life.* New York: Horace Liveright, Inc., 1927.
Dunning, Jennifer. *"But First a School": The First Fifty Years of the School of American Ballet.* New York: Viking-Penguin, 1985.
Emmanuel, Maurice. *The Antique Greek Dance After Sculptured and Painted Figures* (1895), trans. Harriet Jean Beauley. London: John Lane Company, 1916.
Exhibition Paris, 1900, a Practical Guide . . .
Fanchon, Gordon, and Gerald Myers (ed.). *Philosophical Essays on Dance, with Responses by Choreographers, Critics, and Dancers.* Based on a conference at the American Dance Festival. Brooklyn, N.Y.: Dance Horizons, 1981.

Fanny Elssler: Materialen. Catalogue of an exhibit of Elssler memorabilia mounted by the Austrian Theater Museum. Vienna: Heermann Bohlaus Nachf., 1984.

Fellom, Martie. "Joan of Arc and Martha Graham." Paper prepared for a course on Martha Graham, given in New York University's Department of Performance Studies, 1980.

Flaubert, Gustave. *The Complete Works of Gustave Flaubert,* Vols. III and IV (*Salammbô* and *Hérodias*), trans. unknown. London: M. Walter Dunne, 1904.

Flitch, J. E. Crawford. *Modern Dancing and Dancers.* London: Grant Richards Ltd./ Philadelphia: J. B. Lippincott Company, 1912.

Fokine, Mikhail. *Memoirs of a Ballet Master,* trans. Vitale Fokine, ed. Anatole Chujoy. Boston: Little, Brown & Company, 1961.

Foregger, Nikolai. "Experiments in the Art of the Dance," trans. David Miller. *The Drama Review,* Vol. 19, No. 1 (T-65), March 1975, pp. 74–77.

Forti, Simone. *Handbook in Motion.* Halifax: The Press of the Nova Scotia College of Art and Design/New York: New York University Press, 1974.

Frankl, Paul T. *Form and Re-form: A Practical Handbook of Modern Interiors.* New York: Harper & Brothers, 1930.

Friedman, Martin. *Noguchi's Imaginary Landscapes,* catalogue of an exhibit organized by the Walker Art Center, 1978. Minneapolis, Minn.: The Walker Art Center, 1978.

Fuller, Loïe. *Fifteen Years of a Dancer's Life* (1913). Brooklyn, N.Y.: Dance Horizons, n.d.

Garafola, Lynn. "The Travesty Dancer in Nineteenth Century Ballet." *Dance Research Journal,* Vol. 17, No. 2, Fall 1985/ Vol. 18, No. 1, Spring 1986, pp. 35–40.

Garis, Robert. "Balanchine and Stravinsky: Facts and Problems." *Ballet Review,* Vol. 10, No. 3, Fall 1982, pp. 9–24.

Gautier, Théophile. *Mademoiselle de Maupin,* trans.[?]. New York: The Modern Library [1918?].

———. *One of Cleopatra's Nights, and Other Fantastic Romances,* trans. Lafcadio Hearn. New York: Brentano's Publishers, 1927 (1st American ed., 1899).

———. *The Romantic Ballet as Seen by Théophile Gautier,* trans. Cyril Beaumont (1932, revised 1947), Brooklyn, N.Y.: Dance Horizons, 1973.

———, Philarète Chasles, and Jules Janin. *Les Beautés de l'Opéra.* Paris: Soulié, 1845.

Geva, Tamara. *Split Seconds: A Remembrance.* New York: Harper & Row, Publishers, 1972.

Goethe, Johann Wolfgang. *The Eternal Feminine, Selected Poems of Goethe,* ed. Frederick Ungar. New York: Frederick Ungar Publishing Company, 1980.

Goldberg, RoseLee. *Performance: Live Art 1909 to the Present.* New York: Harry N. Abrams, Inc., 1979.

Goldner, Nancy. *The Stravinsky Festival of the New York City Ballet.* New York: The Eakins Press, 1973.

Gordon, David. "It's About Time." *The Drama Review,* Vol. 19, No. 1 (T-65), March 1975, pp. 43–52.

Gordon, Mel. "Foregger and the Dance of the Machines." *The Drama Review,* Vol. 19, No. 1 (T-65), March 1975, pp. 68–73.

Graham, Martha. *The Notebooks of Martha Graham.* New York: Harcourt Brace Jovanovich, 1972.

Grigoriev, Serge Leonidovitch. *The Diaghilev Ballet 1909–1929* (1953). London: Penguin, 1960.

Grove, Mrs. Lilly [Frazer], et al. *Dancing.* The Badminton Library. London: Longmans, Green & Company, 1901 (1st ed., 1895).

Gruen, John. *The Private World of Ballet.* New York: Viking Press, 1975.

Guest, Ivor. "Dandies and Dancers." *Dance Perspectives 37,* Spring 1969.

———. *The Divine Virginia: A Biography of Virginia Zucchi.* New York: Marcel Dekker, 1977.

———. *Fanny Elssler.* Middletown, Conn.: Wesleyan University Press/London: Adam & Charles Black, 1970.

———. *Jules Perrot: Master of the Romantic Ballet.* London: Dance Books Ltd., 1984.

———. *The Romantic Ballet in London: Its Development, Fulfillment and Decline.* London: Pitman Publishing, 1954 (with new introduction, 1972).

———. *The Romantic Ballet in Paris.* London: Dance Books Ltd., 1966 (revised, 1980).

Hambidge, Jay. *The Elements of Dynamic Symmetry* (1919). New York: Brentano's Publishers, 1926.

Harris, Margaret Haile. *Loïe Fuller; Magician of Light.* Catalogue of an exhibition at the Virginia Museum, March 12–April 22, 1979. Richmond: The Virginia Museum, 1979.

Harte, Bret (ed.) *Outcroppings: Being Selections of California Verse.* San Francisco: A. Roman & Cory/New York: W. J. Middleton, 1866.

Haskell, Arnold. *Balletomania: The Story of an Obsession.* New York: Simon & Schuster, 1934. Reprinted with additional material as *Balletomania: Then and Now.* New York: Alfred A. Knopf, 1977.

Haskell, Barbara, with an additional essay by John G. Hanhardt. *BLAM! The Explosion of Pop, Minimalism, and Performance, 1958–1964.* Published in conjunction with an exhibit at the Whitney Museum of American Art, September 20–December 2, 1984. New York: The Whitney Museum and W. W. Norton, 1984.

H'Doubler, Margaret Newell. *The Dance.* New York: Harcourt, Brace & Company, 1925.

Heath, Charles. *The Beauties of the Opera and Ballet* (1845). New York: Da Capo Press, Inc., 1977.

Helpern, Alice J. "The Evolution of Martha Graham's Dance Technique." Ph.D.

thesis, New York University School of Education, Health, Nursing, and Arts Professions, 1981.
Hinman, Mary Wood. *Gymnastic and Folk Dancing* (5 vols.). New York: A. S. Barnes, 1923–1928.
Hitchcock, Henry Russell, and Philip Johnson. *The International Style: Architecture Since 1922* (ca. 1932). New York: W. W. Norton & Company, 1966.
Holmes, Olive (ed.). *Motion Arrested: Dance Reviews of H. T. Parker.* Middletown, Conn.: Wesleyan University Press, 1982.
Honour, Hugh. *Romanticism.* Icon Editions. New York: Harper & Row, Publishers, 1979.
Horwitz, Dawn Lille. *Michel Fokine.* Boston: G. K. Hall & Company, Twayne Publishers, 1985.
Horwitz, Tem, and Susan Kimmelman with H. H. Lui. *Tai Chi Ch'uan: The Technique of Power.* Chicago: The Chicago Review Press, 1976.
Huang, Al Chung-liang. *Embrace Tiger Return to Mountain—The Essence of Tai Chi.* Moab, Utah: Real People Press, 1973.
Huelsenbeck, Richard. *Memoirs of a Dada Drummer.* New York: Viking Press, 1974.
Hugo, Victor. *Oeuvres Complètes,* Vol. 5, *Odes et Ballades, Les Orientales.* Paris: Ernest Flammarion, 1853.
Jackson, Kristin. "Kabuki Theater and 19th Century Ballets: A Comparative Study . . . in terms of Theater Conventions, Performance Style, and Thematic Content." Paper written as partial requirement for an MFA degree, New York University, Tisch School of the Arts, Dance, 1985.
Janin, Jules. *Les Beautés de l'Opéra* (see under Gautier).
Jeans, Sir James. *Physics and Philosophy* (1942). Ann Arbor: University of Michigan Press, 1958.
Jenkyns, Richard. *The Victorians and Ancient Greece.* Cambridge, Mass.: Harvard University Press, 1980.
Johnston, Jill. *Marmalade Me.* New York: E. P. Dutton & Company, Inc., 1971.
Jowitt, Deborah. "A Conversation with Bessie Schönberg." *Ballet Review,* Vol. 9, No. 1, Spring 1981 *(American Modern Dance: The Early Years),* pp. 31–63.
———. *Dance Beat, Selected Views and Reviews.* New York: Marcel Dekker, 1977.
———. *The Dance in Mind: Profiles and Reviews, 1976–1983.* Boston: David R. Godine, Publisher, 1985.
Judson Dance Theater, 1962–1966. Published in conjunction with an exhibit and festival as part of the Bennington College Judson Project. Bennington, Vt.: Bennington College, 1981.
Jullian, Philippe. *Dreamers of Decadence, Symbolist Painters of the 1890s,* trans. Robert Baldick. New York: Praeger Publishers, 1971.
———. *Les Orientalistes: La Vision de l'Orient par les Peintres Européens au XIX Siècle.* Fribourg, Switzerland: Office du Livre, 1977.

Karsavina, Tamara. *Theatre Street.* New York: E. P. Dutton & Company, Inc., 1931.
Keeler, Charles. *The Simple Home* (1904). Santa Barbara, Calif., and Salt Lake City, Utah: Peregrine Smith, Inc., 1979.
Kendall, Elizabeth. *Where She Danced.* New York: Alfred A. Knopf, 1979.
Kermode, Frank. "Loïe Fuller and the Dance Before Diaghilev." Excerpt from "Poet and Dancer Before Diaghilev" (from *Puzzles and Epiphanies,* 1962). *Theatre Arts Magazine,* September 1962, pp. 6–22.
King, Eleanor. *Transformations.* Brooklyn, N.Y.: Dance Horizons, 1978.
Kirby, Michael. *The Art of Time: Essays on the Avant Garde.* New York: E. P. Dutton & Company, Inc., 1969.
Kirkland, Gelsey, with Greg Lawrence. *Dancing on My Grave.* Garden City, N.Y.: Doubleday & Company, Inc., 1986.
Kirstein, Lincoln. "Blast at Ballet: A Corrective for American Audiences" (1938). Reprinted in *Three Pamphlets Collected.* Brooklyn, N.Y.: Dance Horizons, 1967.
———. *Four Centuries of Ballet: Fifty Masterworks* (1970). New York: Dover Publications, 1984.
———. *Nijinsky Dancing.* New York: Alfred A. Knopf, 1975.
———. *Portrait of Mr. B.* New York: Ballet Society/Viking-Penguin, 1984.
———. "What Ballet Is About: An American Glossary." *Dance Perspectives 1,* Winter 1959.
Kleist, Heinrich von. "Puppet Theatre" (see Copeland and Cohen, p. 181).
Klosty, James (ed.). *Merce Cunningham* (1975), 2nd ed., with a new foreword by Klosty. New York: Limelight Editions, 1986.
Kochno, Boris. *Diaghilev and the Ballets Russes.* New York: Harper & Row, Publishers, 1970.
Kostelanetz, Richard (ed.). *Metamorphosis in the Arts: A Critical History of the 1960's* (1967). New York: Assembling Press, 1980.
———. *The New American Arts.* New York: Collier Books/ London: Collier Macmillan, Ltd., 1965.
———. *The Theater of Mixed Means: An Introduction to Happenings, Kinetic Environments, and Other Mixed-Means Performances.* New York: Dial Press, 1968.
Krassovskaya, Vera. "Marius Petipa and the 'Sleeping Beauty'," trans. Cynthia Read. *Dance Perspectives 49,* Spring 1972.
Kriegsman, Sali Ann. *Modern Dance in America; The Bennington Years.* Boston: G. K. Hall & Company, 1981.
Kshessinska, Mathilde. *Dancing in Petersburg: The Memoirs of Kshessinska,* trans. Arnold Haskell. London: Victor Gollancz Ltd., 1960.
Lazzarini, John and Roberta. *Pavlova: Repertoire of a Legend.* A Dance Horizons Book. New York: Schirmer Books/London: Collier Macmillan, Publishers, 1980.
Le Corbusier. *Toward a New Architecture,* trans. Frederick Etchells, from 13th French ed. New York: Brewer & Warren, Inc., 1927.

Lesschaeve, Jacqueline. *Merce Cunningham: The Dancer and the Dance.* New York and London: Marion Boyars, Publishers, 1985.

———. *Merce Cunningham: Le Danseur et la Danse: Entretiens avec Jacqueline Lesschaeve.* Paris: Belfond, 1977.

Levien, Prince Peter. *The Birth of the Ballet-Russe,* trans. L. Zarine (1936). New York: Dover Publications, Inc., 1973.

Levinson, André. *Ballet Old and New* (1918), trans. Susan Cook Summer. Brooklyn, N.Y.: Dance Horizons, 1982.

———. *Marie Taglioni, 1804–1884,* trans. Cyril W. Beaumont (1930). London: Dance Books, 1977.

Lewis, Dio. *The New Gymnastics for Men, Women, and Children,* 10th ed. Boston: James R. Osgood & Company, 1873 (1st ed.: 1867).

Liu, Da. *T'ai Chi Ch'uan and I Ching: A Choreography of Body and Mind.* New York: Harper & Row, Publishers, 1972.

Livet, Anne (ed.). *Contemporary Dance.* New York: Abbeville Press, Inc. (in association with the Fort Worth Art Museum, Texas), 1978.

Lloyd, Margaret. *The Borzoi Book of Modern Dance* (1949). Brooklyn, N.Y.: Dance Horizons, n.d.

Loewenthal, Lillian. "Isadora Duncan in the North." *Dance Chronicle,* Vol. III, No. 3, 1979–80, pp. 227–253.

McDonagh, Don. *Martha Graham* (1973). Paperback edition with revised choreochronicle by Andrew Mark Wentink. New York: Popular Library, 1975.

MacDonald, Nesta. "Isadora Reexamined: Lesser Known Aspects of the Great Dancer's Life, Part I, 1877–1900." *Dance Magazine,* July 1977, pp. 51–66 (series continued through December 1977).

———. *Diaghilev Observed by Critics in England and the United States, 1911–1929.* Brooklyn, N.Y.: Dance Horizons/London: Dance Books Ltd., 1975.

MacDougall, Allan Ross. *Isadora: A Revolutionary in Art and Love.* New York: Thomas Nelson & Sons, 1960.

McLuhan, Marshall. *Understanding Media: The Extensions of Man.* New York: New American Library, 1964.

———, and Quentin Fiore. *The Medium Is the Massage.* New York: Bantam Books, Inc., 1967.

McMahon, Deirdre. "The Feminist Mystique." *Dance Theatre Journal,* No. 3, Winter 1985, pp. 8–10.

Madsen, Stephan Tschudi. *Sources of Art Nouveau* (1956, Oslo), trans. Ragnar Christopherson. New York: Da Capo Press, Inc., 1976.

Magriel, Paul (ed.). *Chronicles of the American Dance from the Shakers to Martha Graham* (1948). New York: Da Capo Press, Inc., 1978. All these monographs originally appeared in issues of *Dance Index.*

——— (ed.). *Nijinsky, Pavlova, Duncan: Three Lives in Dance* (1947). New York: Da Capo Press, Inc., 1977.

Maiorano, Robert, and Valerie Brooks. *Balanchine's Mozartiana: The Making of a Masterpiece.* New York: Freundlich Books, 1985.

Major, A. Hyatt. "Mei-Lan Fang." *Dance Index,* Vol. 1, No. 7, July 1942, pp. 107–109.

Martin, John. *America Dancing: The Background and Personalities of the Modern Dance* (1936). Brooklyn, N.Y.: Dance Horizons, n.d.

———. *The Modern Dance* (1933). Brooklyn, N.Y.: Dance Horizons, n.d.

Martins, Peter. *Far from Denmark.* Boston and Toronto: Little, Brown & Company, 1982.

Matson, Tim. *Pilobolus.* New York: Random House, 1978.

Meikle, Jeffrey L. *Twentieth Century Limited: Industrial Design in America.* Philadelphia: Temple University Press, 1979.

Migel, Parmenia. *The Ballerinas: From the Court of Louis XIV to Pavlova* (1972). New York: Da Capo Press, Inc., n.d.

Mitchell, Margaretta, and Suzanne Riess. *The Temple of the Wings,* Vol. 1 (interviews conducted with Sulgwynn and Charles Quitzow in 1972). Berkeley: Regional Oral History Office, the Bancroft Library, University of California, 1973.

Money, Keith. *Anna Pavlova: Her Life and Art.* New York: Alfred A. Knopf, 1982.

Moore, Lillian. *Artists of the Dance* (1938). Brooklyn, N.Y.: Dance Horizons, 1969.

Moore, Thomas. *Lallah Rookh.* Paris; Baudry's European Library, 1835.

Mørk, Ebbe (ed.). *Salut for Bournonville.* Published in conjunction with an exhibit, November 1979–January 1980. Copenhagen; Statensmuseum for Kunst, 1979.

Morris, R. Anna, *Physical Education in the Public Schools.* New York: American Book Company, 1892.

Morris, Robert. "Notes on Dance." *Tulane Drama Review,* Vol. 10, No. 2, Winter 1965, pp. 179–186.

———. "Notes on Sculpture. Part 1," *Artforum,* Vol. IV, No. 6, 2/66, pp. 42–44; "Notes on Sculpture, Part 2," *Artforum,* Vol. V, No. 2, 10/66, pp. 20–23; "Notes on Sculpture, Part 3," *Artforum,* Vol. V, No. 10, Summer 1967, pp. 24–29.

Murray, Gilbert (ed. and trans.). *The Athenian Drama for English Readers,* Vol. III (Euripides). London: George Allen, 1905–1911.

Neidish, Juliet. "Whose Habitation Is the Air." *Dance Perspectives 61,* Vol. 16, Spring 1975 ("All That Strange and Mysterious Folk: Studies in Ballet Supernaturals," includes articles by Susan Au and Susan Reimer-Sticklor), pp. 4–17.

Neville, Amelia Ransome. *The Fantastic City: Memoirs of the Social and Romantic Life of Old San Francisco,* ed. and revised, Virginia Brastow. Boston and New York: Houghton Mifflin Company, 1932.

New Dance USA. Catalogue containing articles by Sally Banes, Jill Johnston, and Allen Robertson, issued in conjunction with a festival presented by the Walker Arts Center, Minneapolis, October 3–10, 1981.

Newman, Barbara. *Striking a Balance: Dancers Talk About Dancing.* Boston: Houghton Mifflin Company, 1982.

Newton, Stella Mary. *Health, Art & Reason: Dress Reformers of the Nineteenth Century.* London: John Murray, 1974.

Nietzsche, Friedrich. *Ecce Homo* and *The Birth of Tragedy from the Spirit of Music,* trans. Clifton Fadiman. New York: Modern Library, 1927.

Nijinska, Bronislava. *Early Memoirs,* ed. and trans. Irina Nijinska and Jean Rawlinson. New York: Holt, Rinehart, & Winston, 1981.

Nijinsky, Romola, with Lincoln Kirstein. *Nijinsky* (1934). New York: Simon & Schuster, 1934.

Nijinsky, Vaslav. *The Diary of Vaslav Nijinsky,* ed. Romola Nijinsky. Berkeley; University of California Press, 1971.

Noguchi, Isamu. *A Sculptor's World.* New York and Evanston, Ill.: Harper & Row, Publishers, 1968.

Oskar Schlemmer—The Triadic Ballet. Akademie der Kunste, Documentation 5. Berlin: Druckhaus Heinrich, 1985.

Oxford Book of French Poetry. Oxford: The Clarendon Press, 1907.

Padgette, Paul (ed.). *The Dance Writings of Carl Van Vechten.* Brooklyn, N.Y.: Dance Horizons, 1974.

Pater, Walter. *The Renaissance* (1873). New York: New American Library, 1959.

Percival, John. *The World of Diaghilev.* New York: Harmony Books, 1971.

Petipa, Marius. *Russian Ballet Master,* ed. Lillian Moore, trans. Helen Whittalaer. London: Dance Books, 1971.

Petit Courrier des Dames, Journal des Modes, 1839–[41]. Paris: Imprimerie de Dondy, Dupré.

Pierpont, Claudia Roth. "Balanchine: Of Metaphors and Women." *Ballet Review,* Vol. 8, No. 4, 1980, pp. 353–372.

Praz, Mario. *The Romantic Agony* (1933). Cleveland: The World Publishing Company, 1965.

Racster, Olga. *The Master of the Russian Ballet: The Memoirs of Cav. Enrico Cecchetti.* New York: Da Capo Press, Inc., 1978.

Rainer, Yvonne. *Work 1961–73.* Halifax: The Press of the Nova Scotia College of Art and Design/New York: New York University Press, 1974.

Rambert, Marie. *Quicksilver* (1972). London: Macmillan, 1983.

Raskind, Lisa Bonoff. "The Question of Moving Symbol and Meaning in the Romantic Ballet." *Epoché, History of Religion at UCLA Newsletter,* Vol. 5, No. 2, Fall 1977.

Rather, Lois. *Lovely Isadora.* Oakland, Calif.: The Rather Press, 1976.

Reed, John R. *Decadent Style.* Athens, Ohio: Ohio University Press, 1985.

Reich, Charles. *The Greening of America; How the Youth Revolution Is Trying to Make America Livable.* New York: Random House, 1970.

Reynolds, Nancy. *Repertory in Review: Forty Years of the New York City Ballet.* New York: Dial Press, 1977.

Riasanovsky, Nicholas V. *A History of Russia.* New York: Oxford University Press, 1963.

Rogers, Frederick Rand. *Dance: A Basic Educational Technique.* New York: The Macmillan Company, 1941.

Rose, Barbara. *American Art Since 1900, a Critical History.* New York and Washington, D.C.: Praeger Publishers, 1967.

Rose, Bernice. *Jackson Pollock: Drawing into Painting.* New York: Museum of Modern Art, 1980.

Rosen, Charles, and Henry Zerner. *Romanticism and Realism: The Mythology of Nineteenth Century Art.* New York: W. W. Norton & Company, 1984.

Roslaveva, Natalia. *Era of the Russian Ballet* (1966). New York: Da Capo Press, Inc., 1979.

———. "Prechistenka 20: The Isadora Duncan School in Moscow." *Dance Perspectives 64,* Winter 1975.

Rousseau, Jean-Jacques. *Émile, or Treatise on Education,* abridged, translated, and annotated, William H. Payne. New York and London: D. Appleton & Company, 1911.

Ruyter, Nancy Lee Chalfa. *Reformers and Visionaries: The Americanization of the Art of Dance.* Brooklyn, N.Y.: Dance Horizons, 1979.

Said, Edward. *Orientalism.* New York: Pantheon, 1978.

St. Denis, Ruth. *An Unfinished Life.* New York: Harper and Brothers, Publishers, 1939.

Salz, Jonah. "Otojiro Kawakami and Sada Yakko: International Reformers." Master's thesis, New York University, Graduate School of Arts and Science, 1985.

Sand, George. *Lélia,* translated with an introduction by Maria Espinosa. Bloomington: Indiana University Press, 1978.

Sayler, Oliver M. *Revolt in the Arts.* New York: Brentano's, 1930.

Scheyer, Ernst. "The Shapes of Space: The Art of Mary Wigman and Oskar Schlemmer." *Dance Perspectives 41,* Spring 1970.

Schlemmer, Oskar. *The Letters and Diaries of Oskar Schlemmer,* ed. Tut Schlemmer, trans. Kristina Winston. Middletown, Conn.: Wesleyan University Press, 1972.

———, Lazlo Moholy-Nagy, and Farkas Molnar. *The Theatre of the Bauhaus,* trans. Arthur S. Wensinger. Middletown, Conn.: Wesleyan University Press, 1961.

Schlundt, Christena L. "Into the Mystic with Miss Ruth." *Dance Perspectives,* No. 46, Summer 1971.

———. *The Professional Appearances of Ruth St. Denis and Ted Shawn: A Chronology and an Index of Dances, 1906–1932.* New York: The New York Public Library, Astor, Lenox, and Tilden Foundations, 1962.

———. *Tamiris, A Chronicle of Her Dance Career, 1927–1955.* New York: The New York Public Library, Astor, Lenox and Tilden Foundations, 1972.

Schneemann, Carolee. *More Than Meat Joy.* New Paltz, N.Y.: Documentext, 1979.

Schneider, Ilya Ilyitch. *Isadora Duncan: The Russian Years,* trans. David Magarshak. New York: Harcourt Brace & World, Inc., 1968.

Schwab, Raymond. *The Oriental Renaissance: Europe's Rediscovery of India and the*

East, 1680–1880, trans. Gene Patterson-Black and Victor Reinking. New York: Columbia University Press, 1984.

Sears, David. "Graham's Classics in Revival." *Ballet Review,* Vol. 10, No. 2, Summer 1982, pp. 25–34.

Second, Albéric. *Les Petits Mystères de l'Opéra.* Paris; G. Kugelmann/Bernard-Latte, 1844.

Selden, Elizabeth. *The Dancer's Quest: Essays on the Aesthetics of the Contemporary Dance.* Berkeley: University of California Press, 1935.

Seroff, Victor. *The Real Isadora.* New York: Dial Press, 1971.

Serullaz, Maurice. *Eugène Delacroix.* New York: Harry N. Abrams, Inc., 1971.

Shaw, George Bernard. *Music in London, 1890–1894.* London: Constable & Company, Ltd., 1931.

———. *Our Theatres in the Nineties,* 3 vols. London: Constable & Company, Ltd., 1931.

Shawn, Ted, with Gray Poole. *One Thousand and One Night Stands.* New York: Da Capo Press, Inc., 1979 (from the original edition, published Garden City, N.Y.: Doubleday & Company, 1960).

———. *Ruth St. Denis: Pioneer and Prophet.* San Francisco: John Henry Nash, 1920.

Shelton, Suzanne. *Divine Dancer: A Biography of Ruth St. Denis.* Garden City, N.Y.: Doubleday & Company, Inc., 1981.

———. "The Jungian Roots of Martha Graham's Dance Imagery." Paper delivered at the Dance History Scholars Conference, Ohio State University, February 1983.

Sherman, Jane. *Denishawn: The Enduring Influence.* New York: G. K. Hall & Company, Twayne Publishers, 1983.

———. *The Drama of Denishawn Dance.* Middletown, Conn.: Wesleyan University Press, 1979.

———. *Soaring: The Diary and Letters of a Denishawn Dance in the Far East, 1925–1926.* Middletown, Conn.: Wesleyan University Press, 1975.

Siegel, Marcia B. (ed.). "Nik: A Documentary." *Dance Perspectives 48,* Winter 1971.

———. *The Shapes of Change: Images of American Dance.* Boston: Houghton Mifflin Company, 1979.

Slonimsky, Yuri. "Balanchine, The Early Years," ed. Francis Mason, trans. John Andrews. *Ballet Review,* Vol. 5, No. 3, 1975–1976.

———. "Marius Petipa," trans. Anatole Chujoy. *Dance Index,* Vol. 6, Nos. 5 and 6, May–June 1947.

Sokolova, Lydia. *Dancing for Diaghilev: The Memoirs of Lydia Sokolova,* ed. Richard Buckle. New York: Macmillan & Company, 1961.

Sommer, Sally. "Loïe Fuller." *The Drama Review,* Vol. 19, No. 1 (T-65), March 1975.

———. "Loïe Fuller: From the Theater of Popular Entertainment to the Parisian Avant-Garde." Ph.D. thesis, New York University, Graduate School of Arts and Science, 1979.

———. "The Stage Apprenticeship of Loïe Fuller." *Dance Scope,* Vol. 12, No. 1, Fall/Winter, 1977/1978, pp. 23–34.

Sorell, Walter. *Dance in its Time: The Emergence of an Art Form.* Garden City, N.Y.: Anchor Press/Doubleday, 1981.

———. *Hanya Holm: The Biography of an Artist.* Middletown, Conn.: Wesleyan University Press, 1969.

———. (ed. and trans.). *The Mary Wigman Book.* Middletown, Conn.: Wesleyan University Press, 1975.

Souritz, Elizabeth. "Fedor Lopukhov: A Soviet Choreographer in the 1920s," trans. Lynn Visson, revised and ed. Sally Banes. *Dance Research Journal,* Vol. 17, No. 2, and Vol. 18, No. 1, Fall 1985–Spring 1986, pp. 3–20.

Spencer, Charles. *Leon Bakst.* New York: St. Martin's Press, 1973.

———, with contributions by Philip Dyer and Martin Battersby. *The World of Serge Diaghilev* (1974). New York: Penguin Books, 1979.

Stebbins, Genevieve. *Delsarte System of Expression* (1885).

A reprint of the revised edition of 1902. Brooklyn, N.Y.: Dance Horizons, 1977.

———. *Dynamic Breathing and Harmonic Gymnastics: A Complete System of Psychical, Aesthetic and Physical Culture,* 2nd ed. New York: Edgar S. Werner, 1892.

Stecher, W. A. (ed.). *Gymnastics: A Textbook of the German-American System of Gymnastics.* Boston: Lee & Shepard, Publishers, 1895.

Steegmuller, Francis (ed. and connecting text). *"Your Isadora": The Love Story of Isadora Duncan and Gordon Craig.* New York: Random House and the New York Public Library, 1974.

Stewart, Virginia, and Merle Armitage. *The Modern Dance* (1935). Brooklyn, N.Y.: Dance Horizons, 1970.

Sticklor, Susan. "Angel with a Past." *Dance Perspectives 61,* Vol. 16, Spring 1975 ("All That Strange and Mysterious Folk: Studies in Ballet Supernaturals," Includes essays by Susan Au and Juliet Neidish), pp. 18–29.

Stodelle, Ernestine. *Deep Song.* A Dance Horizons Book. New York: Schirmer Books, 1984.

Stokes, Adrien. *Russian Ballets* (1935). New York: E. P. Dutton & Company, 1936.

Swift, Mary Grace. *A Loftier Flight: The Life and Accomplishments of Charles-Louis Didelot, Balletmaster.* Middletown, Conn.: Wesleyan University Press/London Pitman Publishing, 1974.

Sypher, Wylie. *Rococo to Cubism in Art and Literature.* New York: Random House, Vintage Books, 1960.

Taper, Bernard. *Balanchine: A Biography,* revised and updated ed. New York: Macmillan/London: Collier Macmillan, 1974.

Taylor, Paul. *Private Domain.* New York: Alfred A. Knopf, 1987.

Terry, Walter. *Frontiers of Dance: The Life of Martha Graham.* New York: Thomas Y. Crowell Company, 1975.

———. *I Was There: Selected Dance Reviews and Articles, 1936–1976,* compiled and ed. Andrew Mark Wentink. New York: Marcel Dekker, 1978.

———. "The Legacy of Isadora Duncan and Ruth St. Denis." *Dance Perspectives 5,* 1959.

Thiess, Frank. *Der Tanz als Kunstwerk: Studien zu einer Aesthetik der Tanzkunst.* Munich: Delphin-Verlag, 1923.

"Time to Walk in Space: Essays, Stories, and Remarks" (about Merce Cunningham), by Clive Barnes, Carolyn Brown, John Cage, Arlene Croce, Merce Cunningham, Edwin Denby, Jill Johnston, David Vaughan. *Dance Perspectives 34,* Summer 1968.

Tompkins, Calvin. *The Bride and the Bachelors: Five Masters of the Avant-Garde,* expanded ed. New York: Penguin Books, 1968.

———. *Off the Wall: Robert Rauschenberg and the Art World of Our Time.* Garden City, N.Y.: Doubleday & Company, Inc., 1980.

Tracy, Robert, with Sharon DeLano. *Balanchine's Ballerinas: Conversations with the Muses.* New York: Simon & Schuster, Linden Press, 1983.

Vaillat, Léandre. *Marie Taglioni, ou la Vie d'une Danseuse.* Paris: A. Michel, ca. 1942.

Vaughan, David. *Frederick Ashton and His Ballets.* New York: Alfred A. Knopf, 1977.

Véron, Louis. "Behind the Scenes at the Opera in Taglioni's Day," trans. Cyril Beaumont. Extract from *Mémoires d'un Bourgeois de Paris* (1853–1855). *Dancing Times,* Vol. XIV, No. 159, January 1924, pp. 403–407.

Vincent, L. M. *Competing with the Sylph: Dancers and the Pursuit of the Ideal Body Form.* New York: Andrews & McMeel, Inc., 1979.

A Visitor's Guide to Paris and the Exposition of 1900.

Vivekânanda, Swâmi. *Râja Yoga.* New York: Brentano's Publishers, 1929.

Volkov, Simon. *Balanchine's Tchaikovsky,* trans. Antonina W. Bouis. New York: Simon & Schuster, 1985.

Waley, Arthur. *The Nō Plays of Japan* (1921). Rutland, Vt., and Tokyo: Charles E. Tuttle Company, 1976.

Watson, J. Madison. *Handbook of Calisthenics and Gymnastics (with music).* New York and Philadelphia: Schermerhorn, Bancroft, & Company, 1864.

Whitman, Walt. *Leaves of Grass.* New York: Doubleday, Page & Company, 1923.

Wickes, Frances G. *The Inner World of Choice.* New York, Evanston, Ill., and London: Harper & Row, Publishers, 1963.

Wigman, Mary. *The Language of Dance,* trans. Walter Sorell. Middletown, Conn.: Wesleyan University Press, 1966.

Wilder, Thornton. *Three Plays.* New York, Bantam Books, 1958.

Wiley, Roland John. "Alexandre Benois' Commentaries on the first *Saisons Russes.*" *The Dancing Times,* October 1980, pp. 28–30.

———. *Tchaikovsky's Ballets.* Oxford: Clarendon Press, 1985.
Willson, A. Leslie. *A Mythical Image: The Ideal of India in German Romanticism.* Durham, N.C.: Duke University Press, 1964.
Winter, Marian Hannah. *The Pre-Romantic Ballet.* London: Sir Isaac Pitman & Sons Ltd./New York: Pitman Publishing, 1974.
Young, La Monte. "Lecture 1960." *Tulane Drama Review* (now *The Drama Review*), Vol. 10, No 2, Winter 1965, pp. 73–84.
Young, Stark. *Immortal Shadows.* New York: Charles Scribner's Sons, 1948.
Zukav, Gary. *The Dancing Wu Li Masters* (1979). New York: Bantam Books, 1980.

Credits

Grateful acknowledgment is made for permission to reprint from the following individuals: to Sally Banes for a quotation from "Icon and Image in New Dance," copyright © 1981 by Walker Art Center; to Faubion Bowers for the passage from his *Japanese Theatre,* copyright © 1974 (Japan) by Faubion Bowers; to Selma Jeanne Cohen for excerpts from "The Beautiful Danger: Facets of the Romantic Ballet," *Dance Perspectives 58,* by Erik Ashengreen, copyright © 1974 by Dance Perspectives Foundation, and from "Nik: A Documentary," *Dance Perspectives 48,* edited by Marcia B. Siegel, copyright © 1971 by Dance Perspectives Foundation; to Merce Cunningham for a passage from "The Impermanent Art," copyright © 1955 by Merce Cunningham; to Ann Daly for the passage from "The Balanchine Woman: Of Hummingbirds and Channel Swimmers," *The Drama Review,* Vol. 31, No. 1, copyright © 1987 by New York University and the Massachusetts Institute of Technology; to Simone Forti for a passage from *Handbook in Motion,* copyright © 1974 by Simone Forti; to Nancy Goldner for quotations in *The Stravinsky Festival of the New York City Ballet,* copyright © 1973 by Nancy Goldner; to Ivor Guest for quotes from *The Divine Virginia,* copyright © 1977 by Ivor Guest, and *The Romantic Ballet in Paris,* copyright © 1966, 1980, by Ivor Guest; to Olive Holmes for the quotation from *Motion Arrested,* copyright © 1982 by Olive Holmes; to Lincoln Kirstein for the quotation from *Nijinsky Dancing,* copyright © 1975 by Lincoln Kirstein; to Barbara Newman and Tanquil Le Clercq for the quotation from *Striking a Balance,* copyright © 1982 by Barbara Newman; to Carolee Schneemann for a passage from *More Than Meat Joy,* copyright © 1979 by Carolee Schneemann. The editors of *Dance Chronicle* have allowed me to excerpt archival material from "Maud Allan, Part III: Two Years of Triumph," by Felix Cherniavsky, *Dance Chronicle,* Vol. 7, No. 2, copyright © 1984 by Marcel Dekker. Permission to quote from Doris Humphrey's letter was kindly granted by Charles Woodford and the Dance Collection of the New York Public Library, Lenox, Astor, and Tilden Foundations.

Acknowledgment is also made to the publishers who have given me permission to reprint material: to Doubleday & Co., Inc., for the excerpt from *One Thousand and One Night Stands* by Ted Shawn with Gray Poole, copy-

An abridged version of *The Search for Motion* appeared in *Dance Research Journal,* 17/2 and 18/1, Fall 1985/Spring 1986 (double issue).

Index

Note: Numbers in *italics* refer to illustrations.